Environment and Social Theory

While the environment has been a perennial theme in human thought, the environment and how humans value, use and think about it has become an increasingly central and important aspect of recent social theory. It has become clear that the present generation is faced with a series of unique environmental dilemmas, largely unprecedented in human history.

Environment and Social Theory outlines the complex interlinking of the environment, nature and social theory from ancient and pre-modern thinking to contemporary social theorising. It explores the essentially contested character of the environment and nature within social theory, and draws attention to the need for critical analysis whenever the term 'nature' and 'environment' are used in debate and argument. Drawing on a broad understanding of social theory, the book examines the ways major religions such as Judaeo-Christianity have and continue to conceptualise the environment as well as analysing the way the nonhuman environment plays important roles in Western thinkers such as Rousseau, Malthus, Marx, Darwin, Mill to Freud, Horkheimer and the Frankfurt School. It also discusses major contemporary thinkers such as Jurgen Habermas, Anthony Giddens, Richard Dawkins and Jared Diamond, and the controversy around Bjorn Lomborg's *The Skeptical Environmentalist*. The book also explores the relationship between gender and the environment, postmodernism and risk society schools of thought, and the dominance of orthodox economic thinking (which we ought to view as an ideology) in contemporary social theorising about the environment. It concludes with an argument for an explicitly interdisciplinary green social theoretical approach which combines insights from the natural sciences such as evolutionary biology, physics and ecology with social scientific knowledge drawn from social, political and ethical theories and ideas.

Written in an engaging and accessible manner, *Environment and Social Theory* provides the student with an indispensable guide to the way in which the environment and social theory relate to one another.

John Barry is Reader in Politics at Queen's University, Belfast.

Routledge introductions to environment series
Published and forthcoming titles

Titles under Series Editors:
Rita Gardner and A.M. Mannion

Environmental Science texts

Atmospheric Processes and Systems
Natural Environmental Change
Biodiversity and Conservation
Ecosystems
Environmental Biology
Using Statistics to Understand the
  Environment
Coastal Systems
Environmental Physics
Environmental Chemistry
Biodiversity and Conservation, second
  edition
Ecosystems, second edition

Titles under Series Editor:
David Pepper

Environment and Society texts

Environment and Philosophy
Environment and Social Theory
Energy, Society and Environment,
  second edition
Environment and Tourism
Gender and Environment
Environment and Business
Environment and Politics, second edition
Environment and Law
Environment and Society
Environmental Policy
Representing the Environment
Sustainable Development
Environment and Social Theory,
  second edition

Routledge introductions to environment series

Environment and Social Theory

Second edition

John Barry

Routledge
Taylor & Francis Group

LONDON AND NEW YORK

First published 1999
by Routledge
2 Park Square, Milton Park, Abingdon, Oxon OX14 4RN

Simultaneously published in the USA and Canada
by Routledge
270 Madison Ave, New York, NY 10016

Reprinted 2000, 2002

Second edition 2007

Routledge is an imprint of the Taylor & Francis Group, an informa business

© 1999, 2007 John Barry

Typeset in Times New Roman and AG Book by
Keystroke, 28 High Street, Tettenhall, Wolverhampton
Printed and bound in Great Britain by
Antony Rowe Ltd, Chippenham, Wiltshire

British Library Cataloguing in Publication Data
A catalogue record for this book is available from the British Library

Library of Congress Cataloging in Publication Data
Barry, John, 1966–
Environment and social theory / John Barry. – 2nd ed.
p. cm. – (Routledge introductions to the environment)
Includes bibliographical references and index.
ISBN-13: 978–0–415–37617–4 (hardcover : alk. paper)
ISBN-13: 978–0–415–37616–7 (softcover : alk. paper) 1. Human ecology–Philosophy.
2. Human ecology–Religious aspects. 3. Social sciences–Philosophy. 4. Human
beings–Effect of environment on. I. Title.
GF21.B28 2007
304.2–dc22   2006020746

ISBN10: 0–415–37617–3   ISBN13: 978–0–415–37617–4 (hbk)
ISBN10: 0–415–37616–5   ISBN13: 978–0–415–37616–7 (pbk)
ISBN10: 0–203–09922–2   ISBN13: 978–0–203–09922–3 (ebk)

Contents

Illustrations

Figures

Boxes

Series editors' preface
Environment and Society titles

The 1970s and early 1980s constituted a period of intense academic and popular interest in processes of environmental degradation: global, regional and local. However, it soon became increasingly clear that reversing such degradation would not be a purely technical and managerial matter. All the technical knowledge in the world does not necessarily lead societies to change environmentally damaging behaviour. Hence a critical understanding of socio-economic, political and cultural processes and structures has become, it is acknowledged, of central importance in approaching environmental problems. Over the past two decades in particular there has been a mushrooming of research and scholarship on the relationships between social sciences and humanities on the one hand and processes of environmental change on the other. This has lately been reflected in a proliferation of associated courses at undergraduate level.

At the same time, changes in higher education in Europe, which match earlier changes in America, Australasia and elsewhere, mean that an increasing number of such courses are being taught and studied within a framework offering maximum flexibility in the typical undergraduate programme: 'modular' courses or their equivalent.

The volumes in this series will mirror these changes. They will provide short, topic-centred texts on environmentally relevant areas, mainly within social sciences and humanities. They will reflect the fact that students will approach their subject matter from a great variety of different disciplinary backgrounds; not just within social sciences and humanities, but from physical and natural sciences too. And those students may not be familiar with the background to the topic, they may or may not be going on to develop their interest in it, and they cannot automatically be thought of as being at 'first-year level', or second- or third-year: they might need to study the topic in any year of their course.

The authors and editors of this series are mainly established teachers in higher education. Finding that more traditional integrated environmental studies or specialised academic texts do not meet their requirements, they have increasingly met the new challenges caused by structural changes in education by writing their own course materials for their own students. These volumes represent, in modified form which all students can now share, the fruits of their labours.

To achieve the right mix of flexibility, depth and breadth, the volumes, like most modular courses themselves, are designed carefully to create maximum accessibility to readers from a variety of backgrounds. Each leads into its topic by giving an adequate introduction, and each 'leads out' by pointing towards complexities and areas for further development and study. Indeed, much of the integrity and distinctiveness of the Environment and Society titles in the series will come through adopting a characteristic, though not inflexible, structure to the volumes. Each introduces the student to the real-world context of the topic, and the basic concepts and controversies in social sciences/humanities which are most relevant. The core of each volume explores the main issues. Data, case studies, overview diagrams, summary charts and self-check questions and exercises are some of the pedagogic devices that will be found. The last part of each volume will normally show how the themes and issues presented may become more complicated, presenting cognate issues and concepts needing to be explored to gain deeper understanding. Annotated reading lists are important here.

We hope that these concise volumes will provide sufficient depth to maintain the interest of students with relevant backgrounds, and also sketch basic concepts and map out the ground in a stimulating way for students to whom the whole area is new.

The Environment and Society titles in the series complement the Environmental Science titles which deal with natural science-based topics. Together this comprehensive range of volumes which make up the Routledge Introductions to Environment Series will provide modular and other students with an unparalleled range of perspectives on environmental issues, cross-referencing where appropriate.

The main target readership is introductory-level undergraduate students predominantly taking programmes of environmental modules. But we hope that the whole audience will be much wider, perhaps including second- and third-year undergraduates from many disciplines within the social sciences, science/technology and humanities, who might be taking occasional environmental courses. We also hope that sixth-form teachers and the wider public will use these volumes when they feel the need to obtain quick introductory coverage of the subject we present.

David Pepper and Phil O'Keefe
1997

Series International Advisory Board

Australasia: Dr P. Curson and Dr P. Mitchell, Macquarie University

North America: Professor L. Lewis, Clark University; Professor L. Rubinoff, Trent University

Europe: Professor P. Glasbergen, University of Utrecht; Professor van Dam-Mieras, Open University, The Netherlands

Preface to second edition

A lot has happened in the six years or so since the publication of the first edition of *Environment and Social Theory*. The academic study of social theory and the environment has exponentially increased in that time with the ever-growing production of new books, journals, networks and research projects and findings. The area of social theory and the environment is now so vast that it is impossible for anyone to keep abreast of new developments, and thus this current volume will no doubt contain multiple omissions (as well as the usual errors and mistakes). It is however a positive complaint to make, since it demonstrates the vitality of the issues that the intersection of social theory and the environment raise – no doubt in part spurred on by the growing evidence of continuing ecological degradation from global climate change to more local losses of valued landscapes and species. Of particular note is the addition of other books in the Routledge Environment and Society series, such as Erika Cudworth's *Environment and Society* (2003), Susan Buckingham-Hatfield's *Gender and Environment* (2000) and Pratt, Howarth and Brady's *Environment and Philosophy* (2000), all of which were extremely useful in writing the second edition.

There also seems to be a discernible shift within disciplines such as politics and sociology towards mainstreaming the study of the relationship between society, social theory and the environment (less so with economics, as I discuss in Chapter 8), allied to the welcome growth of interdisciplinary programmes of undergraduate study, though more marked in North America than in the UK or Europe which lags behind in this respect. The interdisciplinary focus in research and research funding (discussed in Chapter 11) is also to be welcomed, driven in part by the (belated) recognition that technological and scientific approaches to complex and contested social-environmental issues and problems need the insights of the humanities and social sciences, particularly in terms of policy-making and governance.

In the 'real-world' we have witnessed the horror of 9/11, an event so seared on the minds of the present generation that it is enough to truncate the event to the day it happened and the subsequent birth of the US-led 'war on terror'; the illegal war and occupation in Iraq, motivated in part by the need of Western powers for secure sources of oil; the maturing of the anti-globalisation movement into the global justice movement and the associated development of the World Social Forum as alternative 'global civil society' sites of resistance and opposition to the World Trade Organization and other institutions of the global and globalising economy (discussed in Chapter 6) and of course the now permanent background presence of 'climate change' in most societies; and most recently the associated popularity of 'energy security' and 'peak oil'. All of this means that research into the relationship between the environment and society and social theory is even more important and needed in the current age, and will continue to be needed in the coming decades as human societies come to terms with transforming themselves into 'post-carbon' economies against an unstable and aggressive international relations context in which we cannot rule out future 'resource wars' by powerful states as part of an extension of the 'war on terror'.

I dedicate this book to my daughter Dearbhla, who along with her sister Saoirse, are never-ending sources of inspiration and hope for the future and whose future interests and well-being are motivations behind much of my academic interest in this area.

I would like to thank Zoe Kruze at Taylor & Francis for her patience in waiting over two years beyond the original deadline, and her understanding of the hectic work and family context within which this second edition was written. The three anomyous referees who commented on the proposal for a second edition made some excellent suggestions, and I hope they will forgive me if I have not acted on them all.

Much of the additional material in this second edition and the thinking behind the rewriting of the book is based on the numerous publications I have written since the first edition, and my teaching and supervision of both undergraduate and postgraduate students. I owe a great debt of thanks to all those students, some of whom are now fellow members of the profession, whom I have taught or supervised and from whom I always learn so much. Here I would like to thank Iorweth Griffiths for co-writing the section on Giddens, which draws extensively on his Ph.D. thesis, and another former Ph.D. student Kate Farrell, who introduced me more fully to ecological economics. I also owe a debt of thanks to all those colleagues with whom I have discussed many of the issues in this book and to whom I have presented papers at conferences and workshops. Although too many to name, I would like to particularly thank Derek Bell, Andy Dobson, Brian Doherty, Peter Doran, Robyn Eckerlsey, Geraint Ellis, Mat Paterson, David Schlosberg, Graham Smith, Piers Stephens, Sharon Turner and Marcel

Wissenburg for the various conversations and discussions we have had over the past six years.

It is a curious aspect of the modern academic world that while scholars such as myself talk of the nonhuman environment, we rarely experience it (though this could be just me!), and while we may talk of the importance of community, we find ourselves hermetically hooked to our computers emailing rather than talking to one another face to face. Both of these I intend to rectify between now and the third edition!

Finally, I would like to acknowledge the work and support of my partner Yvonne who deserves credit as a co-worker on this publication in the sense that this book could not have been written without her taking on an unfair (albeit termporary) burden of child-care and domestic work. While hardly in the same league as the products of real care work, I hope this book does at least prove I was doing something productive all that time in the office!

John Barry
Belfast, May 2006

Introduction: the environment and social theory

The 'control of nature' is a phrase conceived in arrogance, born of the Neanderthal age of biology and philosophy, when it was supposed that nature existed for the convenience of man.

(Carson, 1962: 100)

While it may be true that there is nothing new under the sun, when it comes to discussing environmental issues it seems that perhaps this is not the case. The novelty of environmental issues and problems, from global warming, climate change and biodiversity loss, to concerns for animal rights or the intrinsic value of nature, should not blind us to the fact that humans have always thought about their relationship to the environment. As such the environment and our relationship to it is a long-established issue in social theory.

The newness of environmental concerns is more apparent than real in that thinking about the environment, its meaning, significance and value is as old as human society itself. However, it is clear that the present human generation is faced with a series of unique environmental dilemmas, largely unprecedented in human history. The present human generation is the first one, for example, to have the capacity to destroy the planet many times over, while at the same time it is also the first generation for whom the natural environment cannot be taken for granted. So while the environment has been a perennial theme in human thought, the environment and how humans value, use and think about it has become an increasingly central and important aspect of recent social theory and political practice.

The overall aims of this book are the following:

- to introduce and discuss the ways in which the environment has been used and abused in social theory both past and present;
- to trace some of the historical origins of this relationship and to demonstrate the importance of the environment for social theory;

- to examine key concepts such as 'environment', 'human', 'nonhuman' and 'nature' and related concepts in social theory;
- to explore some of the ideological uses of the environment in social, political and moral thought;
- to outline how some central thinkers and forms of social theory have thought about the environment;
- to outline both the 'greening' of recent social theory and the development of a green social theory.

Outline of the book

The book is roughly divided into two parts. The first, historical, part (Chapters 2–4) offers a chronological discussion of past and present uses (and abuses) of the environment in social theory, from Judaeo-Christianity to contemporary social theory. The second part (Chapters 5–8) looks at a variety of contemporary social theories, from economics, to gender, postmodernism, risk society and recent attempts to integrate ecological and biological thinking into social theory.

Chapter 1 defines how social theory is understood and used in the book, and looks at some general conceptual issues of social theorising and the environment. These general introductory issues include how we define what we mean by 'environment', and how this is related to such terms as 'nature' and 'nonhuman'. This chapter also looks at how and why the environment is used and abused in social theory, particularly when environment is understood as 'nature' or the 'natural environment'. Four dominant models or understandings of the environment which are often used in social theory are then outlined, and the chapter ends with a discussion of one of the most common ways in which the environment is used in social theory in terms of 'reading-off' principles or how society should be from observations of how the natural world operates.

Chapter 2 outlines the historical relationship between social theory and the environment. It focuses on exploring the Judaeo-Christian legacy and moves on to examine the Enlightenment as a key turning point in social theorising about the environment. More specifically, it uses reactions to the industrial and democratic revolutions to organise the discussions of how the place, role and power of environment varied in different forms of social theory.

Chapter 3 looks at pre-Enlightenment and Enlightenment social theorising about the environment. After a brief discussion of the different views of classical political philosophers such as Hobbes, Locke and Rousseau, this chapter looks at nineteenth-century social theory and the environment. The centrality of social theory being 'scientific' is examined before moving on to analyse progressive and reactionary social theorising about the environment. The latter proceeds by focusing on four social theorists: Malthus, Darwin, Spencer and Kropotkin. Following on from a discussion of the classical Marxist analysis of the relationship between society and the environment is an examination of the liberal perspective, focusing on the work of John Stuart Mill.

Following on from the previous chapter, twentieth-century social theory and the environment is the topic of Chapter 4. It begins with classical sociology, then moves on to a discussion of some key twentieth-century forms of social theory, such as the work of Sigmund Freud and existentialism. It then discusses the critical theory of the Frankfurt School of social thought and its critique of the 'dark side' of the Enlightenment and modern societies. This section focuses on how the domination of the external environment by modern societies (via the application of science and technology) can lead to the domination and distortion of human social relations and internal human nature. This then leads into a discussion of Herbert Marcuse and his take on the relationship between the domination of the natural environment and human emancipation. The latter half of the chapter is taken up with a critical examination of some recent social theory and the environment, focusing on the work of Jürgen Habermas and Anthony Giddens respectively.

Chapters 5 and 6 look at right- and left-wing social theoretical reactions to and interpretations of the environment and environmental politics. Chapter 5 which looks at right-wing perspectives explores the history of the relationship between environment and conservative thinking, noting the centrality of ideas of a 'natural order' and concern for future generations (based on respect for the past and tradition) to conservative thinking as well as the Malthusian concern with class and population increase. This chapter also looks at the authoritarian reaction to the 'limits to growth' thesis central to the green/ecological movement, and holds that exclusive concern with population often denotes a right-wing, developed world and sometimes racist concern with the population of the Southern world, without acknowledging that the 'ecological footprint' of the smaller populations of the developed North is much greater than the 'majority world' in the South. Hardin's (in)famous 'tragedy of the commons' perspective is analysed and critiqued, as is the 'free market environmentalist' perspective that uses it to justify the privatisation of environmental public goods. The 'Promethean' viewpoint underpinning the right-wing rejection of the idea of ecological limits to economic growth is also outlined and analysed, as is the related right-wing discourse of those who reject the science behind environmental problems such as climate change or ecological degradation. This chapter also looks at the controversy around the publication of *The Skeptical Environmentalist* by Bjorn Lomborg which highlights the politicisation of science and the ideological (mis)uses of science and empirical data (particularly statistics) in a manner which anticipates some of the arguments of Ulrich Beck in Chapter 9. The 'green conservatism' of prominent political theorist John Gray is also examined as an interesting example of the ways in which ecological concerns and thinking have influenced right-wing forms of social theory. Left-wing reactions are the focus of Chapter 6, which looks at the historical and theoretical reaction of Marxism to the environment and ecological crisis, which ranges from a negative rejection of the ecological perspective as a middle-class, reactionary and right-wing phenomenon (largely based on the 'template' of the Marx–Malthus exchange in the nineteenth century), to more recent attempts to develop an eco-Marxist and eco-socialist theory and politics. The environmental justice movement is also discussed as a class- and race-based form of ecological and health-motivated politics of local resistance to development decisions, and represents a grassroots form of ecological activism which has a left-wing, progressive character. The other main left-wing political tradition analysed is that of anarchism in general, and in particular the social ecology thinking associated with Murray Bookchin, including those critical of his conceptualisation of eco-anarchism. The recent emergence of the anti-globalisation/global justice movement and the political potential and aims of the 'World Social Forum' are also discussed as forms of left-wing (though not exclusively so) engagements with central issues of global ecological protection and a radicalised politics of sustainable development and global justice.

The issue of gender is a key issue within modern social theory, and Chapter 7 looks at the insights and necessity of adopting a gendered analysis of the relationship between society and environment. It begins with an analysis of the historical and conceptual set of gendered dualisms within Western culture. These gendered dualisms begin from an idea of culture or society being separate and above nature, and involve the identification of women and female with 'nature' and the 'natural'. The chapter then proceeds to discuss three of the main forms of eco-feminist social theory, essentialist or spiritual eco-feminism, materialist eco-feminism and political economy, and finally resistance eco-feminism.

Taking economics as a form of social theory, Chapter 8 looks at the ways in which economic theory has viewed, valued and conceptualised the environment. Beginning by showing how 'the economic problem' sets up the conceptual relationship between economics and the environment, it then outlines some of the historical connections between them in various economic schools of thought. In particular, the ways in which the emergence of the modern market economy and economics conceptualised and legitimated particular uses of the environment is discussed in terms of the relationship between land, labour and the enclosure movement, and the relationship which existed both historically and theoretically between material progress, poverty and economics. It then examines the relationship between economic theory and environmental policy-making within the contemporary liberal democratic state, before looking at environmental economics as a contemporary form of 'economising the environment'. Finally, it introduces and discusses ecological economics as a recent development within economics which attempts to integrate ecological and social insights into the examination of the economy, and outlines 'green political economy' as a more politically motivated development of the ecological economics perspective.

Chapter 9 explores two recent social theories and how they analyse the environment – namely, 'risk society' and postmodernism. In the first half, Ulrich Beck's 'risk-society' thesis is discussed as a particular approach to the environment and environmental risk. The character of risk is explored before moving on to how the precautionary principle (which is fast becoming a central regulating principle of social–environmental interaction) may be seen as a logical extension of Beck's thesis. 'Reflexive modernisation' (a theme Beck shares with Giddens), stemming in part from a particular way modern societies can cope with increased environmental (and other) risks, is discussed in terms of how it seems to imply a redefinition of progress. A central part of this redefinition involves the extension of democratic forms of decision-making to more areas of social life. The latter half looks at postmodern approaches to the environment and environmental issues. Environmentalism and its commonalities with postmodernism are discussed in terms of a shared rejection of modernity. The insights of postmodern questioning of the concepts of nature and environment are explored, as are the useful and provocative suggestions of Tim Luke on 'eco-governmentality'. However, some problems with postmodern environmentalism are outlined. Primarily, an argument is made that environmentalism in particular, and social–environmental problems in general, do not necessarily call for the rejection of modernity. Instead, and in keeping with one aspect of Habermas' thought, a claim is made that environmentalism can be seen as a critical analysis of and call for the fulfilment or completion of the 'project of modernity'.

Chapter 10 is an exploration of some of the issues involved in the relationship between ecology, biology and social theory. A critical analysis of sociobiology is offered as an example of the way in which biology and particular understandings of the natural world and human nature have been used to advance or support particular political positions. Using the work of Ted Benton and Peter Dickens in particular, the implications of ecology and biology for social theory are discussed in terms of the desirability and necessity for a 'critical naturalistic' form of social theory. Of particular interest in this and the concluding chapter are the consequences for social theory of seeing human beings as both 'biologically embodied' and 'ecologically embedded'.

Chapter 11 outlines some of the main principles of green social theory and uses this as the starting point for an examination of the 'greening' of social theory. Building on the insights of green social theory, and ideas discussed in previous chapters, some suggestions are made concerning the implications of the greening of social theory. A central claim in the greening of social theory is held to include the overcoming of a strict and permanent separation of 'society' and 'environment'. A consequence of this is that the greening of social theory requires an interdisciplinary approach to the study of society and environment in which the insights of the natural sciences are integrated with the insights of the social sciences. This in turn suggests new models of social theory and modes of social theorising.

1 'Nature', 'environment' and social theory

Key issues

- What is social theory?
- Environment, nature and the nonhuman.
- Social theorising and the environment.
- The uses and abuses of the environment in social theory.
- Four environments for humans in social theory.
- The 'reading-off' hypothesis.

Introduction

What does one consider when one thinks about the 'environment'? Is the environment the trees, plants, animals that we see around us? Is it the Amazonian rainforests or the world's climate systems upon which all life on the planet depends? Are genetically modified organisms part of the environment? Is the environment the same as 'nature'? Does the 'environment' have to do with concepts such as 'biodiversity', 'ecosystems' and 'ecological harmony'? Can we say that the room where you are probably reading this book constitutes an 'environment'?

The problem (which can also be an advantage) with the concept 'environment', like many other concepts such as 'democracy', 'justice' or 'equality', is that it can take a number of different meanings, refer to a variety of things, entities and processes, and thus cover a range of issues and be used to justify particular positions and arguments. While of course the environment cannot refer to anything (that is, it refers to some identifiable and determinant set of 'things'), it is an extremely elastic term in that there are many things – the room you are sitting in, the book itself, the chair, the desk, other people, the fly on the window, and the unseen micro-organisms and the air around you – all of which could be considered to constitute your present and immediate 'environment'. Like many things, the environment can mean different things depending on how you define and understand it, or who defines it.

In many respects thinking about and theorising the environment is one of the most enduring aspects of human thought. For example, the question of the proper place of human society within the **natural order** has occupied a central place in philosophy since its beginnings. *Hence, why, how and in what ways the environment, and related concepts such as 'nature' and the 'natural', are used in social theory is not only extremely interesting but absolutely crucial, given the different meanings and power of these terms when used in argument and justification.* For example, calling something 'natural' implies that it is beyond change, immutable, fixed and given. Hence the power of using this term to justify a particular argument, and the need to be aware of how and why the environment and related concepts are employed in social theorising.

At the same time, alongside the theoretical or academic interest, there is a very important practical aspect to thinking about and theorising the environment in relation to society. This has to do with the increasing quantity and quality of environmental problems which face every society on Earth, both nationally and globally. Global warming and climate change, deforestation, desertification, pollution, **biodiversity** loss and the controversies over the benefits and dangers of **genetic engineering and biotechnology** – all are familiar terms in our every-day lives. All of these, and others, seem to suggest that there is an 'environmental crisis' which faces humanity (and the nonhuman world), the like of which is unprecedented in human history. For the first time in history, humanity has at its disposal the capacity radically to alter the environment (primarily through the application of science and technology), and even has the capacity (though thankfully still not the willingness) to completely destroy life on Earth 'as we know it' (as Dr Spock would say) through the collected nuclear, biological and chemical weapons of mass destruction possessed by a minority of the world's nations. At the same time as being the first generation which has this capacity to affect the environment, one could also say that (particularly since the rise of the green or environmental movement) this is the first generation which knows (or at least has some sense) that it is transforming the environment in a way which will affect the state of the environment inherited by future generations. Hence thinking about or analysing the role or place of the environment in social theory (the aim of this book) is not simply of theoretical, but also of great practical interest.

The importance of analysing the environment and social theory can also be seen when one considers that the majority of the world's environmental problems are largely the result of human social action or behaviour. Global warming, for example, is accepted by the vast majority of the world's scientists to be the result of increased carbon emissions by humans, principally through energy production and consumption (the burning of **fossil fuels**, such as coal, gas and petroleum to create electricity) and forms of transport which rely on such fossil fuels. Hence social theory, defined below as the systematic study of how society is and ought to be, has an important role to play in explaining, understanding and providing possible solutions to the 'environmental crisis'.

What is social theory?

'Social theory' as a field of study is particularly difficult to accurately determine or define. As understood here, social theory is the systematic study of human society, including the processes of social change and transformation, involving the formulation of theoretical (and empirical) hypotheses, explanations, justifications and prescriptions. In disciplinary terms 'social theory' is often associated with sociological theory, and modern social theory has its origins in the sociological tradition. This book however, takes a broad rather than a narrow understanding of social theory, in that it encompasses sociological theory but goes beyond it to include other disciplines and intellectual traditions and approaches. As may be seen from the range of authors and disciplinary approaches surveyed in this book, social theory includes the 'social-scientific' approach to the study of society (in terms of the disciplines one finds in the social-scientific approach to studying society and social phenomenon – sociology and anthropology, politics, international relations, economics, legal studies, women's studies, cultural studies). However, social theory may also include the disciplinary approaches of history, philosophy and moral theory and cultural geography. Thus 'social theory' acts as an umbrella under which are gathered a range of approaches to thinking about society, explaining social phenomena, and offering justifications for advocating or resisting social transformation.

The main disciplinary approaches of this book are: sociological theory (including cultural theory), political theory, economics and political economy, but it also includes the history of social thought. In broad terms what may be called an interdisciplinary conception of social theory is used throughout the book.

The historical origins of social theory may be found in **the Enlightenment**, though aspects of modern social theory may also be found in pre-Enlightenment thinkers and schools of thought (particularly in political philosophy and political economy, as outlined in Chapters 2 and 3). And it is in reaction to the Enlightenment, and the emergence of 'modern society', that a large part of past and contemporary social theory finds its subject. It is in the spirit of the early emergence of social theory that a broad understanding of it is adopted here. In its origins, social theory covered the broad field of the systematic or disciplined study of society in all its various aspects: political, economic, cultural, social, legal, philosophical, moral, religious and scientific. Social theory as the systematic or scientific study of society included looking at such social phenomena as the relationship between the individual and society, the origins and character of cultural practices, and the relationships within and between everyday life and social institutions, such as the family, the nation, the state and the economy. As May points out, in the nineteenth century the main trends in social theory were 'First, an interest in the nature or social development and social origins. Second the merging of history and philosophy into a "science of society". Third, the attempt to discover rational-empirical causes for social phenomena in place of metaphysical ones' (May, 1996: 13). In a similar fashion, this book attempts to offer an equally broad and inclusive view of social theory, though of course many issues, writers and ideas are necessarily left out, or only briefly mentioned. At the same time, we can use the Enlightenment as a way to demarcate modern social theory by noting that the 'subject' of modern social theory is 'the analysis of modernity and its impact on the world' (Giddens *et al.*, 1994: 1). In particular, modern social theory analyses the impact of the industrial, liberal-capitalist socio-economic system which has come to shape the modern global and globalising world.

Social theory typically has two dimensions, one descriptive the other prescriptive. In its descriptive aspect, social theory *describes* society and advances particular explanations for social phenomena, events, problems and changes within society. For example, a social theory may involve explaining the emergence of contemporary far-right politics across Europe by reference to a rise in unemployment, the negative economic effects of globalisation and a consequent appeal of populist nationalist politics in response to the erosion of 'national sovereignty' or 'national pride'.

The *prescriptive* dimensions of social theory are the ways in which social theory not only tells a story of the way society is, but also tells how society ought to be. Here social theory advances particular normative or value-based arguments, justifications and principles to support its claims about how society ought to be ordered, changed or whatever. This prescriptive aspect of social theory can broadly take two forms. On the one hand, it can seek to justify the present social order, that is, suggest that the way society is is how it ought to be. This may be described as a 'mainstream' or 'conservative' position in which the aim of social theory is to legitimate, defend and justify the current way society is organised, its principles, institutions and ways of life.

On the other hand, some forms of social theory seek to argue that society ought to be transformed, organised along different principles and with different institutions from those upon which the current social order is based. These forms of social theory may be broadly described as 'critical' in that they are critical of the current way society is organised and seek to provide reasons for why it ought to be changed and organised along different principles or with different institutions. The classical example of a critical form of social theory is Marxism, which criticises the current capitalist, liberal democratic organisation of societies in the 'West' or 'developed' world, suggesting an alternative 'communist' or 'socialist' mode of social organisation. Below are some examples of mainstream and critical forms of social theory which will be looked at in this book.

Mainstream social theory	Critical social theory
Conservatism	Marxism/socialism
Neo-classical economics	Feminism
Sociobiology	Ecologism/green social theory
Social Darwinism	Postmodernism

While it is nearly always an advantage to adopt broad and flexible, rather than narrow and rigid, approaches to the study of society, such an approach is particularly advantageous (indeed, some would say necessary and not just desirable) when it comes to social theory and the environment. The adoption of an explicitly interdisciplinary approach to studying the relation between society and environment is something that has become a dominant perspective in recent work in this area (Barry, 1999b), and will be discussed in greater detail in Chapter 10. In part, this is due to the rather simple fact that there is not just *one* relation between society and environment (as this and other books in the Routledge Environment and Society series seek to demonstrate). Rather, the relation between society and environment denotes a series of relationships: physical, social, economic, political, moral, cultural, epistemological and philosophical, covering a multi-faceted, multi-layered, complex and dynamic interaction between society and environment. Given the multiple relations between society and environment, it is clear that no one discipline or approach can hope to capture the full complexity of the various relations between society and environment. Hence an interdisciplinary approach drawing on a variety of sources is not only useful, but in some ways is necessary in discussing how the environment has been conceptualised, used and abused within social theory.

With regard to social theorising about nature or the environment (and as indicated below, the two are not necessarily the same), one can trace two other approaches alongside critical and mainstream. These are what one may call 'naturalist' and 'social constructionist' approaches. Naturalist social theorising about the environment generally takes the view that the environment is external to society and exists as an independent 'natural order' outside of society. Social-constructionist

	Critical	Conservative/mainstream
Naturalist	e.g. anarchism	e.g. Malthusianism/Sociobiology
Social Constructionist	e.g. Marxism	e.g. neo-classical economics

Figure 1.1   Social theory and the environment schema

Source: Author

approaches, on the other hand, see 'environment' and 'nature' as constructions of society, and therefore focus on analysing the internal relations within society. Combining them together, we get a fourfold schema of social theoretical approaches to the environment. This schema may be used as a rough guide to understanding particular social theories and theorists.

Environment, nature and the nonhuman

One way of starting our exploration of the place or role of environment in social theory is to look at what we mean by 'environment'. First, we can note that the environment is an 'essentially contested' term. The phrase 'essentially contested' simply means that the term has no universally agreed and singular meaning or definition. The importance of these issues should of course be obvious when social theorising about the environment and its relationship to human social concerns. One of the first and most obvious issues about the environment and social theory concerns the **fact/value distinction**. This refers to the way in which the environment, and related terms, are used not just in a descriptive sense, that is dealing with the facts, but how they are also used to express, justify or establish particular values or judgements, courses of action and reaction, policy prescriptions and ways of thinking. Thus while the environment is used to simply describe the world, that is, to tell us how the world is, it is also used to prescribe how the world *ought to be*, or making some normative (value) claim about something. For example, the term 'natural' carries with it a host of different value meanings, sometimes positive ones of 'wholesome' or 'healthy' (as in organic food), sometimes negative ones, 'uncultured' or 'backward' (as in passing judgement on a group's way of life).

A good way to start thinking about the environment is to list its various definitions and understandings. Often when one is trying to define terms or concepts, a good place to start is a dictionary and thesaurus. Here are some definitions of 'environment' that can be found:

> environment: 'surroundings, milieu, atmosphere, condition, climate, circumstances, setting, ambience, scene, decor' (taken from a computer thesaurus).

> environment: 'situation, position, locality, attitude, place, site, bearings, neighbourhood' (*Roget's Thesaurus*, 1988).

> environment: 'surroundings, conditions of life or growth' (*Collins English Dictionary and Thesaurus*, 1992).

Thus while the environment is often taken to mean the nonhuman world, and sometimes used as equivalent to 'nature', it can take on a variety of meanings. The roots of the term 'environment' lie in the French word *environ* which means 'to surround', 'to envelop', 'to enclose'. Another closely related French word is 'milieu', which is often taken to mean the same as environment. An important implication of this idea of environment is that 'An environment as milieu is not something a creature is merely *in*, but something it *has*' (Cooper, 1992: 169). What Cooper means by this is that environment is not just a passive background or context within which something lives or exists. It is also something that is possessed in the sense that to have an environment is an important part of what the creature or entity *is*. That is, to have an environment is a constitutive part of who or what the creature is, so that one cannot identify a creature without referring to its environment. On this reading, anything that surrounds or environs is an environment. But 'to surround' by itself tells us little. We need to know *what* is surrounded in order to know what *the environment* in question is. That is, without some specified thing to refer or relate to (a species such as humans, or a culture or place) the term 'environment' means very little. Or rather without a referent, the environment can mean everything that surrounds everything that exists. In referring to everything, it also refers to nothing in particular and is therefore of little use as an analytical concept for social theory!

Thus it is important to note that the environment is a *relational* concept or idea in that we need to know what or who is the subject of discussion in order to define an environment. While we may often speak of 'the environment', what is usually at issue is a 'particular environment'. Hence often, but not always, the environment within social theory is defined in relation to ourselves and particular human social relations, and particular historical and cultural contexts. For example, when people in the Western world speak of the environment they usually mean the physical nonhuman environment, such as the countryside, forests, animals, rivers and so on. However, in other cultures the environment may include these things but also include non-physical things such as spirits and the ghosts of one's ancestors. It is for this reason that it is misleading to equate the environment with 'nature' in the sense of the nonhuman natural world, though this is often how it is understood. For example, the 'environment' can refer to the non-natural environment, as in the human, social or built environment. At the same time, as a relational concept we can speak of the environments of other animals, organisms or planets. It is an interesting and instructive thought to consider that we are part of the environment of other creatures (and of each other in many respects).

However, 'nature' does not only refer to the nonhuman world, but is, as Raymond Williams noted, 'perhaps the most complex word in the language' (Williams, 1988: 221). This is because 'nature' can and does refer to both 'human nature' and nonhuman nature (understood as natural environment), thus crossing the boundary between that which is human and nonhuman. Indeed, the complexity and power of 'nature' has to do in large part with the fact that it can be used to unite (as well as separate) the human and the nonhuman. Here are some meanings of nature:

> Nature: 'Nature comes from *nature*, OF [Old French] and *natura*, L [Latin], from a root in the past participle of nasci, L [Latin] – to be born (from which also derive *nation, native, innate*, etc.)' (Williams, 1988: 219).

> nature: '1. The essence of something . . . 2. Areas unaltered by human action, i.e. nature as a realm external to humanity and society. 3. The physical world in its entirety, perhaps including humans, i.e. nature as a universal realm of which humans, as a species, are a part' (Castree, forthcoming).

> nature: 'n. inborn or essential character or quality; temperament, disposition; instinct; universe, especially of living things, collectively; unspoilt wild life, scenery, and vegetation; the original unaltered or uncivilised state, especially of man [*sic*] . . . [Latin. *natura*, from *natus*, past participle of *nasci*, to be born]' (*The Children's Dictionary*, 1969: 398).

These definitions point to the way in which 'nature' can refer to both human and nonhuman issues, properties, processes and entities. Thus we can say that every living thing (both human and nonhuman) has its particular 'nature', as in a more or less determinate set of innate dispositions, characteristics and impulses. At the same time, nature can also simply refer to the totality of the nonhuman world, making it synonymous with the natural environment.

However, sometimes nature opposes environment. For example, one of the enduring debates within social theory concerns the 'natural' or 'innate' causes of human behaviour as opposed to its 'environmental' or 'external' causes. This is the common 'nature versus nurture' debate one finds within social theory and everyday discourse. Here 'nature' refers to 'human nature' understood as some 'given' or unalterable *internal* essence of human beings, while 'environment' refers to the *external* social environment within which humans are brought up and socialised. This issue is dealt with in more detail in Chapter 8. Thus both environment and nature are extremely complex, contested as well as very flexible terms.

Social theorising and the environment

In common usage, the environment usually refers to the physical world which environs or surrounds something. Most commonly of all, in modern parlance, the environment is often thought of as synonymous with the 'natural world' or 'nature'. That is, the environment is often thought of as something that is objective rather than subjective. This is another way of understanding the fact/value distinction in that to say the environment is objective means it is a factual reality independent of our subjective value judgements. As objective reality, the environment just is. Closing one's eyes or mind to one's surroundings does not mean that they disappear. This is something most of us learn as we grow older; young children often believe that simply closing one's eyes is sufficient to make their environment (and all it contains, such as angry adults!) go away. Now while I do not wish to suggest that the environment does not or cannot refer to 'nature' (meaning the nonhuman world and its processes and entities), a less restrictive understanding of the environment is a more fruitful approach to take when relating the environment to social theory. That is, thinking about the environment as something that can and does mean more than the 'natural world' can both help us in thinking about the natural world as well as revealing the complexity of social theorising about the environment.

One of the problems in social theorising about the environment has been that the latter has been viewed by the former as essentially something that is both nonhuman and also beyond human society and culture. So, for example, the environment has been understood as the 'natural world' or nonhuman nature, something which surrounds us and is also beyond human culture. This is the view of the environment which one gets from popular nature programmes on television, such as the excellent natural history programmes produced by the BBC (e.g. David Attenborough's 'Life on Earth' or 'Planet Earth' series). The point is not to reject these understandings but to widen how we think about the environment so as to incorporate these and other possible meanings. Using the term 'environment' as simply another way of speaking about 'nature' or the 'natural world' within social theory is understandable, but one needs to be aware of the danger of missing something important about the environment if we define (and thus confine) it so narrowly.

Particularly in modern everyday language and in modern social theory (Soper, 1995), there is a marked tendency simply to equate the environment with the 'natural'. Often one finds the two terms used interchangeably. An example is O'Brien and Cahn's statement that 'the study of nature, and the relationship between human civilization and the environment, have always held a prominent position in social and political inquiry. Humans have long been interested in discovering our place in the hierarchy of nature' (1996: 5). The point is not that we should never equate the two concepts – indeed it is very difficult to consistently distinguish 'nature' from 'environment' – but rather we should be aware that distinguishing between them is required in critically analysing the concept of environment within social theory. As in many forms of human inquiry (particularly in the humanities and social sciences) part of the process of theorising about something involves making distinctions between different concepts, terms, relations and processes.

One important distinction which may be drawn is between 'nature' as conveying an abstract, almost neutral sense of the nonhuman world, and 'environment' as associated with a more local or determinate sense of a nonhuman (or human) milieu or surrounding. That is, 'nature' is often understood as referring to the conditions of life (for both human and nonhuman species) and all that exists on this planet as a whole, while 'environment' is often associated with a particular subset of these conditions, a subset defined in relation to a particular organism or entity. Thus we can speak of 'nature' without referring to any particular organism or entity, but 'environment' implies the environment of some particular organism, species or set of these. As Ingold puts it, nature is the 'reality *of* the physical world of neutral objects apparent only to the detached, indifferent observer' while the environment is the 'reality *for* the world constituted in relation to the organism or person whose environment it is' (1992: 44). Or as Cooper expresses it, 'an environment [is] a field of significance' (1992: 170), that is, significant for someone or something. Even when both nature and environment are used in reference to the nonhuman world, 'nature' is often associated with an abstract, universal sense of the nonhuman world, referring to the totality of the latter. In contrast, 'environment' refers to a particular, less abstract and more local and determinate part of the natural world.

Like many of our concepts and terms, environment and nature are formed in contrast to their opposites. As well as consulting a dictionary or thesaurus, another good way to get a sense of what terms mean is to seek out their opposites. At least at an initial stage of inquiry, one can find out quite a lot about a concept by seeing with what it is contrasted. This dualistic form of analytical inquiry simply means that we compare a particular term with its opposite. It is important to point out that this form of inquiry will not capture the full complexity of an issue, since thinking about something cannot be reduced to simply specifying something and then discussing its opposite, but it is a useful way to start. So to what do these terms 'non-environment' and 'non-natural' refer? What do they mean?

Since a whole book could be taken up in exploring the full range of issues involved in this task (see Soper, 1995), what I intend to do is highlight some of the more obvious ways in which our understanding of these terms can be advanced by comparing them with their antonyms. In the binary set of concepts below, we can find out quite a lot about the meaning of the environment (qua 'nature').

environment/nature	*opposite of*	human society/culture
nature/nonhuman	*opposite of*	human
naturally occurring	*opposite of*	human-made/artificial
nature	*opposite of*	nurture

One of the first things we should note about 'nature' and 'environment' (when used as referring to the nonhuman world or processes) is that they are viewed in opposition to human society and culture. In this respect, whatever is environ- mental or natural is something which is separate from and independent of human society. And in some respects this seems to be, at least intuitively, true. For example, trees grow and ecosystems function independently of human society and culture. At this very basic level nature or the 'natural environment' does not depend on humanity. Indeed, the opposite would seem to be the case: that is, humans in common with every other living species depend on their environment to survive and flourish. So on this first analysis: the environment is something that is separate from human society. *However, this separation does not mean that humans do not have a relation with their environment.* Since they depend on their environment, and exist within the environment, they are obviously related to their environment. But to say that humans are related to and depend upon the environment is not to say that they are the same as the environment. Like any other species, humans exist in a condition of separation from but at the same time a relationship to and with their environment.

Second, there is also another dimension to the relationship: that between 'nature' and the 'natural environment' as 'nonhuman' in contrast to 'human'. 'Nature' as 'nonhuman' may thus be used to define what is 'human' or what is properly human. In this way, nature as nonhuman is an extremely important concept, one might say a foundational concept, in social theory, in that it defines what is the human, or properly human.

Third, we see that the 'environment' can refer to that which naturally occurs, in contrast to that which is human-made or artificial. Indeed, this final set of opposing concepts – between the 'natural' and the 'artificial' – is one of the most central ways in which humans have and do think about the environment. We commonly think of the environment as entities (rocks, rivers), species (bears, lions, foxes) and processes (carbon cycles, hydrological cycles) which are emphatically not the products of human society. Thus the environment here is that which occurs without human intervention, and many natural processes and entities pre-date humanity and human society. Here we have a notion of the environment qua nature which is one of the oldest and most enduring conceptions humans have of the environment. This particular conception of the environment resonates with the idea of the environment as something nonhuman, the external and eternal natural and naturally occurring surroundings which envelopes both humans and nonhuman entities.

In some ways, it finds an echo in the Christian doctrine of the environment as 'God's Creation', that is, the environment as something which is not of human origin or design. One can also appreciate the distinction between the 'natural' and the 'artificial' when one considers the difference people perceive between certain foods and goods which are 'natural' and those that are 'processed'. Added to a factual distinction between what is 'natural' and what is 'artificial' are a whole range of evaluative positions in which one or the other is seen as 'superior' or 'better' than the other. For example, Goodin (1992) suggests that a 'green' theory of value rests on the claim that naturally occurring processes have a particular value precisely because they are not the work of human hands. As he puts it, in answer to his question 'What is so especially valuable about something having come about through natural rather than through artificial human processes?' is that 'naturalness [is] a source of value' (Goodin, 1992: 30). That something is 'natural', of nonhuman origin, and existing independently of human actions or interests, is held by many people to be something of value. According to Goodin, a 'natural' landscape is more valuable than a 'humanised' landscape, in the same way as a 'fake' or reproduction is never as valuable as the original (Elliott, 1997). Placing such stress and importance on the value of 'natural' and naturalness is a distinctly 'green' position, though one which many non-greens may share.

However, on the other hand, there are those for whom the 'artificial' is superior to the 'natural'. Here we can think of arguments in which whatever is human-made or produced is viewed as more valuable than whatever is naturally created. An extreme example of this is what can be called the **technocentric** position which holds that human creations are vastly superior to natural ones, so that it can not only ask 'What's wrong with plastic trees?' (Krieger, 2000), but answer that there is nothing wrong with them, and indeed they are superior to natural ones. One can see some of the origins of this view of the superiority of the human over the natural in the 'perfectionist' justifications for human transformation of the natural environment that were prevalent in pre-Enlightenment Western Christianity, as discussed in Chapter 2.

However, it also needs to be remembered that there is a continuum between the two poles of the 'natural' and the 'artificial'; there are of course many intermediate positions between them. As will be seen throughout this book, this distinction between the 'natural' and the 'artificial' and their relative evaluative weightings (superior/inferior) is something that shadows much social theorising about the environment and our relationship to it. One can trace many of the origins of the debates about the relationship between society and the environment through looking at how, at different times and places, different values are attached to the 'natural' and the 'artificial'. For example, whereas nowadays there is a premium attached to things 'natural' (and not just for health reasons, as in organic food), not so long ago, natural produce was regarded as 'backward', 'uncivilised' or not advanced; a sign of socio-economic and cultural inferiority. For example, in the last century, to live 'close to nature' (either in hunter-gatherer communities or rural-agricultural settings) or consume natural produce, meant one was not as advanced or cultured as those who did not live close to nature (but in urban areas and cities) and who enjoyed 'artificial' and 'processed' goods and services. Thus there is no determinate or singular, agreed or fixed reading of the natural and the artificial; they mean different things and are given different evaluations in different social and cultural settings and in different historical periods. The point of social theory is to make us aware of these evaluative distinctions, to try and understand them, and if possible suggest explanations for them and to critically interrogate and perhaps challenge them. In this way we can say that there are no 'value-neutral' readings of the environment as nonhuman nature. That is, when one describes the environment as nonhuman nature, implicit in those descriptions are certain value judgements and normative positions. This is partly because 'nature' and the 'natural' carry with them various meanings and express a variety of evaluative judgements (ranging from the good/positive to the inferior/negative). And as will be seen later on, in discussing the 'reading-off' hypothesis, when social theories 'read' the 'environment' they often project or map particular ways of thinking and values on to the environment rather than simply offering a 'neutral' or 'objective' account.

At a very basic level one can intuitively grasp what it means to say that the environment is **socially constructed** by noting how different societies, different ways of thinking and social theorising display distinct ways of thinking about and perceiving the environment. For example, while the environment for a typical city-dweller may mean the houses, buildings, waste spaces, parks as well as 'nature' (meaning the nonhuman natural world), for a country-dweller the environment may mean fields, domestic and wild animals, hedgerows, stone walls and the seasons, as well as 'nature'. Thus environments differ and depend upon that to which one is relating the environment in question. At the same time, this example shows that while there may be different conceptions of the environment, they do not necessarily have to be contradictory. This is due in part to the relational character of the environment – that is, the environment is that which surrounds something, some entity or someone (including collections and groups). My environment (however I construct this) does not necessarily have to contradict your environment (however you construct it). The map (the representation), after all, is not the territory (the physical reality), but a particular 'reading' or representation of the territory. As Foster notes, '"The environment" . . . is something upon which very many frames of reference converge. But there is no frame of reference which is as it were "naturally given", and which does not have to be contended for in environmental debate' (1997: 10).

Alongside this discursive or conceptual sense of the 'social construction' of the environment, there is a *material* dimension to the 'construction' of the environment which refers to the real, material, physical production and transformation of the environment by the human species. Such transformations of the environment by humans include agriculture, the creation of particular landscapes by human practices different from the environment if left in its 'natural' (i.e. untransformed) state, the creation of hybrid species of plants and animals as a result of human intentional selection and cross-breeding (which includes modern biotechnological techniques and the human manipulation of genetic information).

The uses and abuses of the environment in social theory

Conceptions of the environment differ, sometimes dramatically. In some cultures, or within particular worldviews (ways of thinking), the environment can include the dead, one's ancestors and/or other entities from the 'supernatural' realm, such as gods, goddesses, spirits, angels, ghosts and so on. Thus the environment, as that which environs, depends not only on something to environ, but what constitutes the surrounding environment. Hence the environment does not necessarily refer to the physical environment (whether natural or human-made).

The full complexity of the social construction of the environment can be seen if we examine how we think about the environment. 'The environment' as a term of social discourse (that is, as a part of human language, thinking and acting) is of course a human concept. It is difficult to imagine that other species see or construct their environment using the conceptual tools which humans do. Indeed, the vast majority of other species do not 'conceptualise' their environment at all (at least as far as we know), they simply get on with it and live within their particular environment. To complicate matters, when we focus on human societies we find that there is often a difficulty in translating what one culture refers to as 'environment' into another cultural context or language.

One of the first things we can say about the environment is that while it refers to something 'out there' in the 'real world', this does not mean that it cannot be a social construction. Even in scientific discourse (the theory and practice of the scientific community, with its internal principles, standards and rules for what constitutes 'scientific knowledge'), which is a discourse about 'facts' telling us 'how things are' in the environment, the latter may still be regarded as a social construction. The facts of science about the external world are human social facts, particular interpretations of the way the world seems to us. It is more or less accepted within the scientific community that there are no observer-independent facts about the world. The world we see around us, and the scientific knowledge we have of that world, is as it is because of the types of beings we are. That is, if humans could not see colours apart from black and white, other colours would not exist for us. Rather like Adam in the Garden of Eden (discussed below), humans 'name' things and in naming them they exist, or 'come into being' for us, which is not the same as saying we create them. For example, while entities such as trees would doubtless exist without humans, the concept or category of a 'tree' would not. In a world in which humans did not exist, 'trees' would not exist, though the vegetative physical entities to which the term refers would.

When we think about the environment and when we apprehend it, we do so from particular perspectives and in more or less distinct ways. One way of explaining this is to say that humans (and nonhumans) have particular modes of apprehending the environment, that is, distinct and different ways of seeing, feeling and thus 'constructing' the world (the world as it seems to them). For example, we may say that given the type of beings humans are, with our particular physiological and sensory make-up, the environment we see and appreciate will be different from, say, how a fly or a bat perceives the same 'environment'. However, this is not to say that all humans have the same environment. Clearly they do not, since humans live in a wide (though not unlimited) variety of different environments, ranging from the harsh arctic environment to the tropical environment of the Equator. At the same time, different cultural or other value-based views of the environment mean that not all humans (who share the same senses) will necessarily have the same meaning of 'environment'. For example, conceptualising the environment as an 'inhospitable wilderness' carries with it different meanings and origins and will have different implications and effects from conceptualising it as a 'garden', or 'God's creation'. Some of the implications of this are bought out in the next section.

Four environments for humans in social theory

Environment as wilderness

According to Rennie-Short, 'Wilderness is a word whose first use marks the transition from a hunter-gathering economy to an agricultural society' (1991: 5). That is, 'wilderness' is a view of the natural environment from a 'civilised' or 'cultured' perspective. Rennie-Short suggests that there are two general responses to wilderness. The first is a negative reaction, in which the dominant aim is to 'pacify', 'tame' or 'conquer' wild nature, turning it into a 'garden' for human enjoyment and in line with human aims.

This has been the dominant view of wilderness for most of human history: wild nature as dangerous, uncontrollable and unstable, a permanent threat to the human social order (Oelschlaeger, 1992). Thus in European history, wild areas, such as dark forests, mountains and swamps, were regarded as dangerous places, and people looked upon them with fear. A good example of this is the status and view of forests in European folk-tales and fairy-tales as places where humans were not welcome, which were inhabited by wild animals, such as wolves and bears, as well as by 'supernatural' creatures such as witches, goblins, evil spirits and so on. Indeed, the root of the term 'bewilder', meaning confusion, disorder and over-whelmed, may be found in the idea of wilderness. People were in danger of becoming bewildered when they left the security of the town and entered the disorder and anarchy of the wilderness, which could then threaten the order of human society. This negative view of 'raw' nature is something which may be seen in the Christian idea of 'perfecting nature', taking raw nature and improving it, transforming it into a more productive, tamed and aesthetically pleasing environment. It is also something that may be seen in the history of America with its myth of the 'Wild West' and the frontier society it created, an historical experience which has left its mark on American national identity, as many authors have pointed out (Cronon, 1995; Arnold, 1996; Nash, 1967). From this frontier society sprang a society and mentality of 'rugged, free, independent individuals' pitting themselves against an untamed wilderness which they had to dominate and control in order to survive and prosper. Thus how one conceives of the environment may be part of how one conceives of collective, especially national, identities.

The second, more positive view of wilderness is what one may call the 'romantic' or 'green' view. Here wilderness is celebrated, something to be cherished and valued in the face of a world in which the natural environment is increasingly 'developed' or destroyed by humans. As Zimmerman puts it, 'Wilderness is a direct reminder that not every thing can be reduced to the status of a human product, project or construct; wilderness is the "other" which reminds humanity of its own dependency on the powers at work in wilderness, but also in humanity itself' (1992: 247). This view can trace its roots to the romantic reaction to the processes of industrialisation in the late eighteenth and early nineteenth centuries as discussed in Chapter 3. This positive view of wilderness may be seen in contemporary popular culture, in which 'nature wild and free' is not only viewed as something to be valued, but also under threat from the bureaucratic, mundane and stultifying processes of **modernisation**. Popular films such as *Free Willy*, about the 'freeing' of a tamed killer whale into its rightful 'wild environment', or others such as *Finding Nemo*, also express this contemporary sentiment.

Environment as countryside/garden

Environment as countryside represents a different sense of environment from wilderness. In opposition to 'wilderness' and wild nature, the natural environment as countryside may be seen as a 'garden', a 'tamed' or humanised natural environment. Here the environment denotes nature transformed by and in the service of human needs, aims and intentions. Historically and conceptually, one can situate environment as countryside between 'wilderness' and the urban environment. That is, the idea of the environment as countryside lies between the natural environment untouched by human hands (wilderness) and the created, artificial environment created by humans (the urban environment). The idea of the environment as countryside or garden is often used to explain differences in national culture and environmental attitudes between European countries and other 'industrialised' countries. Unlike other countries such as America and Australia, countries like Britain do not have areas of wilderness, and the natural environments of these countries are humanised environments in the sense that they are the product of past human transformation. They are more like worked gardens than wilderness.

Just as with wilderness, Rennie-Short (1991) suggests that there is a connection between environment as countryside/garden and national identity. As he puts it, 'In most countries the countryside has become the embodiment of the nation, idealized as the ideal middle landscape between the rough wilderness of nature and the smooth artificiality of the town, a combination of nature and culture which best represents the nation-state' (1991: 35). Echoing some themes which will be taken up in later chapters, he also points out that 'the countryside is seen as the last remnant of a golden age . . . the nostalgic past, providing a glimpse of a simpler, purer age . . . [a] refuge from modernity' (1991: 31, 34). For example, in Britain (as elsewhere) the countryside, its inhabitants and particular ways of life are often held to best preserve how life used to be, before crime, stress, competitiveness and materialism became daily features of life. Here the 'countryside' acts as a 'living reminder' of a 'better' age in which things such as patriotism, hard work, loyalty to the land and monarch, respect for authority, piety and community formed the substance of daily life and identity.

At the same time, the ideas and experiences of those who wish to escape the 'rat race' associated with life in the 'big city', 'drop out' of the 'mainstream' and 'return to the land' is something that has been a periodic feature of Western social theory and historical experience for the past 200 years. A good example of this is the 'counter-cultural' movement of the 1960s in Western societies which, as part of its alternative to modern life in industrialised, urban societies, advocated a rural, agricultural setting, living communally, frugally and in 'tune with nature'. The popular 1970s BBC series *The Good Life* expressed some of these themes.

The urban environment

By the urban environment is meant the human-made spaces, buildings, developments and structures one finds in towns and cities, as opposed to either wilderness (where there is little or no human trace) or the rural countryside (where the natural environment and its processes are 'managed' but not created by humans). The city and town represent the artificial environment that humans make for themselves and others, and the emergence of the urban environment, both historically and conceptually, is also the most modern of environments which humans (and nonhumans) inhabit. Indeed, the connection between the urban environment and modernisation is such that standard accounts of the latter have urbanisation as a constitutive aspect. That is, to be 'modern' is to live in an urban as opposed to a rural environment.

The creation and development of the urban environment over the past 200 years has profoundly affected how people viewed and thought about the natural environment. As more and more people moved from the land to the cities (as part of the 'agricultural revolution' which was a condition for the industrial revolution) they became increasingly removed from direct contact with nature. This removal from the natural environment led in part to a heightened sense of the symbolic status of the natural world (wilderness and countryside) and a concern for its preservation. That is, the less direct and daily contact people had with the natural world, the more symbolically powerful it became (as a symbol of 'a golden age', as in 'countryside' or a reminder of the 'disorder' which surrounded society, as in wilderness). As outlined in Chapter 4, in discussing Giddens, the experience of urbanisation and separation from the natural environment is a key aspect in the rise of environmental consciousness, the aesthetic appreciation of nature and concern with the preservation of the natural world. The positive value placed on all things 'natural' and 'environmental' which we see today is due in part to the fact that most people do not have direct experience of nature, and the natural carries with it a sense of purity, goodness, wholeness and wholesomeness which many find lacking in the polluted, overcrowded, unsafe and artificial environments of the city and urban areas. In times of crisis or when people experience life as crowded, polluted, unsafe and so on, it is understandable why a return to a simpler, purer and less complicated way of life seems attractive.

The global environment

The idea of the 'global environment' is the most recent conceptualisation of 'environment', and has its origins in the debates about the 'environmental crisis' in the 1960s and 1970s. It is thus significant to note that the concept of the global environment originates against a background of vulnerability, threat and risk (some of which will be taken up in Chapter 9 in discussing Ulrich Beck's 'risk society' thesis). While there has always been a global environment or biosphere, with the emergence of threats to it, it became an object of public and political interest. Of particular importance in the development of the global environment was the pictures of the Earth taken from space in the late 1960s (see Figure 1.2). It was these images more than anything else which really brought home to people that we all live on and share the same planet Earth. From this developed an environmentally based 'planetary consciousness', expressed not only in the emerging green and environmental movements but also in events such as Earth Day in 1970, and the concept of 'spaceship earth' (Boulding, 1966).

However, it was in the 1980s and 1990s that the idea of the 'global environment' gathered momentum. With the emergence of 'global environmental problems' such as biodiversity loss, global warming, ozone depletion – that is environmental problems which are global in scope, though not necessarily global in origin – the idea of a global environment made perfect sense. The idea of the global environment has become a central aspect of social theorising about **globalisation**, with the global environment now taking its place alongside the 'global economy', 'global communications' and the 'global village'. Just as the global economy creates a global network of socio-economic relations between distant places and people, likewise the global environment expresses another level of the interdependence of distant people and places. Environmental problems, like the modern global free market economy, do not respect national boundaries.

The global environment idea carries with it some important messages. The first is interconnectedness; that is, we are all in the same boat when it comes to threats

Figure 1.2 View of the Earth from the Moon

to the global environment. The second is that at its heart (and reflecting its origins) the global environment is an environment under threat from humans. The third, and following on from the last is that talk of threats to the '*global* environment' often carries with it the implicit idea that since these threats are global, faced by everyone on the planet, and we are all in the same boat together, then somehow we are all to blame. Yet, according to many greens and developing world activists, the latter is not the case, since the main cause of these global problems is in the advanced, industrialised world. However, the idea of a 'shared earth' which is related to the idea of the 'global environment' is used to convey the idea of shared fate and common responsibility of humanity. A fourth and final issue is that the idea of a global environment that is threatened leads to the need for global co-operation between nation-states and peoples to solve or cope with these problems, most pressing of which is **climate change**.

The 'reading-off' hypothesis

Perhaps the oldest and most dominant way in which the environment has been used in social theory is what may be called 'the reading-off hypothesis' (Barry, 1994). This hypothesis, as its name suggests, proposes that if we wish to find out about human society, there are important lessons to be learned from a close examination of the natural world in general, and animal behaviour in particular. This hypothesis is based on the power of the notion of 'naturalness' in social theory and argument. An example of its power is how calling something 'unnatural' is a common way of ending a discussion or argument. To say something or someone is 'unnatural' is a pejorative, negative judgement, meaning they or their behaviour are objectively and simply wrong. If I do not like what you do, calling you or your behaviour 'unnatural' is a powerful put-down. Its power lies in that I do not have to argue or 'prove' how or why you or your behaviour is unnatural.

The reading-off hypothesis claims that we can both better describe or explain human society by applying the knowledge derived from the study of the non-human world, since we too are part of nature, a particular species of animal. It is thus common in naturalist forms of social theorising about the environment. But alongside this 'descriptive' claim, it also purports that the study of the nonhuman world can have prescriptive power. In short, the reading-off hypothesis within social theory states that we can describe the human social world as it is and prescribe how it ought to be from the application of knowledge gained from the study of the natural world. We can 'read off' how human society is and ought to be from looking at the nonhuman world.

The appeal of reading off from nature lies largely in the idea of the 'givenness' of nature and natural processes: that is, whatever is 'natural' or 'part of nature' is simply the way things are and ought to be, and there is nothing we can do to alter it. As Smith points out, 'The authority of "nature" as a source of social norms derives from its assumed externality to human interference, the givenness and unalterability of natural events and processes and behaviours' (1996: 41). Whatever is deemed 'natural' and 'unnatural' simply is, that is an unalterable 'fact' of the 'way things are' (and ought to be) and cannot (or should not) be altered by human intervention. Thus if one says that women *are* 'naturally' – i.e. by their natures – mothers and home-makers, then one has a powerful argument for saying women *ought* to be mothers and home-makers and not seek employ-ment outside of the home. Or, if one says that homosexuality is 'unnatural', one is in effect saying that it is a transgression of or 'going against' nature and the natural order of things, and hence it is simply wrong and/or harmful. These uses of 'natural' and 'unnatural' are common in everyday and academic argument.

It is important to note that this 'reading-off' hypothesis is a two-way process. 'Reading off' how human society or social relations ought to be from an examination of the natural world inevitably involves the projection of social claims/aims/positions on to the natural world. Rather than 'read off' we 'read into'. That is, we 'read off' from the natural world what we project into it. Thus, for some, the natural world is a place of harmony, co-operation and balance between different species and environments, while for others 'nature is red in tooth and claw', a place of competition and 'survival of the fittest'. In social theory, there are no determinate readings of the environment, because there are no value-free readings.

However, an important point that must be raised with regard to the reading-off hypothesis is not concerned with particular readings or interpretations of the natural world. Rather, we need to ask why this particular device or strategy is used within social theory in the first place, and ask why any reading of the natural environment should be seen as telling us something important about how human society is and ought to be. This theme of reading off from the natural environment will be explored in greater detail in Chapters 2 and 3.

Conclusion

The environment, and its related terms, their meanings, status and significance for humans, and the relationship between them and humanity, constitutes one of the oldest themes in human thought. Particularly when environment is equated with 'nature' or 'natural' it is a powerful form of argument, and thus one must be aware of how it is used (and who by) in social theory. In Chapter 2 we explore some of the ways in which early forms of social theory used the environment and related concepts in justifying or grounding particular positions and arguments.

Summary points

- A broad, flexible and interdisciplinary understanding of social theory is used in this book. Social theory, as the systematic study of society, covers the following disciplines (with particular emphasis on the first three): sociology, politics, political economy, cultural theory, philosophy, cultural geography, legal studies and history.

- Such an interdisciplinary approach is argued to be particularly appropriate when dealing with the matrix of relations between society and environment.
- The environment and its connected terms, such as nature and the natural, are relational concepts.
- There are many meanings and understandings of the environment, but one of the most common is the equation of the environment with 'nature'.
- While this equation of the environment with nature can be useful, it is necessary to expand the concept of environment beyond a simplistic equation with nature.
- Within social theory, this idea of the environment as nature may be seen to express itself in the distinction between 'nonhuman' and 'human', and 'natural' and 'artificial'.
- In thinking about the environment one needs to be aware of the social and cultural meanings attached to the environment. There are no 'value-neutral' readings of the environment.
- Some important conceptualisations of the environment include: environment as wilderness, environment as countryside/garden, the urban environment and the global environment.
- One of the most common uses of environment in social theory is the reading-off hypothesis in which how the human social world ought to be is derived or read off from how the natural world is organised.

Further reading

On social theory, both Derek Layder's *Understanding Social Theory* (2nd edn), London: Sage, 2005, and Tim May's *Situating Social Theory*, Milton Keynes: Open University Press, 1996, offer good introductory accounts of the historical origins and main traditions or currents of Western social theory. A more advanced set of readings of contemporary social theory may be found in Anthony Giddens *et al.* (eds), *The Polity Reader in Social Theory*, Cambridge: Polity Press, 1994, and for more recent work in social theory a good source is S. Siedman and J. Alexander's *The New Social Theory Reader*, London: Routledge, 2001, and Pip Jones' *Introducing Social Theory*, Cambridge: Polity Press, 2003.

A difficult but worthwhile read on the various complexities of the idea of nature and the environment may be found in Kate Soper's *What Is Nature?*, Oxford: Blackwell, 1995. In addition, my own *Rethinking Green Politics: Nature, Virtue and Progress*, London: Sage, 1999, contains a fuller discussion of the distinction between 'nature' and 'environment' as well as the relationship between 'culture' 'society' and the former, and the 'reading-off' hypothesis. An excellent book on the 'reading-off' issue is John Meyer's *Political Nature: Environmentalism and the Interpretation of Western Thought*,

Boston, MA: MIT Press, 2001, which distinguishes 'derivative' from 'dualist' accounts of nature within the history of Western political thought.

See also David Cooper's 'The Idea of Environment' in David Cooper and Joy Palmer (eds), *The Environment in Question: Ethics and Global Issues*, London: Routledge, 1992, and Tim Ingold's 'Culture and the Perception of the Environment', in E. Croll and D. Parkin (eds), *Bush Base, Forest Farm: Culture, Environment and Development*, London: Routledge, 1992. Other good texts include Robert Boardman's *The Political Economy of Nature: Environmental Debates and the Social Sciences*, Basingstoke: Palgrave, 2001, and M. Redclift and T. Benton (eds), *Social Theory and the Global Environment*, London: Routledge, 1994.

For an examination of the cultural dimensions of social–environmental relations see Kay Milton's readable and informative book *Environmentalism and Cultural Theory*, London: Routledge, 1996. For an excellent and readable account of different ideas of the environment (such as wilderness, countryside and city) see John Rennie-Short, *Imagined Country: Society, Culture and Environment*, London: Routledge, 1991, and Elizabeth Croll and David Parkin's edited volume, *Bush Base, Forest Farm: Culture, Environment and Development*, London: Routledge, 1992. Other good texts include Neil Evernden's *The Social Creation of Nature*, Baltimore, MD, and London: The Johns Hopkins University Press, 1992, and David Golblatt *Social Theory and the Environment*, Cambridge: Polity Press, 1996.

2 The role of the environment historically within social theory

Key issues

- **Non-Western views of the environment.**
- **The Judaeo-Christian legacy.**
- **The Enlightenment, environment and social theory.**
- **The industrial revolution.**
- **The democratic revolution.**

Introduction

The aim of this chapter is to set the scene for the later discussions of the role of the environment in social theory by looking at how social theory has historically viewed and used the environment. This chapter traces some of the historical antecedents of how previous human generations at different times, places and within different cultures have conceptualised and thought about the environment and social–environmental relations. A second aim will be to look at some of the historical roots of Western social theorising about the environment in general, and at the legacy of Judaeo-Christianity and the Enlightenment in particular. Finally, a third aim is to look at some of the historical origins of the 'green' social theoretical perspective, focusing on certain antecedents of green thought in two broad reactions to the Enlightenment: namely, the reactions to the industrial revolution and the French and American 'democratic revolutions' of the late eighteenth century.

Historically, social theory has been largely concerned with reflecting on human society, critically analysing it, proposing the best arrangement of society for *human beings*. While there have been some notable exceptions, as will be discussed below, social theory has historically been overwhelmingly **anthropocentric**; that is, largely concerned with humans, human interests and human social relations.

One has only to examine some of the great texts of social theory (covering such disciplines as politics, philosophy, sociology, history, economics) to quickly see that 'the environment' as an explicit object of examination is either absent, or else is seen as a natural 'backdrop' against which human history, politics and social development takes place. Thus the environment (largely viewed as 'nature') is often a mute or a passive object of human manipulation within the history of social thought. It is rarely at the forefront of social theory historically, being seen as something that just is, standing over and above human affairs and the enduring natural context within which those affairs occur. However, this is not to say that the history of Western social thought has little to say about the environment. For the most part the environment has been regarded as a necessary collection of resources or means to human ends, an attitude towards the natural environment which still predominates today, but which is being challenged by greens and others who suggest that this attitude is both morally objectionable and results in environmental problems for society.

Non-Western views of the environment

While most of this book will concentrate on the relationship between Western social theory and the environment, where appropriate references will also be made to non-Western perspectives and insights. This is particularly important as there is a strong argument to suggest that many of the environmental and social problems that we see around the world today, at least in part, may have to do with the predominance of a particular Western way of thinking about and interacting with the environment and a distinctly Western or European set of values, institutions and principles around 'modernity' and 'development', which will be dealt with later in this chapter. Here all I wish to do is indicate that there is and has been a variety of non-Western social theorising about the environment.

Historically, as in the Western world, most non-Western social theorising about the environment took religious and 'traditional' cultural forms (meaning that how people thought about the environment was largely governed by myths and stories which were handed down from one generation to the next as tradition). In the Middle Eastern civilisations of Egypt and Mesopotamia we can find a wide variety of ways of thinking about the environment. According to Hughes, 'The attitude of the peoples of Mesopotamia toward nature . . . is marked by a strong sense of battle. Nature herself was represented in Mesopotamian mythology as monstrous chaos, and it was only by the constant labours of people and the patron gods that chaos could be overcome and order established' (1994: 34). This idea of the 'struggle against nature' is something that has framed much of the debate about the relationship between human society and the environment, and is something that will be taken up in later chapters. It serves as an opposing view to the idea of human harmony with a bountiful nature which many suggest is typical of some aboriginal and 'hunter-gatherer' worldviews, which are discussed briefly below.

The oriental religious teachings of Confucianism, Shintoism and Buddhism each had their particular views on the proper place of the environment in their particular worldview, and all had their own rules and principles concerning the treatment and use of the environment. Generally speaking, Buddhism displayed a marked respect for the natural environment and a basic Buddhist belief is that all forms of life (human and nonhuman) are interdependent, which includes the principle of *ahimsa* or avoiding harm to other living beings. Hindu religious thought denoted particular ways of treating domestic animals, and forbade the eating of beef. Islam had its own particular set of rules, laid down in the Koran, about the proper way of thinking and relating to the environment. As Morgan puts it, 'Muslims have a strong sense that the whole universe, sun, moon, stars, trees, birds and flowers are God's creation and "signs" of His being, and that humans are *khalifa*, vice-regents under God with responsibility to care for what God has made' (2001: 394). This idea of taking care of God's creation is similar to the 'stewardship' tradition within Judaeo-Christianity which is discussed below.

One of the things which all the 'great religions' of the world share (Buddhism, Islam, Judaeo-Christianity and Hinduism), and which is important to note, is a common character as 'agricultural' religions. That is, these particular religions can trace their historical roots back to the period after (the majority of) humans had left the 'hunter-gatherer' stage of human social evolution and were prominent in societies and empires which were overwhelmingly agricultural civilisations. A second issue of note is that most of the civilisations in which these religions originated were also civilisations in which cities and towns were important places of political, economic, religious and military organisation and power.

Aboriginal peoples in Africa, Australia, the Americas and Asia also had their own, usually spiritually informed, traditional ways of thinking about and treating their environments. The forms these traditional ways of thinking and acting took ranged from animism, a belief in spirits of the forest or of particular animals, nature worship and sun worship. In general, in marked contrast to the Judaeo-Christian religious worldview (and also to the other formal religions outlined above), these traditional aboriginal cultures were less anthropocentric and more inclined to emphasise the continuity rather than the separation between the human and the nonhuman worlds. These more 'eco-friendly' worldviews have been a constant source of inspiration for green forms of social theory and action. As Wall notes, 'Greens and fellow travellers have used existing hunter-gatherer groups and their ancient ancestors as an example of ecological good conduct' (1994a: 20). However, this stratergy is not without its problems and critics.

For example, the 'ecological wisdom' of the American Indians has long been a (contested and sought after) point of reference for many environmental arguments and groups, on the grounds that the philosophy, worldview and associated ways of living of American Indian cultures represent real world examples of 'living lighter on the earth' and how to live in harmony and balance with the environment. Equally, as Guha (1989) has correctly pointed out there is a tendency amongst some Western social theorising about the environment to (selectively) read non-Western (usually Eastern) philosophical values and positions into Western ecological thinking, such as the association of 'deep ecology' with Buddhist or aspects of Hindu thinking, which does not pay attention to or respect the cultural specificity of those non-Western forms of thinking and ways of life. Those who wish, for example, to make a case for the harmony between Chinese, Indian or other Asian cultures and nature would do well to go to the original sources rather than to popular and selective Western interpretations of the religious, cultural or folk wisdom of these non-Western sources (Elvin, 1998). However, as Milton points out:

> The myth of primitive ecological wisdom, however misleading it may be, is useful in drawing attention to the fact that a concern to protect the environment from the effects of human activity need not be part of an oppositional ideology. It may be part of the cultural status quo, part of the way in which the members of a particular society has always understood their place in the world.
>
> (Milton, 1996: 33)

The Judaeo-Christian legacy

To the extent that theological debates about spiritual and worldly matters may be said to constitute a form of social theorising, we can say that Judaeo-Christianity was a limited, though none the less significant, reflection on the relationship between human society and the natural environment. The importance of beginning our analysis of the relationship between Western social theory and the environment with an examination of the Judaeo-Christian legacy cannot be overstated. For while many see Western societies and social theory as 'secular' or non-religious, it remains the case that exploring their Judaeo-Christian origins and contexts can be extremely illuminating. Generally speaking, it is more accurate and useful to describe Western societies as 'post-Christian'. What is meant by this is that, although it is no longer the case that these societies are deeply Christian in the way they were in the past, it is still the case that Christianity and

Christian values and perspectives have shaped and continue to be reflected in many of the practices, institutions and cultures of these 'secular' societies. This, as I hope to show below, is particularly the case with regard to Western social theorising about the environment.

For example, proof of the continuing power of Christian thinking could be seen in the anti-war arguments made by some Christian groups in America who ran publicity campaigns around the theme of 'What would Jesus drive?', which attempted to provoke Christians into reflecting on the connection between the US-led war and invasion of Iraq (and the environmental and human costs of that act), the oil resereves of Iraq as one of the main reasons for going to war, and how, by reducing their dependence on oil via buying smaller, more fuel-efficient cars, this could be seen as living according to their Christian principles. In particular, this Christian environmental campaign criticised the rise of fuel-inefficient SUV and light truck ownership and use in America. According to its website, 'The Lordship of Christ extends throughout every area of life. Nothing is excluded from his Lordship. This includes our transportation choice. This is why the question "What would Jesus drive?" is one that all Christians should ponder seriously. Obeying Jesus in our transportation choices is one of the great Christian obligations.' The campaign generated a great deal of publicity both in America and Europe but is also significant as a reminder of the continuing mobilising capacity of religious sentiment in Western societies.

In the Jewish tradition, the 'natural environment' was generally seen as something akin to 'wilderness' and against which human society had to struggle. However, at the same time there are more harmonious views on the interaction of humanity and nature. It is worth noting that Judaism in particular has much to say on the proper treatment of domesticated animals, and forbade the needless destruction of the environment even to subdue one's enemies (Swartz, 1996). According to Ives:

> The Torah orders the creation of green belts around cities (Numbers 35:4), and the laws against grafting diverse seeds and cross breeding animal species (Leviticus 19:19) can be understood in modern terms as concern for bio-diversity. Shabbat is a weekly rest for humans, animals and the natural world. We are called upon in Jewish law to offer blessings for all manner of natural phenomena (rainbow, lightning, shooting stars, the first blossoms of a tree, etc.). A most dramatic ecological gesture is *Shemita*, the seventh year rest for the environment, when all fields lie fallow. Maimonides declares that meditating on nature is one of the key ways a person can fulfil the commandment to 'love God with all your heart' (Mishne Torah, Yesodei Hatorah 2:2).
>
> (Ives, 2004: 2)

In contrast to some hunter-gatherer views of the environment, and more in keeping with the Mesopotamian view mentioned above, both the Jewish and Christian views of the environment were not of a 'giving environment' (Milton, 1996: 116–18). The idea of the 'giving environment' denotes the (often mis-leading) positive conception of the typical environment of hunter-gatherer peoples, in which people simply picked or procured what they needed from the abundant resources of their immediate environment, without much effort. In marked contrast the Judaeo-Christian attitude to the environment is a combination of a negative view of 'wilderness' (viewed as chaos and a threat to human social order), coupled with a deep sense of how the environment required intensive human labour and effort, such as agricultural and animal husbandry practices, in order that humans could survive and prosper from 'ungiving' and often hostile natural environments.

The latter idea of having to labour for a living in the world is directly related to the biblical story of Adam and Eve being banished from the Garden of Eden (which was similar to the 'giving environment' of some hunter-gatherer peoples) for having defied God and eaten fruit from the tree of knowledge. In banishing Adam and Eve from this comfortable environment in which all their needs were met without having to work (and in which they along with the beasts were vegetarian), God curses Adam and his descendants (i.e. all humans) to have to 'work by the sweat of his brow'. The importance of the Garden of Eden story, as the Christian creation story, is not whether it is 'true' or not. Rather its signifi-cance lies in its being one of the first systematic and most powerful stories or narratives about the relationship between humans and the environment. As such, we can say that this story constitutes an important attempt to theorise the environment and our proper relationship to it. It contains many of the elements that surface later in social theorising about the environment. These include: a particular conception of 'environment' and its status as the 'home' or 'proper place' of humans; the role of knowledge in how we think and ought to think about and interact with the environment; the distinction between a 'giving' and a 'non-giving' environment; the crucial role of human labour in our relationship to the environment; and finally, the dangers inherent in particular forms of thinking about and using the environment for humans. All of these, and others, are issues which arise in different forms of social theorising about the environment and will be discussed in later chapters.

Aspects of the roots of this Judaeo-Christian attitude towards the environment may be traced in part to ancient Greece and the story of Prometheus, the Greek hero who stole fire from the Gods and which symbolises humanity's triumph over nature viewed as an enemy or a denying force for humanity. According to Dryzek, 'In Greek mythology Prometheus stole fire from Zeus, and so vastly increased the human capacity to manipulate the world for human ends. Prometheans have an unlimited confidence in the ability of humans and their technologies to overcome any problems presented to them – including what can now be styled environmental problems' (1997: 45).

In the Christian Bible one can trace some of the roots of how the environment has been viewed and treated within Western society and social theory. Typically, people point to the passage in Genesis in which God orders Adam and Eve to 'dominate and subdue' the Earth and 'go forth and multiply', which demonstrates the extremely anthropocentric character of Christianity (see Box 2.1). This anthropocentrism within Christianity is an attitude to the nonhuman world in which the environment is viewed and valued instrumentally; that is, on this particular reading of the Bible, humans are permitted and indeed encouraged to use the environment and value it only insofar as it is useful to human ends or purposes. In other words, the environment has no value in itself (**intrinsic value**), but has **instrumental value**; that is, its value or worth is given by how useful or instrumental it is in fulfilling some purpose other than its own, or the needs or ends of some other entity. According to Lynn White Jr. (1967) in an influential essay entitled 'The Historical Roots of our Ecologic Crisis', on his reading of Christianity, what he calls the 'domination of nature' story and imperative in Genesis and biblical teaching makes Christianity the most anthropocentric of all religions. Indicative of the superiority of humans over the nonhuman world is not just that 'man' [*sic*] is created in the image of God, but God also gives Adam the power to name each creature. For White, 'Christianity . . . not only established a dualism of man and nature but also insisted that it is God's will that man exploit nature for his proper ends' (1967: 1205) and he concluded that 'we shall continue to have a worsening ecologic crisis until we reject the Christian axiom that nature has no reason for existence save to serve man' (1967: 1206). However, for White this did not mean the abandoning of a religious approach to thinking about our relationship to nature in general or Christianity in particular. White proposed that there were resources within Christianity which could remedy the 'domination' narrative. In particular he pointed to the teachings of St Francis of Assisi:

> The greatest spiritual revolutionary in Western history, Saint Francis, proposed what he thought was an alternative Christian view of nature and man's relation to it; he tried to substitute the idea of the equality of all creatures, including man, for the idea of man's limitless rule of creation. He failed. Both our present science and our present technology are so tinctured with orthodox Christian arrogance toward nature that no solution for our ecologic crisis can be expected from them alone. Since the roots of our trouble are so largely religious, the remedy must also be essentially religious, whether we call it that or not. We

must rethink and re-feel our nature and destiny. The profoundly religious, but
heretical, sense of the primitive Franciscans for the spiritual autonomy of all
parts of nature may point a direction. I propose Francis as a patron saint for
ecologists.

(White, 1967: 1207)

In opposition to the 'domination' view, Passmore (1980) also suggests an
alternative interpretation of Christian teaching about the environment. According
to him, there is a 'stewardship tradition' derived from Christian thought in which
the natural environment was 'God's Creation'. This stewardship tradition
pre-dates Christianity, and its origins may be found in the Greek philosopher
Plato who in the *Phaedrus* wrote that '"It is everywhere the responsibility of
the animate to look after the inanimate". Man . . . is sent to earth by God "to
administer earthly things", to care for them in God's name' (quoted in Passmore,
1980: 28; see Box 2.1).

This was also the Jewish position. As Swartz points out, 'And though their efforts
to tame the land, to make it more productive and more dependable, were often
marvels of ingenuity, they understood, as well, the limits to their mastery – for
they knew God as Sovereign of the Land, and . . . they acknowledged God's
ownership' (1996: 88). Since the natural world had not been made by humans, it
was not their exclusive property to treat and use as they wished. Within the
stewardship tradition, rather than the nonhuman world being made for humans
(a position which did eventually come to dominate Western views of the environment), as stewards of God's creation humans were in a sense made for the
nonhuman world, or rather they were God's 'managers' or stewards holding
responsibility for God's property. This meant that there were certain rules
governing how the environment, its plants, animals and so on were to be treated.
One implication of the stewardship view, and one which will be picked up in the
conclusion, is that as stewards of creation, humans had an obligation to pass
on the natural environment to future generations (see Box 2.1). An important
point here is whether this obligation meant passing on the environment in
the same state as they found it, or whether humans were obliged to 'improve' the
natural environment. A good example of this sustainability injunction (long
before the concept of 'sustainability' was conceived) is the following from the
chancellor of Reichenhall, an old Bavarian salt-works city in 1661, who stated:
'God created the woodlands for the salt-water spring, in order that the woodlands
might continue eternally like the spring. Accordingly shall the men behave: They
shall not cut down the old trees before the young trees have grown up' (von Bulow,
1962: 159).

As Passmore (1980) and others (e.g. Pepper, 1984) have pointed out, this Christian
stewardship view, in which social–environmental interaction was governed by

Box 2.1 Judaeo-Christian theory and the environment

'Then God said, "Let us make man in our image, after our likeness and let them have dominion over the fish of the sea, and over the birds of the air, and over the cattle, and over all the earth, and over every creeping thing that creeps upon the earth." So God created man in his own image, in the image of God he created him; male and female he created them. And God blessed them, and God said to them, "Be fruitful and multiply, fill the earth and subdue it; and have dominion over the fish of the sea and over the birds of the air and over every living thing that moves upon the earth." And God said, "Behold, I have given you every plant yielding seed upon the face of all the earth, and every tree with seed in its fruit; you shall have them for food"' (Genesis 1: 26-9).

The domination of nature interpretation

'Christianity in opposing and destroying pagan animism made it possible to exploit nature in a mood of indifference to the feelings of natural objects' (White, 1967: 1206).

'[I]n the Christian separation of man from the animals and the Christian view that nature was made for man, there lie the seeds of an attitude to nature far more properly describable as "arrogant"' (Passmore, 1980: 12).

The stewardship tradition

'Genesis . . . makes this duty [of 'man' towards all nature and all life] clear when it tells us that God put Adam into the Garden of Eden "to dress and to keep it", i.e. to manage and protect it' (Passmore, 1980: 29).

'The tradition of stewardship legitimates the reordering of the non-human world in the interests of human welfare provided this is balanced with a sufficient regard for obligations to conserve the natural world, to protect the moral interests of wild and domesticated animals, and to regard the interests of future generations as well as those of presently existing persons' (Northcott, 1996: 129).

'The end of man's creation was, that he should be the viceroy of the great God of heaven and earth in this inferior world; his steward . . . bailiff or farmer of this goodly farm of the lower world' (seventeenth-century Chief Justice, Sir Matthew Hale, quoted in in Passmore, 1980: 30).

The perfection of nature thesis

'The word "nature" derives . . . from the Latin *nascere*, with such meanings as "to be born", "to come into being". Its etymology suggests, that is, the embryonic, the potential rather than the actual. We speak, in this spirit, of an area still in something like its original condition as "not yet developed". To "develop land", on this way

continued

of looking at man's relationship to nature, is to actualise its potentialities, to bring to light what it has in itself to become, and this means to perfect it' (Passmore, 1980: 32).

'The view that man has responsibility for handing over to his descendants a nature made more fruitful by his efforts is not . . . an entirely contemporary innovation, or an attempt to appeal to moral feelings which simply do not exist: it has deeper roots in Western civilisation' (Passmore, 1980: 32).

religious considerations, began to give way to a more interventionist and anthropocentric viewpoint in which humans could use their God-given powers of creativity and ingenuity to 'perfect' nature for 'the Glory of God' (see Box 2.1). The 'perfection of nature' idea resulted in providing a religious justification for what we would now call the 'development' of the environment. This transformation of the natural world by humans in the West took many forms, from the creation of geometrically symmetrical landscape gardens of the famous eighteenth-century landscaper 'Capability' Brown (in contrast to the messy, irregular patterns of 'wild' or 'natural' environments), to the straightening of rivers and the draining of swamps.

Another extremely important contribution which Christian thinking made to theorising the environment is the 'Great Chain of Being' (though strictly speaking this pre-dates Christianity and may be found in other religions and non-religious thought). The essence of this view, as the name suggests, was that the world was made up of a hierarchical set of relationships with God at the top of the chain and clay/dirt at the bottom, with angels, men, women, animals and plants in between (see Figure 2.1). Thomas Aquinas gave clear expression to this idea:

> As we observe . . . imperfect beings serve the needs of more noble beings; plants draw their nutrients from the earth, animals feed on plants, and these in turn serve man's use. We conclude, then, that lifeless beings exist for living beings, plants for animals, and the latter for man. . . . The whole of material nature exists for man, inasmuch as he is a rational animal. . . . We believe all corporeal things to have been made for man's sake.
>
> (Quoted in Kinsley, 1996: 110)

However, it is interesting to note that in both Jewish and Christian theology there have been those who have rejected this hierarchical view. For example, the Jewish kabbalist (mystic) Maimonides declared: 'It should not be believed that all the beings exist for the sake of the existence of humanity. On the contrary, all the beings too have been intended for their own sakes, and not for the sake of something else' (in Swartz, 1996: 93). A similar argument was advanced by

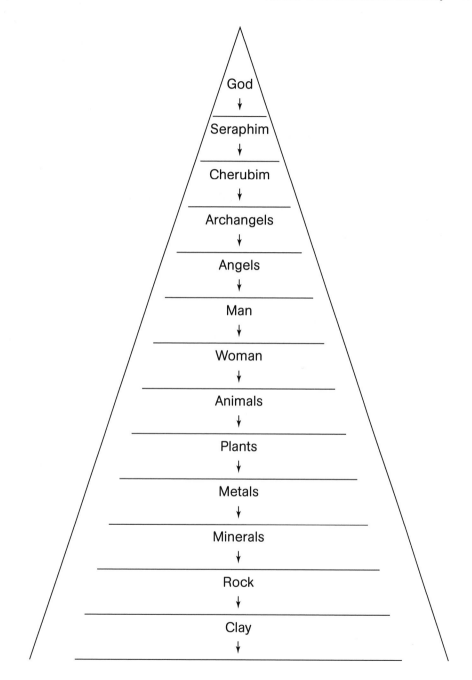

Figure 2.1 The Christian 'Great Chain of Being'

Source: Adapted from Pepper (1984)

St Francis of Assisi who famously preached to the animals and developed a Christian pantheism in which the natural environment partook of the divinity and grace of God, and was not simply a set of spiritually meaningless or empty resources to be used. For his ecological awareness and earth-sensitive theology in 1979 the Catholic Church made St Francis the patron saint of animals and ecology.

Despite these and other counter-currents, the dominant attitude of Judaeo-Christianity to the environment has been one based on the Great Chain of Being. The hierarchical arrangement of entities implied different grades of value or importance such that those above were more valuable/important than those below. Often there are also gradations within these broad categories, such that certain metals, for example, were more valuable than others (gold was higher than copper), or within the animal category (cows were higher than rats). This divine order, in which there is a place for every living and non-living thing (both natural and supernatural), is something which still frames how many people view and think about the natural environment. Both in everyday life and in social theory we find that this Great Chain of Being idea operates, such that humans are regarded as 'higher' or 'more important' than animals or plants, and indeed this also led to views in which certain human beings were 'higher' than others (white European males being higher or superior to all others). More popular representations of this idea is the commonplace designation of certain animals as superior to others such as the lion as the 'king of the jungle', as in the popular children's animation, *The Lion King*. Or as the pigs in George Orwell's *Animal Farm* might put it, 'All animals are equal, but some animals are more equal than others.'

The Great Chain of Being idea linked with the 'perfection of nature' view in that those higher up the chain could legitimately transform and 'improve', 'perfect' or manage those entities below. Thus human transformation of the nonhuman environment was permissible within the framework of the Great Chain of Being. As the eighteenth century drew near the Christian legitimacy associated with this interventionist-instrumental view of the environment became increasingly difficult to sustain in the face of growing intellectual and practical challenges to the Christian worldview. Thus by the time of the Enlightenment and the beginnings of the industrial revolution in Europe, human use of the environment, particularly in agriculture, rudimentary commercial manufacturing, landscaping, the scientific harvesting of forests and the creation of waterways, had largely ceased to be legitimated by the idea of 'God's creation' which implied that there were moral or normative limits to what humans could do to the environment. Human transformation and use of the natural environment became increasingly divorced from a strict medieval Christian framework from the sixteenth to the eighteenth centuries. If the environment was becoming increasingly vulnerable to human manipulation and transformation from the sixteenth century on, this vulnerability became outright exploitation with the coming of the Enlightenment era in the eighteenth century. By exploitation is meant that the use of the environment was less and less regulated by moral (religiously based) considerations, and was viewed increasingly in non-moral, economic terms. A consequence of this development was an erosion of the boundary between legitimate 'use' and illegitimate 'abuse', such that the critria for decisions about how humans interact with the environment were 'profit' or 'usefulness' rather than ethical notions of 'right' and 'wrong'.

The Enlightenment, environment and social theory

'The Enlightenment' (which may also be termed 'modernity') is often understood as the series of interconnected and sometimes radical changes that took place within Europe in the mid- to late eighteenth century, across numerous fields of human thought and action. There is no one exact date to which we can point as the dawning of the 'age of reason', nor is there one writer or school of social thought to which we can trace the exact origins of the momentous changes in European intellectual, political, economic and social life which occurred at this time. In terms of tracing and understanding the historical (and contemporary) relationship between social theory and the environment, the Enlightenment is of central significance. Not only do the origins of many current environmental problems lie in the Enlightenment (particularly the industrial revolution), but some of the roots of 'green' critiques of and alternatives to industrialism also lie in the various reactions to the Enlightenment. Hence the Enlightenment represents an important turning point in the place of the environment within social theory. As Porter points out, 'The Enlightenment believed people could improve themselves by improving nature, offering a programme of progress through science, technology and industry' (1994: 174). A typical example of the profound belief in progress and in the improvement of humanity by the application of reason (particularly scientific knowledge and its technological application) in Enlightenment thinking is the following passage from the French philosopher Condorcet:

> A very small amount of ground will be able to produce a great quantity of supplies of greater utility or higher quality; more goods will be obtained for a smaller outlay; the manufacture of articles will be achieved with less wastage in raw materials and will make better use of them. . . . The improvement of medical practice, which will become more efficacious with the progress of reason and social order, will mean the end of infectious and hereditary diseases and illnesses brought on by climate, food, or working conditions. It is

reasonable to hope that all other diseases may likewise disappear as their
distant causes are discovered. Would it be absurd, then, to suppose that this
perfection of the humanity species might be capable of infinite progress?

(Condorcet, 1995: 35–7)

The important point to note about the Enlightenment is that human progress and
improvement is premised on the more effective exploitation of the natural
environment. As the passage from Condorcet above shows, Enlightenment social
theory had at its heart the exploitation of the natural environment by the use
of scientific knowledge and the application of technology to industrial production. For purposes of exposition, what I intend to do in this section is simplify the
Enlightenment into two component aspects, namely the industrial revolution and
the democratic revolution.

The industrial revolution

By the term 'industrial revolution' is meant the various changes that took place
in European economic life both in terms of concepts, theories and ideas and
in terms of actual practice during the period from about the sixteenth to the
nineteenth centuries, which laid the basis for the emergence and development of
modern industrial society and a capitalist organisation of the economy. Britain
is often seen as the cradle of the industrial revolution, in that it was the first
country to transform itself along industrial lines. As the 'workshop of the world',
Britain exhibited many of the features which later industrial societies would
develop, and Britain became the model for industrialisation.

Central to the industrial revolution is a particularly instrumental attitude towards
the natural environment. The environment was seen as a collection of means for
human ends, raw materials for the factories, machines and new productive technologies which were being invented. Science was seen as unlocking the secrets
of nature, developing new insights into its inner workings, and in conjunction
with technology provided more effective ways in which humans could exploit
it. In this way the natural environment became 'disenchanted', where once it was
a meaningful order now under the cold, scientific light of reason, it simply became
a collection of means (Barry, 1993). That is, whereas once the environment was
variously seen as 'enchanted' (as in folk legends) or imbued with spiritual significance (as in Christianity where the natural environment was 'God's creation'),
with the industrial revolution, the environment was transformed and reduced
to being a store of raw materials for human economic purposes. By 'disenchant-
ing' the natural environment is meant the draining or eroding of meaning or
significance from it, other than its status as a set of means for human ends. Other
new forms of knowledge included the emergence of 'political economy' as the

systematic study of the new capitalist economic system, with its novel free market, private property-based economy. At the same time, the industrial revolution denoted radical changes in the type of economy and form of social organisation. The shift from a largely rural, agricultural economy to an urban, industrial and class-based economy created a new type of society, a 'modern' one, based on manufacturing, technological innovations, machinery and a complex specialisation and division of labour, in comparison to the feudal social order which preceded it.

Like all social change, the industrial revolution created 'winners' and 'losers', and was accompanied by social upheaval, unrest, suffering and pain. Chief among the 'losers' were the peasantry or 'commoners' who as a result of the enclosure movement (the privatisation of land to which 'commoners' once had rights of access) were forced into the emerging industrial urban areas and became the industrial working class, who resisted the erosion of their way of life, status and relative independence. As a popular anti-enclosure rhymn had it, 'The Law doth punish man or woman who steals the goose from off the common, but lets the greater felon loose, who steals the common from the goose'. Another form of resistance to the industrial revolution were the 'Luddites' in the early nineteenth century (1811–16) who smashed machinery, the introduction of which was causing unemployment and thus great social hardship. This will be discussed in Chapter 6.

As the industrial revolution continued there was also what one may call a Romantic and negative reaction, one that is particularly important for later social theorising about the environment in general, and 'green' social theory in particular. This Romantic backlash against the industrial revolution was motivated by how the latter was destroying and disfiguring the natural environment, turning once beautiful landscapes into ugly, overcrowded cities, polluting factories and mining operations. 'Was Jerusalem builded here, among these dark, satanic mills?' as the poet William Blake graphically put it in reference to the new mines, mills and factories which were disfiguring the English countryside, and which were at the heart of this new and polluting economic system. From the Romantic perspective there was also something arrogant about human domination of the natural world. As Thomas Carlyle put it, 'For all earthly, and some unearthly purposes, we have machines and mechanical furtherances. . . . We remove mountains, and make seas our smooth highway; nothing can resist us. We war with rude Nature; and, by our resistless engines, come off always victorious, and loaded with spoils' (in Clayre, 1977: 229). This theme of a 'war' against nature is something that will also be discussed in Chapter 4.

Industrial capitalism launched the Modern Age, and ripped into the available raw materials with no regard for environmental consequences. By the 19th century Britain had become the 'workshop of the world'.

Having exhausted the supply of fresh trees the industrialists solved the fuel shortage with *fossilised* trees – coal.

The steam-engine churned. The cotton-mills hummed. The iron industry boomed. New coal-pits were opened. Towns grew into cities and village workshops into factories.

Canals, roads and railway lines criss-crossed the country. Britain throbbed with industrial activity, to quote the school history books. But what else happened?

Figure 2.2 'Ecology and Industrial Capitalism'

Source: Croall and Rankin (1981)

The democratic revolution

By the democratic revolution is meant the radical changes that took place politically during the late eighteenth and nineteenth centuries in theory and practice. The two key historical events here are the American Revolution (1775) and the French Revolution (1789). The main aspects of the democratic revolution concerned the principle of popular government – that is, government by the people, of the people and for the people – to replace rule by unelected monarchies, aristocrats and the Church. The slogan of the French Revolution ('liberty, equality, fraternity') neatly sums up the essence of the democratic imperative of the Enlightenment, banishing the divine right of kings and the authority of organised religion. Other salient aspects of the democratic imperative of the Enlightenment included the use of the vocabulary of rights in political, social and moral thought, the increasing emphasis on the individual (both as citizen and producer/ consumer); the emergence of representative government and liberal democracy; the establishment of constitutions, the separation of state powers and the rule of law rather than of man, and finally the creation of nation-states. Writers such as Thomas Paine, Jean-Jacques Rousseau, Voltaire, William Godwin and Montesquieu wrote about, justified, supported and/or took part in this democratic revolution.

While less obvious than in the case of the aims of the industrial revolution, the democratic revolution was equally based on a particularly instrumental attitude towards and use of the natural environment. In the first case most supporters and theorists of democracy argued that this new form of government required material wealth (based on the exploitation of the natural world), and in this sense the industrial revolution was a necessary condition for the flowering of democratic politics. As Tocqueville noted:

> General prosperity is favourable to the stability of all governments, but more particularly of a democratic one, which depends upon the will of the majority, and especially upon the will of that portion of the community which is most exposed to want. When the people rule, they must be rendered happy or they will overturn the state: and misery stimulates them to those excesses to which ambition rouses kings.
>
> (1956: 129–30)

Second, the democratic revolution was a property-owning democracy in the sense that democratic rights were not extended to everyone. Only men with property were permitted to vote, and an important implication of this is that it served to further legitimate the idea of private property in land. That is, extending the right to vote implied extending private property in land to more and more people. This, of course, meant not only regarding the natural environment as raw materials but also as private and transferable property, which could be traded, bought and sold like any commodity in the emerging market economy. This property-based view of democracy was especially clear in the American case, largely because the American democratic revolution was strongly grounded in the political philosophy of John Locke for whom the goal of government was primarily to protect life, individual liberty and private property. Locke's ideas will be discussed in Chapter 3. As the American Declaration of Independence puts it, 'We hold these Truths to be self-evident, that all Men are created equal, that they are endowed by their Creator with certain inalienable Right, that among these are Life, Liberty, and the Pursuit of Happiness'. According to Kramnick, in the context of the Enlightenment, 'Government's purpose was to serve self-interest, to enable individuals to enjoy peacefully their rights to life, liberty and property, not to serve the glory of God or dynasties' (1995: xvi).

Another dimension to this land-ownership issue was the special status of agricultural life and those who work the land. According to William Jefferson, one of the founding fathers of America, 'cultivators of the earth are the most virtuous and independent of citizens', adding, in a statement which echoes the anti-urbanism and anti-commercialism of some green thinking, that 'Merchants have no country. The mere spot they stand on does not constitute so strong an attachment as that from which they draw their gains' (quoted in Miller, 1988: 207, 210–11). Echoing some of the issues raised in the previous chapter concerning the status of countryside and town/city, Jefferson thought that to dwell in the country was to dwell in virtue, while living in the city, separated from nature, was to risk corruption (Rennie-Short, 1991).

Thus the Enlightenment or modernity is an absolutely key moment in the relationship between social theory and the environment, since it represented a radical change both in theory and practice about how the natural environment was viewed, valued, used and conceptualised.

Conclusion

This chapter has explored the historical relationship between social theory and the environment by outlining two opposing ways in which the environment has been theorised in Western social thought. On the one hand, we have the religious approach of Judaeo-Christianity, and the various ways in which it has theorised the environment and the proper relation between humans and the environment. These include the narrative of the Garden of Eden, the competing interpretations of the Christian attitude to the environment (namely the domination of nature view and the stewardship tradition), and finally the idea of the 'Great Chain of Being'. On the other, we have the profoundly secular approaches to theorising the environment offered by the Enlightenment, viewed as a combination of two revolutions: the industrial and democratic. An awareness of both Judaeo-Christianity and the Enlightenment is necessary as historical and conceptual legacies and frameworks to understand the character of social theory and the environment.

Summary points

- Up until the Enlightenment, most social theorising about the environment took the form of religious, mythical or 'traditional' accounts of the origins of the natural world, humans, and the 'proper' relationship between the two.
- Analysing the Judaeo-Christian worldview and its teachings is important in examining the historical origins of the relationship between Western social theory and the environment.
- There are competing views about the 'ecological' character of Christianity. On the one hand, there is the 'domination of nature' interpretation of Genesis in the Bible. On the other, there is the 'stewardship' view where humans are stewards or caretakers (not owners) of 'God's creation', and also the 'perfection of nature' thesis, where humans are obliged or encouraged to 'perfect' or 'develop' nature.
- The Enlightenment (or 'modernity') marks a decisive change in how European civilisation thought about and used the natural environment. With the advent of the industrial revolution, nature became 'disenchanted'.
- The historical origins of social theorising about the environment may be traced to the industrial and democratic revolutions, and reactions to them.

Further reading

For authoritative and scholarly accounts of the theoretical history of Western society, social theory and the environment see John Passmore's excellent *Man's Responsibility for Nature*, London: Duckworth (2nd edn), 1980, and Clarence Glacken's magisterial (and extremely long!) *Traces on the Rhodian Shore*, Berkeley: University of California Press, 1967. A more focused account of the history of thinking about the environment may be found in Donald Worster's *Nature's Economy: A History of Ecological Ideas*, Cambridge: Cambridge University Press (2nd edn), 1994. Keith Thomas' *Man and the Natural World: Changing Attitudes in England 1500–1800*, Harmondsworth: Penguin, 1983, is a very readable account which looks at some of the historical changes in England that preceded the Enlightenment.

A useful overview of the historical place of the environment within Western thought is Derek Wall's reader *Green History*, London: Routledge, 1994, which contains many edited original articles, and Alasdair Clayre's edited volume *Nature and Industrialization*, Oxford: Oxford University Press, 1977. David Pepper has written two books which cover some of the issues discussed in this chapter: *The Roots of Modern Environmentalism*, London: Croom Helm, 1984, and *Modern Environmentalism: An Introduction*, London: Routledge, 1996.

An excellent anthology exploring the relationship between religion and nature is Roger Gottlieb's edited volume, *This Sacred Earth: Religion, Nature, Environment*, London: Routledge, 1996, while for a more in-depth analysis of the Christian perspective see Michael Northcott's *The Environment and Christian Ethics*, Cambridge: Cambridge University Press, 1996, and Michael Barnes (ed.), *An Ecology of the Spirit*, Lanham, MD: University Press of America, 1994. On Islam and its attitudes to the environment see F. Khalid and J. O'Brien (eds), *Islam and Ecology*, London: Cassell, 1992; for Hinduism see R. Prime, *Hinduism and Ecology*, London: Cassell, 1992; on Buddhism see M.E. Tucker and D.R. Williams (eds), *Buddhism and Ecology*, Cambridge, MA: Harvard University Press, 1999.

For an introduction to some of the non-Western approaches to theorising the environment, see Peter Marshall's *Nature's Web: Rethinking Our Place on Earth*, London: Cassell, 1995; Baird Callicott's *Earth's Insights: A Multicultural Survey of Ecological Ethics from the Mediterranean Basin to the Australian Outback*, Berkeley: University of California Press, 1994, and Helaine Selin's edited volume *Nature Across Cultures: Views of Nature and the Environment in Non-Western Cultures*, Boston, MA: Kluwer Academic, 2003.

3 The uses of 'nature' and the nonhuman world in social theory

Pre-Enlightenment and Enlightenment accounts

Key issues

- Pre-nineteenth-century social readings of 'nature'.
- Forerunners and critics of modernity: environment, 'the state of nature' and social theory.
- Nineteenth-century social theory and the nonhuman world.
- Progressive and reactionary social theorising about nature.
- Social theory and natural evolution: Darwin, Spencer and 'Social Darwinism'.
- Natural harmony: Kropotkin and the anarchist reading of the environment.
- Marxist social theory and the environment.

Introduction

The aim of this chapter is to outline some of the ways in which the nonhuman world, its entities, processes and principles have been used (and abused) within the history of social theory. Of particular importance is the ways in which social theorists have appealed to some notion of the 'natural order' or 'nature' to justify, legitimate or illustrate their theories about and prescriptions for the social order. Just as religious thought both in Europe and elsewhere looked to nature for metaphors and lessons for illustrations of God's laws or plans for humans, modern social theory since the Enlightenment has also made both positive and negative references to the nonhuman world. The idea of nature as a 'text', such as 'God's book', or like a book from which we can read, if we know (through science or mystical experience) the language of nature or source of meaning (as opposed to simply a store of means to human ends), is something that has a very long history and is an enduring feature of human thought.

Pre-nineteenth-century social readings of 'nature'

The 'scientific revolution' of the sixteenth to eighteenth centuries constitutes one of the significant backdrops against which social theorising about the nonhuman world took place. This period of European history is also concurrent with the beginnings of modern industrial capitalism, as Pepper (1996: 124) notes. The revolution in how the natural world was understood, its codification and use of mathematical methodology, rigorous modes of inquiry, and what one may call the scientific project of 'dominating' nature and finding out *her* secrets (in Francis Bacon's terms), constitutes the genesis both of 'modern science' and the modern 'worldview'. Unlike the pre-modern, medieval worldview which was based on Christian cosmology, the idea of a 'Great Chain of Being' discussed earlier, and an organistic view of the Earth, the 'modern' worldview was that the Earth is like a clock which is understandable and intelligible to human reason by the use of scientific modes of inquiry. This *mechanistic* conception of nature was at the heart of the emerging modern worldview, and laid the grounds for the Enlightenment and the industrial revolution. Indeed, this mechanistic view is still alive today, and may be seen in advertisements such as BUPA's (a British private medical insurance company) which states that 'Your body is a wonderful machine', itself a modern echo of Rene Descartes' view of the body (human or nonhuman) as a machine (Synnott, 1992). The important point to note is how attitudes towards the nonhuman world were intimately connected to particular ideas about progress, social and historical evolution, and arguments about the best organisation of society.

Forerunners and critics of modernity: environment, 'the state of nature' and social theory

Within social and political theory from the sixteenth century on, the idea of a 'state of nature' was a common device used to illustrate arguments about how the social order ought to be arranged, what the 'good society' is, and how political and social change should be viewed. This section looks at three important political thinkers to illustrate some of the different ways in which the environment was theorised in early social theory.

The 'state of nature' was a device used by various theorists to illustrate, explain and justify their particular social theories. Typically, the state of nature referred to a stage of human evolution as a 'pre-social' state; that is, how humans were prior to the creation of society, the state, social institutions, rules, principles and regulations. Generally speaking this stage of human evolution was regarded as inferior to and depicted a lower level preceding the 'civilised' stage at which human society is created and institutions such as the state, the monarchy (and the market) are founded.

It implied a number of related propositions:

- there is a natural order, with a corresponding 'natural law';
- this natural order can be seen in the rest of nature;
- humans ought to follow this natural order.

Thomas Hobbes

For the seventeenth-century political philosopher Thomas Hobbes (1588–1679), later nineteenth- and twentieth-century anarchist views of humans as *naturally* harmonious, co-operative and not requiring a coercive state to impose social order, based on a reading of nature as co-operative, would have been heresy. For Hobbes, life in the 'state of nature' was 'solitary, poore, nasty, brutish and short', not the natural harmony and co-operation that anarchists thought existed within human society prior to the creation of the state. In this pre-state stage of social development, argued Hobbes, 'human society' as such did not exist, property was not secure, and individuals unfortunate enough to exist in such a state were continually in fear of their lives, not able to make plans for the future, and completely lacking security. Security and insecurity were foundational for Hobbes' thought and the basis upon which he outlined and defended the need for a centralised, authoritive 'soverign' power and nation-state.

In this way, for Hobbes the idea of 'natural society' was not something he regarded as a positive stage in human evolution (as it was for Rousseau), but rather was a primitive and what later social theorists in the eighteenth century would call 'rude state of society' (as in rudimentary or basic). That is, humans living in a natural state were living in an uncivilised and backward stage of social development.

However, as John Meyer has pointed out in his authoritive analysis of Hobbes, 'Hobbes presents the state of nature as "natural", yet he incorporates a number of interpretations and judgements about politics into this "natural condition"' (2001: 58, 83). In other words, as with much of social theorising about the environment, Hobbes 'reads into' nature what he wants to 'read out' of it, thus offering a reading of the natural environment which is ideological and value-based and not 'objective' or 'neutral'. In other words, Hobbes' defence of the need for a strong state and sovereign is based not on some quasi-scientific or objective acccount of the 'state of nature' but on certain values and principles which are not 'natural' but social and political.

John Locke

John Locke (1632–1704), another English social and political thinker, had a more benign view of human social life within a 'state of nature'. However, he did suggest that such a rudimentary state of society was one which, while not perhaps 'nasty, brutish and short', as Hobbes had so graphically suggested, was very basic, poor and in need of improvement and advancement. The way he suggested this be done had profound effects on how the nonhuman environment was viewed by later social theorists. Without going into the complexities of Locke's theory, one of his main arguments was that humans could claim parts of the nonhuman environment as their own, private and exclusive property, with the proviso that there was 'enough and as good' left for others. As he puts it in his famous book, *Two Treatises on Government*:

> Nor was this appropriation of any parcel of *land*, by improving it, any prejudice to any other man, *since there was still enough, and as good left*; and more than the as yet unprovided could use. So that, in effect, there was never the less for others because of his inclosure for himself: for he that leaves as much as another can make use of, does as good as take nothing at all. No body could consider himself injured by the drinking of another man, though he took a good draught, who had a whole river of the same water left to quench his thirst: and the case of land and water, where there is enough, is perfectly the same.
>
> (Locke, 1959: Book II, v. 33; emphasis added)

In this way, Locke is one of the first theorists to rationally justify an instrumental and possessive attitude towards and valuation of the nonhuman environment. In his view, the nonhuman environment, untouched by humans, is 'valueless'. It is also important to point out that Locke's defence of private property in the environment (initially land rights but also private ownership of other produced goods and services) was defended in the name of liberty as well as productivity in the sense that by possessing a natural right to private property, 'man' had a sphere of freedom from fellow citizens and the soverign or the public authorities.

Thus, coupled with Hobbes' view of the 'state of nature' as something to be viewed with horror, Locke's view of nature as worthless in its untouched state, and proposing private property as the most productive way for humans to relate to and use nature, together produced the dominant attitude of social and political theory to the nonhuman environment up until the modern era.

Jean-Jacques Rousseau

However, in Jean-Jacques Rousseau (1712–78), we find a social theorist who differs from both Hobbes and Locke in his assessment of the state of nature; he is also one of the first and most powerful critics of the modern worldview, its social arrangements, aims, principles and animating goal of progress. In his *Discourse on the Origin of Inequality* he explicitly criticised Locke and Hobbes for depicting a 'state of natute' which did not examine the effects of the society which they held developed from that state of nature on human character. As he put it, Hobbes and Locke 'spoke about savage man, and it was civil man that they depicted' (Rousseau, 1997: 38), but agreed with them about the foundational character of private property for modern 'civilised society' but, unlike them, he also points out that private property (and thus the society it supports) is unequal and based on power diffentials and class-based rivalries (in this he anticipates Marx and Kropotkin discussed below).

Going against the grain of Enlightenment thought, Rousseau argued that 'man' is naturally co-operative, and that 'primitive societies like those of the indigenous Americans and Africans were the "best for man"; civilization, far from being a boon, is always accompanied by costs that are greater than the benefits' (Masters, 1991: 456). Rousseau's idea of the 'noble savage', his positive view of life in the state of nature, made him an early critic of the Enlightenment, and he may be seen as a precursor of later Romantic and green ideas of primitive peoples as paragons of ecological wisdom. By the 'noble savage', Rousseau, going against the received view of the status of 'savage' non-civilised peoples, suggested that this pre-civilised stage of human development, and human character, was in fact more virtuous, morally good and admirable than the so-called 'advanced' civilised and cultured stage of social advancement. Writing in 1851, Rousseau declared that 'Before art had new moulded our behaviours, and taught our passions to talk an affected language, our manners were indeed rustic, but sincere and natural' (1995: 365), and also that 'our minds have been corrupted in proportion as our arts and sciences have made advances towards their perfection' (1995: 367).

Rousseau was one of the first in this period to reverse, or at least challenge, the view that the 'Natural' was inferior to the 'Artificial', and that a less complex society was necessarily inferior to an advanced civilised society such as those which existed in eighteenth-century Europe. Going against the self-styled 'progressive' spirit of the age, he viewed the emerging modern European capitalist societies in a rather different light from those social theorists, economists, philosophers and political theorists who viewed the Enlightenment, science, technology and industrial development as unqualifiedly a positive step forward in human social evolution. Rousseau questioned the 'progressive' character of civilised society as representing an advance over previous stages of human social evolution. In this he anticipated a key aspect of the Romantic critical reaction to the industrialism in the nineteenth century, and his critique of modernity and its conception of progress is something that is at the heart of the later emergence of green social and political thought in the twentieth century. At the same time, his use of an evolutionary framework anticipated the work of Charles Darwin in the nineteenth century, discussed below.

Rousseau's critique of the 'artificial' as the opposite of the 'natural', and his laying the blame for social ills on the corrupting effects of 'civilisation', constitutes one of the first critiques of the Enlightenment from a 'green' perspective. For Rousseau, nature and the natural environment represented innocence, authenticity and wholesomeness against the corrupting effects of urban, sophisticated, civilised life (Wingrove, 2000). As Andrew Biro has put it, 'Rousseau from his earliest writings notes that alienation from nature is a chief source of human misery' (2005: 60). In this he was going against the prevailing notion of social evolution from primitive to agricultural, to city-states and on to the eighteenth-century commercial society and modern nation-states, as expressed by the dominant forms of Enlightenment social theory.

In Rousseau's thought we find echoes of the different values attached to different forms of environment in social theory, as outlined in Chapter 2. In depicting a largely rural, agricultural society of 'rough equality' and democratic republican government, one can see that he was using an idea of the environment as countryside/garden as morally (and politically) superior to the large, urbanised cities and forms of government of his time. Again in contrast to Locke and Hobbes, Rousseau questioned both inevitability and desirability of socio-economic inequalities as a necessary 'cost' of social evolution from the 'state of nature' to 'civilised society', preferring a state of 'rough equality' instead where no one was rich enough to buy another person, nor no one poor enough that they were forced to sell themselves. At the same time, his praise of the 'noble savage' implied a positive assessment of 'wilderness' and those who lived there, and may also be seen as a positive endorsement of 'non-Western' cultures and ways of life.

Nineteenth-century social theory and the nonhuman world

This section looks at the ways in which some of the main proponents of nineteenth-century social theory theorised about the nonhuman world. However, before proceeding it would be useful to briefly 'set the scene' by describing the intellectual and social context of the nineteenth century.

While there had been a Romantic 'backlash' or reaction against the industrial revolution, as discussed in the introduction, this constituted a minority opinion. This Romantic reaction against the industrial revolution took the form of an aesthetic-cultural critique which highlighted the devastating effects of this revolution on the natural world and settled, traditional, rural ways of life. Leading proponents of this Romantic backlash included poets, writers and artists such as William Blake, Percy Shelley and William Wordsworth. Part of the rejection of the new industrial age rested on a suspicion and outright mistrust of the application of science and technology to the natural and social worlds. Like any great social change, the industrial revolution in Europe challenged old ways of thinking as well as of doing things, and equally like any change produced winners and losers, those who agreed and those who disagreed with the changes. By and large the nineteenth century could be called the first 'modern' century on account of the widespread changes in manufacturing, social life, culture, politics, economic organisation and above all else in the relationship between human society and the natural environment increasingly understood and mediated by new developments in scientific knowledge and technological innovations.

The application of scientific knowledge did not stop at the level of how humans could more productively exploit the natural world. In keeping with the Enlightenment belief in the ability of human reason to explain and solve almost everything, the nineteenth century also witnessed the rapid (and often indis-criminate) application of knowledge gained from the study of the physical world to the study and organisation of human society. The self-evident success and explanatory power of scientific knowledge convinced many social theorists that one had only to apply the methods of inquiry used in natural science to the study of society. In this way the desire for the study of society to be 'scientific' led to the birth of the 'social sciences', and also led to the first attempts to 'read off' how society is, and ought to be, from observations of the natural world. By 'scientific' is meant the systematic study of phenomena, seeking universal-isations, generalisable and law-like principles of explanation of cause and effect, using testable hypotheses which can be verified or falsified empirically by observation and/or experimentation.

In the nineteenth century, for a theory or theorist to be considered 'unscientific' was to be deemed as falling below the basic standards of social inquiry. When looking at the nineteenth century one must remember that most of the disciplines into which human knowledge and study are divided either emerged or took their present form at this time.

This was particularly the case with the 'social sciences'. Modern economic science, for example, emerged from the older tradition of classical political economy; sociology, 'the science of human beings and their behaviour', began to take shape towards the end of the century; while other disciplines such as law and history began to model themselves as 'scientific disciplines' and separated themselves from older disciplines such as philosophy and theology, which themselves began to develop into their own distinct academic disciplines. All in all the nineteenth century was a remarkable period at every level in Europe in ways of thinking about the natural and social worlds. In every area of social and individual life there were tremendous developments. In almost every aspect or dimension of human life we can think of – economic, social, political, cultural, religious, domestic, legal and personal – the nineteenth century in Europe was the birthplace of a new industrial society, which was both quantitatively and qualitatively different from any previous stage of human social development. And at the base of this industrial society, the ultimate source of its material success and self-understanding lay the domination of this society over the natural environment (and not just its local environment), principally as a result of the application of science and technology within new forms of socio-economic organisation combined with political and military power. Of particular importance was the emergence of the 'self-regulating' market as the principal means of organising the economy, which, together with the centrality of private property and production for profit, gives industrial society its 'capitalist' character (Polanyi, 1947; Goldblatt, 1996). This is discussed in more detail in Chapter 6.

While social theory was in the main tied to the aim of being 'scientific', what also unites different schools of European social thought and individual social theorists of the nineteenth century is a belief in the idea of social progress and the application of scientific knowledge as the means to achieve that end. With the 'death of God' as the philosopher Friedrich Nietzsche (1844–1900) announced towards the end of the century, humanity (or more specifically that portion of it resident in Europe and North America) and its progress became the ultimate aim or dominant goal of the new industrial society. Where once nature was God's creation (not humanity's), and had rules and principles governing human use of it, the growing secularisation of society in Europe meant that nature was viewed increasingly as belonging to humans (or as in the case of European colonisation, the Earth rightly belonged to 'some' humans) and the older, religious-based rules limiting and regulating human use of nature withered away.

While Europe was experiencing the fruits of the industrial revolution, the rest of the world had its part to play in the unfolding drama through colonisation, imperial expansion, and conquest of territory, peoples and cultures (Grove, 1995). Although previous civilisations and societies had used 'non-local' environmental resources, often by force – that is, they were not self-sufficient in terms of the resources they required – the industrial society of the nineteenth century was the first human society to emerge that was truly 'global' in its reach and in its environmental impact. In other words, the industrial societies developing in Europe and North America, through the process of imperialism and colonisation of other parts of the world, Central and South America, Africa, India and South-East Asia, meant that this European industrial civilisation had at its disposal the environmental resources of almost the entire planet. In modern terms the 'ecological footprint' of European societies far exceeded their own 'ecological space'. From minerals and precious ores such as gold, silver and diamonds, to tropical hardwood timber, to spices, exotic animals, tea, coffee, bananas, and in the case of slavery, other human beings, Western industrial societies in the nineteenth century required and depended upon more resources than their immediate European environments could provide. This is a point that will become important later on in discussing the green critique of industrial society. At the same time the 'discovery' of new lands, environments, climates, species and peoples had a direct effect on how the environment and environmental factors were articulated within social theory.

One way of understanding the colonialisation of many parts of the world by European societies lies in the distinction between 'ecosphere' and 'biosphere' forms of social organisation. According to Dasmann, 'Biosphere people draw their support, not from the resources of any one ecosystem, but from the biosphere. . . . [They] can exert incredible pressure upon an ecosystem they wish to exploit . . . something that would be impossible or unthinkable for people who were dependent upon that particular ecosystem' (1993: 121). For Milton, 'Biosphere people are those whose way of life is tied in with the global technological system . . . a biosphere economy is more likely to engender a cavalier exploitative attitude than a sense of environmental responsibility' (1996: 29–30). She goes on to add that 'The history of colonial expansion and industrial progress can be seen as a process in which ecosystem peoples have been transformed into biosphere peoples, often unwillingly, often forcibly, but often (and perhaps increasingly in recent decades) with their enthusiastic co-operation' (1996: 30). In this way the essential 'unsustainable' character of European industrial societies is revealed in the sense that these societies cannot survive on their own ecological space, but need the resources of the rest of the world. In addition, in so needing these resources, such 'unsustainable' social and economic systems are also 'unjust' in that they depend on the consumption of 'unfair' shares of the world's resources, which will be discussed later in Chapter 11.

A corollary of this distinction between 'ecosphere' and 'biosphere' is that ecosystem is unique, and thus demands a particular way of life adapted to and congruent with that particular ecosystem, rather than the universalisation of one way of life as is the case with Enlightenment thought and colonial historical practice. In this way a community's distinctiveness is intimately related to how it interacts with its local ecosystem or bioregion. This interaction, the metabolism between community and environment, and thus the community's identity, is co-determined by the bioregional ecological context. Defending biodiversity and cultural diversity thus goes hand in hand in the bioregional scheme of things, particularly with regard to wilderness conservation and indigenous people as 'authentic inhabitants' or 'ecological guardians' of wilderness areas.

It is important to note the part played by ideas of environment and nature in justifying nineteenth-century colonisation and imperialism. For many European thinkers, indigenous peoples were seen as simply part of the environment; that is, they were not recognised as equal human beings. This inclusion of indigenous peoples as part of the environment is at the root of the *terra nullis* claim used to legitimate the colonisation of Australia. The logic of this argument is brutally concise. Since the Australian aboriginal peoples were part of the Australian environment, like other species, the land of Australia was uninhabited, *terra nullis*, a land empty of 'people' (where what or who counts as 'people' is understood in a Eurocentric and essentially racist manner) and thus could be legitimately claimed as property.

Alongside these racist ideas were other equally objectionable views, such as the idea that peoples of 'hot environments' were lazy, backward and incapable of being industrious. Arnold, discussing the colonisation of India by Britain, notes how 'Environmental forces – climate and disease above all – were repeatedly invoked to demonstrate and explain Indians' moral and physical weakness and to justify the indefinite continuance of imperial rule' (Arnold, 1996: 174). Unlike the temperate climates and environment of Europe, tropical environments produced work-shy, over-relaxed and non-energetic cultures and peoples. According to some social theorists of the time, 'nature' decreed despotism in Asia while the harshness of the Northern European climate made for hardy, self-starting industrious societies. Such environmentally determinist ideas (that environment determines culture and psychological character) were quite common in the twentieth century and indeed may be found in racist discourses and ideas today.

Describing the dynamics of this industrial mode of socio-economic life became the primary subject of analysis for social theory, which for the most part thought of itself as 'scientific'. However, as well as *explaining* the origins and principles of industrial society, how it functioned and developed, using scientific methods and modes of inquiry, social theory was also concerned with *prescribing* how this society ought to be. In keeping with the great explosion in human thought and knowledge that characterises this period, utopian plans, revolutionary critiques of and alternatives to the prevailing industrial social order, radical suggestions for social development and improvement, as well as fewer defences of the status quo, flourished within social theory. It is to an examination of how the prescriptive and descriptive claims of some of these social theories rested upon or used particular understandings of the nonhuman world, that we turn next.

Progressive and reactionary social theorising about nature

Thomas Malthus

One of the most important areas of overlap between the environment and social theory in the nineteenth century begins with Thomas Malthus' theory of population. Malthus (1766–1834) criticised progressive Enlightenment thinkers such as William Godwin (1756–1836) and the Marquis de Condorcet (1743–94) at the end of the eighteenth century for thinking that the future of humanity was destined to be one of improved social, political and economic conditions. In the first edition of his infamous book *An Essay on the Principles of Population as It Affects the Future Improvement of Society, with Remarks on the Speculations of Mr. Godwin, Mr. Condorcet and Other Writers*, published in 1798, Malthus takes the two named theorists and other Enlightenment thinkers to task for foolishly and unscientifically speculating that the 'era of reason', expressed politically in the success of the French Revolution in destroying government by unelected monarchy, gave any grounds for forecasting a society 'devoid of war, crime, government, disease, anguish, melancholy, and resentment, where every man unflinchingly sought the good of all' (Eklund and Herbert, 1975: 81). Anticipating the logic of the green 'limits to growth' debate of the 1970s (discussed in Chapter 9), Malthus argued that Enlightenment views on the future progress of society were unfounded. While he did think that social progress such as less socioeconomic inequality between classes was *undesirable* (something for which Marx was later vehemently to criticise him), Malthus' main claim was that the utopian visions of social progress suggested by Godwin (an anarchist) and Condorcet (a leading French philosopher of the Enlightenment) were *impossible* on 'natural' grounds ('natural' here understood to refer both to the external natural environment and 'internal' human nature, both viewed ahistorically as incapable of change).

Malthus 'had argued that the prospects for progress were continually threatened by population growth and the fact that food production could in no way match such growth' (Dickens, 1992: 22). The root of the problem for Malthus lay in the brute fact that while population increases geometrically (2,4,6,8), food supply increases arithmetically (1,2,3,4). The different rates of growth of the two 'proved' that social progress of the sort favoured and proposed by Enlightenment thinkers was impossible to achieve for what we can now call 'ecological' and 'biological' reasons. Because there were definite limits to the ability of the land to provide food without strict population control (and controlling what Malthus saw as the 'passion between the sexes' was something he thought was extremely difficult to achieve, particularly among what he referred to as the 'lower social orders'), social improvement would come to grief against these non-negotiable, nonhuman ecological limits. Because of this he argued that giving more resources to the poor would only increase their numbers (since he thought the poor were least likely to exercise sexual self-restraint), and thus only add to their misery. Hence he suggested that the state should not intervene to help the poor; rather, if state aid were removed they would be 'motivated' to find gainful employment (instead of being dependent upon the public purse) and encouraged to have fewer children. In articulating this attitude, Malthus may be said to offer an essentially right-wing and conservative view of nature and society in that he held an essentially negative view of human nature against the usual positive views about the perfectability of humanity found in left-wing thinking and also, in keeping with right-wing thought, defended the inevitablity and desirability of inequality within society. These differences between left- and right-wing thinking will be dealt with later in Chapters 5 and 6.

Malthus' theory of population is not significant for its attention to 'ecological' considerations alone, but also for its attempt to be 'scientific'. While in the first edition of his *Essay* (1798), Malthus was largely offering some counter-speculations to those of Godwin and Condorcet, in subsequent editions (finally resulting in *A Summary View of the Principle of Population* in 1830), he attempted to give his theory some scientific credibility by using empirical evidence to prove the validity of his theory. This empirical evidence consisted of demographic, agricultural and other statistics and empirical data from different parts of the world, including travellers' diaries and other written accounts of journeys to foreign lands, and was flawed, patchy and would not pass statistical standards today. However, by the standards of early nineteenth-century social theory, his views were considered to be 'scientific'.

Malthus' social theory, in which he both described one aspect of modern industrial society and also prescribed particular policies, as well as basing it on 'scientific' principles of inquiry, may be said to constitute the model for social theory in the rest of the century. In particular, whether one was for or against industrial society, to be taken seriously, one's theory had to be scientific, rigorous and if possible backed up by empirical evidence. For example, a clear indication of how many social sciences sought to base themselves on the natural sciences is August Comte (1798–1857), one of the founders of modern sociology, who stated in 1853 that 'The subordination of social science to biology is so evident that nobody denies it in statement however it may be neglected in practice' (quoted in Dickens, 1992: 20). The implications of this for social theory are enormous: if the study of society could be reduced to biology, then we could both predict social behaviour as well as establish a set of scientific criteria by which we could judge what sort of social behaviour and social order we ought to have in accordance with scientific principles. The relationship between social theory and biology will be discussed in Chapter 10.

Social theory and natural evolution: Darwin, Spencer and 'Social Darwinism'

Charles Darwin's (1809–82) theory of evolution by natural selection and Herbert Spencer's (1820–1903) extension of this principle to social evolution constituted the next significant juncture in the history of the relationship between the nonhuman environment and social theory. In particular there are three aspects of Darwin's theory of evolution which are worth noting. The first is his theory that humans were evolved from primates. This was an incredibly radical and controversial theory in its day, although, extreme Christian creationists aside, it is now the received scientific wisdom concerning human evolution. This theory not only went against the biblical story of the creation of the Earth and humans by God, but, as will become clear later on, also served to undermine any strict separation between humans and the nonhuman world. Humans were not just *like* animals (somehow like them but actually 'higher' or 'superior' beings) and hence knowledge gained from the study of nonhuman animals might shed some light on human behaviour; after Darwin one could say that humans *were* animals, a particular subspecies of primate, namely *Homo sapiens*.

The second important aspect is Darwin's theory of natural selection which held that biological organisms were adapted to their environments. Due to a 'struggle for survival' between organisms it is those organisms which are best adapted to their environment which will survive and have more offspring and pass on their advantageous biological characteristics to their descendants. The final, and perhaps most significant for social theory, is that Darwin's evolutionary thought demonstrated the foundational continuity between humanity and nature, re-inforcing the fact that humans were not just like animals, we *are* animals. In doing this Darwin not only challenged any strict separation of humanity and nature and the 'social' and the 'natural', but also challenged religion-based views of the 'uniqueness' of humanity. This challenge continues today in attempts by fundamentalist Christians to debunk Darwin's evolutionary ideas and/or propose 'creationist' thinking as on a par with Darwinian evolutionary thinking.

Both of these aspects of Darwinian scientific theory were to have a great impact on social theory. In the history of social theory, it is Herbert Spencer who is seen as the main instigator of applying the 'Darwinian' principle of 'natural selection' and 'survival of the fittest' to the investigation of society. However, while Spencer is often credited with developing a theory of 'Social Darwinism', Dickens points out that 'Spencer started developing his theories some time before the emergence of Darwin's *Origins of Species*' (1992: 20). On the other hand, Spencer took an extreme organic view of society (in that he saw society as a 'super-organism' and therefore subject to the same laws and patterns of development as any other organism), and used Darwin's discoveries regarding the evolution of species to explain, predict and prescribe social relations between human beings.

Social Darwinism as it developed from the mid-nineteenth century on was a particularly harsh social theory in that, echoing the arguments of Malthus, it held that helping the poor only served to enable the 'unfit' to 'artificially' survive, and thus held back the course of social evolution which was premised on the survival of the fittest. In this way, Social Darwinism could be used to justify and legitimate a view of society in which there was little state interference in the 'natural' struggle for survival among human beings. It was, and sometimes still is, used to justify *laissez-faire* capitalist or classical free-market forms of an unequal social organisation of the economy and society. Like Malthus before him, Spencer expressed a version of **classical liberalism** in rejecting state interference in society and economy, most explicitly articulated in his *The Man Versus the State* published in 1884. Spencer and Social Darwinism propounded what we would now call an extremely right-wing libertarian social theory. It proposed a particularly individualistic view of human freedom (the central principle or value for such social theories), based on economic competition and the free market. Social Darwinism also grounded its normative and prescriptive view of society and the proper relations of individual within it on a particular organic conceptualisation of human society and of natural evolution by selection.

The study of natural selection and the interaction between and within species gave us a picture of 'nature, red in tooth and claw'. By analogy, social evolution and the interaction between individuals and groups in society, especially within the economic sphere, is and ought to be governed by the same laws of nature. Hence competition, self-interestedness and ruthlessness in economic life could be justified, as this simply conformed to the principles of social evolution and a particular conception of 'human nature' in which co-operation and solidarity were largely missing. It is perhaps no contingent fact that while Social Darwinism did flourish in Britain in the latter half of the nineteenth and early twentieth centuries, it was in North America, the land of 'rugged individualism', limited government and an extremely entrepreneurial culture, that Spencer's ideas and

Social Darwinism took deepest root, particularly in the work of social theorists such as William Sumner (1840–1910) who stated that 'The millionaires are a product of natural selection' (quoted in Dickens, 1992: 27). Like most Social Darwinists, Sumner's central political and social concern was with defending a particular, classical liberal view of the relationship between state, society and economy, in the name of a particular understanding of human liberty (Sumner, 1992). As Dickens notes, 'Early forms of social theory were largely constructed using analogies between societies and nature. Thus societies were seen as if they were developing like live organisms, or people were seen as struggling for survival in their environment, much in the same way as Darwin had specified in his theories' (1992: 56). As Anderson (2006) points out, and which was also central to Marx, there is a direct connection between Malthus' theory of over-population and Social Darwinism:

> Malthus' deterministic doctrine of overpopulation and his notion that people always breed faster than the growth of food supplies unless checked by poverty, famine, and disease shaped Darwinism; and then it was applied back to human behavior as Social Darwinism, back to the realm of social science but now imbued with all the authority of natural science and even more deterministic than in the original Malthus version. It claims that human beings are by nature competitive, suggesting that human nature is unchanging and that the 'dog-eat-dog' competitiveness of modern capitalism is somehow natural and inevitable (forgetting most of human history in the process and the role of cooperation in human survival).
>
> (Anderson, 2006: 272)

Mary Midgley points out that part of the explanation for the development of Social Darwinism may be found in the ways in which Darwin used metaphors from the worlds of business and warfare to explain natural selection. According to her:

> Theorists, including Darwin, who discussed conflicts of interest in the rest of nature constantly used images drawn from two particular human institutions – war and commerce – and in that non-human context it was not too hard to remember that these were only metaphors. But when the discussion turned back to human affairs, it became much harder to be clear that these were not literal descriptions of human life, new truths about its characteristic motives and intentions, truths which showed that it could all be reduced to these two simple models. The drama shaped by those models was then projected back, in its turn, onto the cosmos, producing a picture of the universe in which commercial rivalry provided the key guiding principle for everything – from gas to genius. Spencer and his followers thus saw competition as the all-explaining pattern both for human life and (somewhat casually, for they were not scientists) for the rest of nature, though Darwin himself, always anxiously aware of the limits of our knowledge, carefully avoided these extensions.
>
> (Midgley, 2001: 48)

Many authors have argued that the application of Darwinian theory (or any knowledge of the natural world) to social affairs is not only misleading but, more importantly, serves an *ideological* function of legitimating particular forms of social arrangements, patterns of power and distribution of wealth. One such example is Social Darwinism. However, there have also been other social theorists who while following the same basic logic of Spencer (that is, reading off from the nonhuman world about how the human social world is and ought to be), have read a different story and drawn different lessons for human society from the natural world. At the other end of the spectrum from Spencer's view of nature as 'red in tooth and claw' and his harsh message of the 'survival of the fittest' which underwrites the centrality of competition and struggle between humans within society (as well as between human society and the nonhuman environment) as necessary and desirable for social evolution and progress, we have the Russian aristocrat, social theorist and revolutionary anarchist Peter Kropotkin.

Natural harmony: Kropotkin and the anarchist reading of the environment

In 1902 Prince Peter Kropotkin (1842–1921) published *Mutual Aid: A Factor of Evolution*, the main message of which was that co-operation was just as important as competition in both human and nonhuman evolution. For Kropotkin, and others who shared his political perspective, this of course led to a different set of principles and prescriptions for human society from those outlined by Social Darwinists, conservatives or liberals. For Kropotkin, as for most left-wing or communist anarchists, humans are not naturally selfish and engaged in a brutal struggle for survival; rather it was only under the particular prevailing social order that humans behaved in this manner. But more than that, Kropotkin's reading of natural selection, in contrast to Spencer's, was premised on the idea that humans were naturally cooperative and that if the 'artificial' institutions of the state and the capitalist organisation of the economy were abolished, humans could enjoy a more harmonious, co-operative and egalitarian social order. And co-operation not competition was superior both in explaining human evolution but also in prescibing social reform and organising society for human improvement. As Miller notes:

> Kropotkin . . . tried to show that Darwinian ideas, properly understood, could be invoked in aid of libertarian communism (thereby rebutting the normal implications of Social Darwinism). Beginning with animals and moving on to human societies, he argued that those groups which have proved most successful in evolutionary terms had done so by developing practices of mutual aid – practices whereby each member came to the help of others in need.
>
> (1991: 271)

In this way, Kropotkin and other communist anarchists echoed one of the main arguments that Rousseau had raised against the Enlightenment and the social and political arrangements of modern industrial society more than a century before. Rather than supporting a *laissez-faire*, competitive capitalist and unequal social order, social theory could find in nature a model of human society based on mutual aid, solidarity, equality and harmony. As Nisbet puts it:

> For Peter Kropotkin . . . the problem of community, that is, genuine and lasting community, resolved itself into a rediscovery of nature: not merely the protection and proper development of nature in the external, physical and biological, sense, but also in the sense of seeking to build community – and in the long run society as a whole – on the most natural of interdependence among men.
>
> (1982: 205–6)

This connection between external nature and internal human nature is something common to most social theorising about the environment which is in the form of 'reading off' social principles from nature. As de Geus, in his study of Kropotkin, puts it, 'in his graphic picture of an ecological society he takes into consideration a set of fundamental principles *which he believes can be derived from nature*, such as mutual aid, solidarity, cooperation, self-government, harmony, balance and community' (1999: 88; emphasis added). Thus, whereas Spencer looked into nature and saw competition and inequality, and a 'natural' defence of *laissez-faire* capitalism, Kropotkin saw mutual aid and solidarity, and proof of the naturalness and superiority of a stateless, anarchist social order. As Thiele puts it, Kropotkin echoed Rousseau in his critique of modern industrial capitalism and his proposal for the integration of older and non-capitalist principles of social organisation:

> It is the pervasiveness of mutual aid, Kropotkin believed, that makes anarchism a viable option for humankind. Kropotkin looked to the Middle Ages as a period of relative well-being whose virtues were trampled under by the rising bourgeois society and the centralized state. Kropotkin suggested that the ideal society of the future might go beyond the decentralized and federative structures of the Middle Ages to foster lives of liberty and harmony, such that cooperation would supplant coercion and the preservation of nature would replace efforts to master it.
>
> (Thiele, 2000: 18)

Marxist social theory and the environment

Karl Marx (1818–83) and Friedrich Engels (1820–95) and their political and social theory of Marxism is the next significant body of nineteenth-century social theory in which the environment played a particular (though ambiguous) role. In their famous work *The Communist Manifesto* (first published in Germany in 1848), after stating that the bourgeoisie (the owners of capital, the ruling class in Marxist terms) has subjected 'the country to the rule of the towns' and 'rescued a considerable part of the population from the idiocy of rural life' (1967: 84), Marx and Engels go on to recognise the great achievements of industrial capitalism. Summing up the spirit of early social theory in the mid-nineteenth century in terms of its acceptance of the domination of the nonhuman world as the basic premise of the industrial society, they state:

> The bourgeoisie, during its rule of scarce one hundred years, has created more massive and more colossal productive forces than have all preceding generations together. Subjection of Nature's forces to man, machinery, application of chemistry to industry and agriculture, steam-navigation, railways, electric telegraphs, clearing of whole continents for cultivation, canalization of rivers, whole populations conjured out of the ground – what earlier century had even a presentiment that such productive forces slumbered in the lap of social labour?
>
> (Marx and Engels, 1967: 85)

It is fair to say that historically classical Marxism, being a product of its time, did not address the range and significance of ecological issues that have come to play such an important part in late twentieth-century political and ethical discourse. Indeed, insofar as ecology stresses natural or absolute limits to economic development, early Marxist theory was vehemently anti-ecological. It is in the Marxist attack on Malthus' theory of population and his argument for subsistence wages that we can trace aspects of the attitude of Marxist social theory to the nonhuman world (Barry, 1998c). For Marx, Malthus' theory was a piece of ideological justification masked as scientific inquiry, and he labelled Mathus' essay as a 'libel on humanity'. As Marx put it, 'Malthus' work . . . is a pamphlet directed against the French Revolution and the ideas of reform which were springing up at that time in England; it served to justify the misery of the working class' (Marx, 1971: 368). Malthus represented the interests of the landed classes, and his theory of population was an attack on the urban poor and a conservative defence of the political status quo. Accoding to Bellemy-Foster, Malthus' greatest fear, which he instilled into the English aristocracy and middle classes, was that unchecked population increase coupled with egalitarian ideas would lead to the middle classes blending with and becoming part of the poor, a possibility which had to be avoided and prevented at all costs, specifically by denying public welfare relief to the poor (Bellemy-Foster, 2000:100).

There is also a link between Malthus, Hobbes and Darwin according to Marx. For Marx, 'Darwin rediscovers in animals and plants his own English society with its social division of labour, competition . . . and the Malthusian struggle for existence', which in turn had come from Hobbes' 'bellum omnium contra omnes' – the 'war of all against all', a theory to justify a strong monarchical state to keep the peace (in Pepper 1984: 134–6).

Marx's attack on Malthus' ideas set the tone, and often the parameters, within which the interaction between Marxism and concerns about the nonhuman world took place. For example, this encounter contains all the main ingredients which mark, and continue to mark, the relationship between ecological social thought and Marxism. First, there is the Marxist claim that stressing non-social ecological limits and conditions is anti-Enlightenment in general and anti-industrial in particular. Second, and following on from the latter, is the equation of anti-industrial with anti-working class, something which, for Marx, was abundantly clear in Malthus' work, and its main ideological aim. Third, we have the importance of science and technology on both sides. On the one hand, we have Marx completely optimistic in the ability of technology, once free of capitalist relations, to transcend so-called 'natural limits'. On the other, we have Malthus' claim that his theory was fully supported by scientific and statistical data, which led to the opposite conclusion from that of Marx, and indeed was at odds with the dominant belief in progress that characterised the early development of industrialisation under capitalism, and which was reflected in the social theory of this period.

In this way ecological concerns from a classical Marxist perspective was another fetter holding back the onward and inexorable rise of the revolutionary proletariat. In many ways ecological considerations were worse than bourgeois political economy because, unlike the latter, an excessive concern with the nonhuman world was held to be anti-industrial and anti-modern, and was argued to be driven by a desire to return to a pre-modern, agrarian, social order. When social theorists discussed the nonhuman world in terms of a Romantic defence of the natural world against industrialisation, against the 'disenchantment of nature' in Max Weber's (1864–1920) famous phrase, as in the Romantic poetry of William Wordsworth or Percy Shelley, or the social theories of Thomas Carlyle (1795–1881) or John Stuart Mill (1806–73), this merely confirmed the regressive and deeply conservative character of these social theories for Marxists. To be 'modern' and 'progressive' meant that nature had to be dominated; as Marx and Engels noted above, nature's forces had to be subjected to man. Those who objected to this domination and purely instrumental view of the nonhuman natural world were either simple-minded sentimentalists (poets such as Wordsworth) or reactionaries who were really motivated by a defence of a feudal, aristocratic social order based on a 'pre-industrial' and 'pre-modern' system of land ownership (Malthus and Carlyle).

At the same time, it is not simply the case that Marx rejected the importance of the relationship between human society and the nonhuman world. Far from it. Marxist social theory is at root a materialist theory of human society, its dynamics and historical evolution. Unlike idealistic social theorists such as the German philosopher G.W.F. Hegel (1770–1831), the basic tenet of Marxism is that it is the material conditions and relations within a society which determine its character. It is not ideas that cause societies to change but rather the material conditions under which the society organises its economic life which rather determine (or influence) ideas in society. This leads contemporary Marxist writers, such as John Bellemy-Foster, to conclude that 'Marx's work cannot be fully comprehended without an understanding of his materialist conception of nature, and its relation to the materialist conception of history. Marx's social thought, in other words, is inextricably bound to an ecological world-view' (2000: 20).

The starting point for Marx is the brute fact that humans have to produce their own means of subsistence. What he means by this is that humans have to use their labour power, skills and creativity to transform the nonhuman world into the things, goods and services they need to survive. This, according to Marx and Engels, is what distinguishes humans from the nonhuman world. As they put it, 'Men can be distinguished from animals by consciousness, by religion or anything else you like. They themselves begin to distinguish themselves from animals as soon as they begin to produce their means of subsistence. . . . The nature of individuals thus depends on the material conditions determining their production' (quoted in Parsons, 1977: 137). Although, like all other species, humans are dependent upon their environment for resources in order to survive, Marx held that humans were different from the rest of nature (and here he was simply stating the prevailing dominant view of the matter) because they did not simply take from nature whatever their natural environment afforded. Except in the 'primitive' hunter-gatherer stage of human evolution, the story of humanity was one where by their collective actions they transformed their environment, and by their labour power transformed the 'raw materials' of the nonhuman environment into usable and valuable artefacts: such as dwellings from simple huts to large cities, clothing from animal furs to designed fashions, and a whole range of goods, things and commodities (see Figure 3.1).

Marx and Engels held Darwin in high esteem, and early Marxist thought agreed with the idea of evolution in history (as in nature) as being based on 'competition'. As Marx wrote, 'Darwin's book is very important and serves me as a basis in natural science for the class struggle in history' (quoted in Dickens, 1992: 45). That is, the Marxist model in which historical change, the evolution of society from one historical stage to the next, which was based on the idea of 'class struggle being the motor force of history', could find support in Darwin's ideas of natural evolution and competition between species.

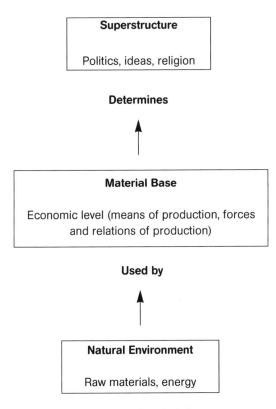

Figure 3.1 The Marxist model of socio-historical change
Source: Author

Following Locke in many respects (though of course drawing completely different political and social lessons), Marxist social theory was premised on the idea that the nonhuman world if left to itself, unused and untouched by human hands, was 'valueless'. Whatever was of value in the world was the product of human labour and creativity. This was the essence of the Marxist labour theory of value. The problem with the capitalist organisation of industrial society for Marxists was that the social organisation of this society was such that the vast majority of the people were denied the full fruits of their labour.

Because under capitalism, those who owned the capital, the factories, the machinery and so on were in a stronger economic position than those who had simply their labour power to sell, the latter were, according to Marx, exploited. Thus, while the industrial **mode of production** (including the factory system, the extensive use of science and technology, and a complex division of labour), was premised on the intensive exploitation of the nonhuman world, its capitalist character also meant that the workers or the proletariat in Marxist terminology were also exploited. And while there have been attempts to claim that there are aspects of Marx's work in which there is a concern with overcoming the exploitation of the nonhuman world (Parsons, 1977), by and large we can say that the primary concern of Marxism was liberating human society from capitalist exploitation, inequality and oppression. In other words – and this is an important point for understanding much of the history and present state of the relationship between Enlightenment social theory and the nonhuman environment – the exploitation or abuse of the nonhuman world was not considered as a significant ethical or political issue for Marx.

As Bellemy-Foster puts it, Marx, while aware of ecological issues and the metabolism between human society and the nonhuman environment, cannot be said to have prioritised this aspect of his analysis:

> Marx and Engels did not generally treat environmental destruction (apart from the role that it played in the direct life of the proletariat – that is, the lack of air, of cleanliness, of the prerequisites for health, and so on) as a major factor in the revolutionary movement against capitalism that they saw as imminent. Where they emphasized ecological contradictions, they did not seem to believe that they were developed to such an extent that they were to play a central role in the transition to socialism. Rather such considerations with regard to the creation of a sustainable relation to nature were part of – even a distinguishing feature of – the later dialectic of the construction of communism.
>
> (Bellemy-Foster, 2000: 140)

Marxism expresses the thoroughly 'modern' view that human social progress (and in this case 'liberation' from a capitalist exploitative social order) is dependent upon the exploitation and domination of the nonhuman environment. As will be seen later in discussing critical theory (Chapter 4) and green social thought (Chapter 11), this is a position which has come under sustained attack in the latter half of this century.

The basic argument of Marxism as a representative modern social theory of industrial society was not about the great strides capitalism had made in finding the 'secrets of nature', and new ways of exploiting 'natural resources' which produced great amounts of material wealth, food, shelter and goods, as well as decreasing illness and increasing the average human life-span, and creating the most affluent social order the world had ever seen. Marx's problems with capitalism were not that he objected to the wealth-producing process, which was based on the exploitation of the nonhuman world; rather he argued that the fruits of this remarkable social order were not distributed equally because a few (the bourgeoisie or the owners of capital) enjoyed the gains while the many (the proletariat or workers and colonised peoples and places) had to bear the costs and reaped few rewards. Marxism argued that the fruits of social labour should be shared equally among everyone, and not be the preserve of a privileged few who happened to own and control the means of production through the capitalist organisation of the economy and the institution of private property.

If nature was exploited under capitalism, Marx's views of a 'post-capitalist' or communist organisation of society was based on the hyper-exploitation of nature (Benton, 1989; Soper, 1996; Barry, 1999a) – though there is a debate about this charge of Prometheanism and also his concern with overcoming human alienation from nature, or what he described as 'man's inorganic body' (Bellemy-Foster, 2000: 135; Pepper, 1993; Biro, 2005: ch. 4). Marx's basic critique of capitalist industrial society was that its relations of production (particularly those relating to private property) were holding back what he called the forces of production (technology, science, the division of labour) from producing even greater levels of material wealth and affluence (see Figure 3.2). In other words, capitalism was not exploiting the nonhuman world efficiently enough to the fullest extent possible; nor did it distribute the wealth created in a just manner. As Soper points out, 'there is nothing overt in Marx's argument that associates socialism with a restraint on material and resource-intensive forms of production or with a use of surplus/free time that would be obedient to environmental limitations on the expansion of certain forms of consumption' (1996: 93).

For Marxists, it is one of the great contradictions of capitalism that the social order which has put a man on the moon is unable to eradicate poverty, homelessness and many other socio-economic problems. Marx's vision of a post-capitalist society is premised on the existence of 'material abundance': that is, communist society is one which has transcended material scarcity. Unlike all previous human societies in which there have never been enough material goods for everyone, and hence every social order has to have some principles or institutions for distributing what goods are available, Marx's communist society was one in which the principle of distribution was 'to each according to their needs'. In other words, Marx was envisaging that under communism the exploitation of nature and the production of wealth, goods and services would be so efficient and productive that all human material needs would be met. Whereas under capitalism only a few can afford an expensive car, for example, under communism everyone could have a car if they needed one because the economic exploitation of nature would be so efficient. Free of constraining, inefficient as well as illegitimate capitalist property relations (and the centralised, capitalist state, which would 'wither away' after the revolution), human beings would be in a position to use the world around them as they saw fit and to meet their material needs and wants. Nobody would go hungry or homeless in this vision of a future post-capitalist society, and more than this, getting rid of capitalist relations of production would enable a more rational, planned, intensive and ultimately more productive exploitation of the nonhuman

The Marxist Boomerang

Marx thought you could fight capitalism by allowing it to grow, thus creating a strong class of industrial workers who would eventually 'break their chains' by kicking out the owners and taking over production themselves. He called such growth the development of the *productive forces*.

Most modern technology, however, by plundering nature is in conflict with the *conditions of production*.

Marx left the impression that natural resources were in principle unlimited – that humanity would always find new ones. In the Communist Manifesto he declared that one of the aims of socialism would be 'to increase the total of productive forces as rapidly as possible'.

Figure 3.2 'The Marxist Boomerang'

Source: Croall and Rankin (1981)

environment. Human liberation and emancipation were to be achieved at the price of the greater exploitation and intensification of the instrumental use of nature.

A more 'environmentally sensitive' interpretation of Marxism may be found in some of the works of Engels, though the emphasis is on the urban environment rather than the natural one. In his *The Condition of the Working Class in England*, Engels described the degrading, filthy and unhealthy urban and working environments of the emerging urban working class, and suggested that the coming communist society would create less unhealthy, unsafe and more aesthetically pleasing urban, living and working environments. At the same time some recent theorists, such as Benton (1989, 1993) and Pepper (1993), working within the broadly Marxist tradition, have suggested more nuanced and less unecological forms of Marxist social theory (for an overview of some of these recent developments between Marxism and ecology see Barry (1998c) and Benton (1996)).

As Giddens has commented, 'in general it is the case that Marx was not a critic of industrialism. Rather for him industrialism holds out the promise of a life of abundance, through turning the forces of nature to human purposes. It is a particular mode of organizing industrial production – capitalism – that needs to be combated, not the industrial order itself' (quoted in Cassell, 1993: 329). Hence Marx may be seen as wanting to intensify the exploitation of the natural environment which capitalism had begun, but to end the exploitation of humans by humans and to distribute the fruits of the exploitation of the environment more equally than under capitalism, as well as viewing his vision of a socialist society as motivated, in part, by a desire to overcome the alienation between humanity and nature. In this way we may say that the legacy of Marx as a social theorist in relation to the environment is at best ambigious, as Soper (1996) contends in that while he cannot be described as a 'green' thinker or have a view of nature beyond its instrumental value for humanity, his thought also contains the potential for a more sustainable relationship between human society and nature which has been developed by various 'eco-socialist' thinkers, as discussed in Chapter 6.

Liberalism, utilitarianism and J.S. Mill: the first 'green' social theorist?

The place of the environment in the work of John Stuart Mill (1806–73), one of the greatest liberal political thinkers of the nineteenth century, may be traced to his recovery from a mental breakdown while he was aided by reading the nature-loving Romantic poetry of Wordsworth and Coleridge. His highly original views about what we would today call 'ecological issues' were atypical of his time and anticipated many issues that would later come to be termed 'green' or 'ecological'.

For example, in an essay entitled 'Nature', Mill demonstrated an awareness of the foundational importance of nature and related ideas in social theory. Particularly interesting for our purposes, in this essay Mill also highlighted some of the dangers of the 'reading-off' hypothesis outlined in Chapter 2. As he puts it:

> the doctrine that man ought to follow nature, or in other words, ought to make the spontaneous course of things the model of his voluntary actions, is equally irrational and immoral. Irrational, because all human action whatever, consists in altering, and all useful action in improving, the spontaneous course of nature. Immoral, because the course of natural phenomena being replete with everything which when committed by human beings is most worthy of abhorrence, any one who endeavoured in his own actions to imitate the natural course of things would be universally seen and acknowledged to be the wickedest of men.
>
> (Mill, 1977: 311)

Thus for Mill, 'reading off' or following nature in how we ought to live and organise society is dangerous, irrational and likely to lead to negative moral and practical consequences. This is a theme also taken up by later social theorists such as Jürgen Habermas, discussed in Chapter 3.

While this essay illustrates that Mill did not have a benign view of nature (like Kropotkin) and tended towards seeing it as close to the image of nature being 'red in tooth and claw', this does not mean that Mill is dismissive of the importance of the natural world for humanity and social theory. As Stephens points out, 'Mill's vision of the natural world was not benevolent . . . but it was nonetheless mitigated by appreciation of its aesthetic and regenerative qualities, as his devotion to the poetry of Wordsworth and Coleridge testifies' (1996: 378).

The appreciation of the therapeutic and aesthetic qualities of the natural environment, which has its roots in Romanticism, was a view shared and extended by American transcendentalism of the late nineteenth century, and indeed continues to this day in both 'deep ecological' thinking and also the use of nature and animals in coping with trauma and mental illness in psychology. This American movement in social and literary theory and practice was a form of 'nature/religion/spirituality' in that it saw God and spirituality as immanent in nature (especially the American wilderness). According to the transcendentalist view, the direct experience and appreciation of nature was a way to enter a 'higher' or 'transcendental' realm of eternal truth, beauty and happiness, away from the mundane distractions of our everyday, urban world. Thinkers associated with this movement, which was directly connected with the debates about the relationship between American national identity and wilderness, discussed in the last chapter, included Henry Thoreau (Buell, 1996) and Ralph Waldo Emerson (Buell, 2004). Emerson, for example, believed that the personal experience of nature (as 'other') could bring human beings towards an appreciation of the prelapsarian conditions enjoyed by humanity in the Garden of Eden and help overcome the pain (unconscious or conscious) engendered through the separation of humanity from God and his creation, nature (Oelschlaeger, 1992: 275).

In 'Of the Stationary State', Book IV, ch. VI of his *Principles of Political Economy*, first published in 1848, Mill suggested a radical critique of received wisdom concerning the progress of society. He suggested that the desire for more and more material goods and services, based on the domination of nature and the more intensive use and application of science and technology, was perhaps too narrow a view of 'social progress'. What Mill had to say is worth quoting at length:

> I cannot, therefore, regard the stationary state of capital and wealth with the unaffected aversion to it so generally manifested towards it by political economists of the old school. I am inclined to believe that it would be, on the whole, a very considerable improvement on our present condition. . . . It may be a necessary stage in the progress of civilization.
>
> . . . But the best state for human nature is that in which, while no one is poor, no one desires to be richer, nor has any reason to fear being thrust back, by the efforts of others to push themselves forward. . . . It is only in the backward countries of the world that increased production is still an important object: in those most advanced, what is economically needed is better distribution, of which one indispensable means is a stricter restraint on population.
>
> (Mill, 1900: 453)

As Eklund and Herbert note, 'alone among the classical economists, Mill did not believe the stationary state was undesirable, since . . . it provided the necessary condition for his program of social reform' (1975: 120). In articulating both a critique of the dominant view of social progress (as the growth in material goods and services and undifferentiated economic growth), based on a notion of natural limits, and in proposing an alternative view of human social development which stressed the non-material or qualitative dimensions of development, Mill, in proposing the stationary state economy, anticipated some of the key concerns of green social thought of the 1960s and 1970s, particularly in relation to green approaches to economics and a sustainable economy, and influential theorists of green political economy such as Herman Daly. These ideas will be discussed in Chapter 6.

In many respects, taking together Mill's social theory, his view of progress and the need for social reform via distributive policies, coupled with what Macpherson (1973) has termed his 'developmental' strand of liberalism (to be distinguished from the 'classical' or *laissez-faire* liberalism which was more dominant at the time), a strong case can be made for seeing him as a prototype 'green' social theorist (Stephens, 1996; de-Shalit, 1997; Wissenburg, 1998; de Geus, 1999; Barry, 2001).

It is in Mill's eloquent defence of the 'stationary state' in which wealth, capital and population are held as constant as possible, that is, a non-growing economy, that the first outline of a 'green' or sustainable society and view of the 'good life' may be found. To those who suggest that a non-growing economy would be a social disaster, Mill replies:

> It is scarcely necessary to remark that a stationary condition of capital and population implies no stationary state of human improvement. There would be as much scope as ever for all kinds of mental cultures, and moral and social progress; as much room for improving the Art of Living, and much more likelihood of its being improved when minds are ceased to be engrossed by the art of getting on.
>
> (Mill, 1900: 455)

Indeed, Mill suggests that in such a non-growing or 'steady-state economy' (Daly, 1973), technological progress could and should be used to lessen the work that has to be done, rather than simply be used to employ less labour and thus enhance profits. This theme, which is at the heart of green political economy (discussed below), again makes Mill something of a green visionary. On the subject of technological improvements, Mill expresses the view that in the stationary state:

> Even the industrial arts might be as earnestly and as successfully cultivated, with this sole difference, that instead of serving no purpose but the increase in wealth, industrial improvements would produce their legitimate effect, that of abridging labour. Hitherto, it is questionable if all the mechanical inventions yet made have lightened the day's toil of any human being. They have enabled a greater population to live the same life of drudgery and impoverishment, and an increased number of manufacturers and others to make fortunes.
>
> (1990: 455)

Another 'green' aspect of Mill's thought is his defence of anti-cruelty to animals legislation (Mill, 1900: 578), something which had a long tradition in English **utilitarian** social thought. Jeremy Bentham (1748–1832), one of the founders of English utilitarianism, for example, was one of the first to give a philosophical justification for limiting cruelty to sentient animals. According to him:

> The day may come when the rest of the animal creation may acquire those rights which could have never been witholden from them but by the hand of tyranny. . . . The question is not, Can they *reason*? nor Can they *talk*? but Can they *suffer*?
>
> (1970: 311)

Utilitarianism, in basing its social and moral arguments on considerations of pleasure, pain, welfare and happiness, was perhaps the first modern social theory to be applicable beyond the species barrier. If one was concerned with creating the 'greatest happiness of the greatest number' (one of the common expressions of utilitarian thought), then to restrict one's concerns to humans only would be arbitrary and unjustifiable, since like humans, animals are sentient, can feel pleasure and pain, experience happiness and sadness and so on. This same utilitarian argument since first articulated by Bentham, then Mill, and also put into practice as the animating principle of a popular 'animal welfare' movement of the late nineteenth century, in which figures such as Henry Salt were prominent theorists and activists, may still be found today in the arguments of philosophers such as Peter Singer (1990) for 'animal rights', and within the politics of the animal liberation movement. From a non-anthropocentric utilitarian perspective much of Western industrilaised farming methods, for example, would be unjustifiable given the amount of pain and suffering industrialised meat production causes to sentient animals (Pratt *et al.*, 2000: 91).

This is not to say that there are other aspects of Mill's social, political and economic thought which would be at odds with green thinking. For example, though the general aim of Mill's remarks on 'The Stationary State' would be welcomed by many greens, his seemingly anthropocentric tone and language, and his somewhat overdependence on population control, as well as his (admittedly ambiguous) defence of colonialisation, would alienate some of them. Mill notes that, 'Only when, in addition to just institutions, the increase of mankind shall be under the deliberate guidance of judicious foresight, can *the conquests made from the power of nature* by the intellect and the energy of scientific discoverers, become the common property of the species, and the means of improving and elevating the universal lot' (1900: 455; emphasis added). Yet, despite (and indeed perhaps because of, as discussed below) the anthropocentric tone, the instrumental view of nature, and the almost exclusive concern with human progress and well-being, Mill can still be counted as one of the first green social theorists.

Conclusion

It is with the Enlightenment or 'modernity' that we can see the first engagements between social theory and the environment, both as a precondition for society and when translated as 'nature' or 'natural' how it was put to a variety of ideological uses throughout the late eighteenth and nineteenth centuries. With the advent of modernity, 'nature' and the natural environment were, according to Weber, 'disenchanted'; that is, the natural world was reduced to a set of 'natural resources' or 'raw materials' for human productive use.

Starting with Rousseau, some social theorists questioned the 'artificiality' of the modern world, and yearned for a return to a simpler, more 'natural' form of social order. This went against the prevailing view that life in the 'state of nature' was, in the infamous words of Hobbes, 'nasty, brutish and short'.

While for some, human nature was 'naturally' competitive and indeed nature had been shown scientifically to be a struggle for survival (classical liberalism), for others, nature was a harmonious web of relations in which co-operation and mutual aid were just as important for evolution and progress (anarchism).

Within dominant nineteenth-century schools of social theory, such as liberalism and socialism, we can discern a common view of 'social progress'. Progress and 'development' were understood as a linear historical path from pre-industrial society to the complex and more advanced stage of the industrial socio-economic order. In many ways this particular linear view of human historical development is an onward and ever upward progression from ignorance, poverty, squalor and backwardness. While socialism criticised the capitalist character of industrial society, it did not criticise the industrialisation process in general or how the latter meant that the nonhuman world was 'dominated' and exploited, and viewed and used primarily as a means to human ends. While the other dominant nineteenth-century social theory, liberalism, generally agreed with socialism on the relationship between society and nature and the pursuit of material progress, J.S. Mill stands out against this dominant shared view of industrial development. As well as his defence of the extension of moral concern to nonhuman animals, his views on the desirability of a 'stationary state' mark him out as an early green social theorist.

Summary points

- In social theory from the nineteenth to the twentieth centuries, an appeal to the 'naturalness' of principles or view of social relations has often been seen as indicating that such relations are 'given', that is, cannot be changed by human will.
- The idea of the 'state of nature' was the dominant way in which pre-Enlightenment social theory conceived of the environment, and it was regarded, by thinkers such as Hobbes and Locke as inferior to civilised society.
- Rousseau went against this negative view of the environment and the state of nature with his ideas of the 'noble savage' and critique of the artificiality of civilisation.
- From 'Social Darwinism' to 'eco-anarchism' there have been ideological uses of the nonhuman world to justify, persuade and legitimate particular forms of social order.

- An early precursor of 'green' social thought is the anarchist reading of nature in which co-operation was seen as just as important as competition for natural, and thus social, evolution and development.
- Within dominant schools of social theory, such as liberalism and socialism, there was a common understanding of 'social progress'. Progress and 'development' were understood as a linear historical path from pre-industrial society to the complex and more advanced stage of the industrial socio-economic order.
- Marxist social theory criticised the capitalist character of industrial society, but not the industrial stage of social historical development. It accepted the necessity and desirability of 'subjugating' nature to humanity, or exploiting the natural environment so as to end human exploitation under capitalism.
- Mill and his brand of 'developmental' liberalism, to be distinguished from 'classical' or libertarian liberalism, anticipated many of the main concerns and aims of green social and political theory, particularly his defence of the 'stationary state' or non-growing 'steady-state economy'.

Further reading

For a general overview of some of the theorists and theories outlined here (and others not discussed) see Derek Wall's *Green History*, London: Routledge, 1994a, and Parts II and III of Marius de Geus' *Ecological Utopias: Envisioning the Sustainable Society*, Utrecht: International Books, 1999.

On pre-Enlightenment social thought and nature, see David Pepper, *The Roots of Modern Environmentalism*, London: Croom Helm, 1984. For a discussion of Hobbes and Locke see Keekok Lee, *Social Philosophy and Ecological Scarcity*, London: Routledge, 1989. Rousseau's thought and the environment are discussed in Shane Phelan's article (1992) 'Intimate Distance: The Dislocation of Nature in Modernity', *The Western Political Quarterly*, 45: 2, and Andrew Biro's *Denaturalizing Ecological Politics: Alienation from Nature from Rousseau to the Frankfurt School and Beyond*, Toronto: University of Toronto Press, 2005: ch. 3. On the Enlightenment and environmental concerns see Alasdair Clayre (ed.), *Nature and Industrialization*, Oxford: Oxford University Press, 1977.

The relationship between anarchist social theory and the environment is discussed in Murray Bookchin, *The Ecology of Freedom: The Emergence and Dissolution of Hierarchy*, Montreal and New York: Black Rose Books (rev. edn), 1991; Marius de Geus, *Ecological Utopias: Envisioning the Sustainable Society*, Utrecht: International Books, 1999, and Graham Purchase, *Anarchism & Ecology: The Historical Relationship*, Edinburgh: AK Press, 1998.

For a discussion of Darwin and Malthus, see Donald Worster, *Nature's Economy*, Cambridge: Cambridge University Press (2nd edn), 1994; for a discussion of Malthus and neo-Malthusan ideas see Eric B. Ross, *The Malthus Factor: Poverty, Politics and Pop in Capitalist Development*, London: Corner House Briefing, 2000.

On Marx, Darwin and environmental issues see Peter Dickens, *Society and Nature: Towards a Green Social Theory*, Hemel Hempstead: Harvester Wheatsheaf, 1992. On Marx and Engels on environmental issues see Henry Parsons (ed.), *Marx and Engels on Ecology*, Wesport, CT: Greenwood Press, 1977; John Bellemy-Foster, *Marx's Ecology: Materialism and Nature*, New York: Monthly Review Press, 2000; Paul Burkett, *Marx and Nature: A Red and Green Perspective*, New York: St Martin's Press, 1999; Jonathan Hughes, *Ecology and Historical Materialism*, Cambridge: Cambridge University Press, 2000. For an assessment of the historical and contemporary relationship between Marxism and the environment see John Barry, 'Marxism and Ecology', in A. Gamble *et al.* (eds), *Marxism and Social Science*, Basingstoke: Macmillan, 1999b. See also David Pepper, *Eco-Socialism: From Deep Ecology to Social Justice*, London: Routledge, 1993; Ted Benton, *Natural Relations: Ecology, Animals and Social Justice*, London: Verso, 1993, and Benton (ed.), *The Greening of Marxism*, New York: Guilford Press, 1996.

On liberalism, John Locke, J.S. Mill and environmental concerns see Marcel Wissenburg, *The Free and the Green Society: Green Liberalism*, London: UCL Press, 1998, and Piers Stephens, 'Plural Pluralisms: Towards a More Liberal Green Political Theory', in Iain Hampsher-Monk and Jeffrey Stanyer (eds), *Contemporary Political Studies 1996*, Belfast: Political Studies Association, 1996, and the debate between Wissenburg and Stephens in *Environmental Politics*, 10: 3, 2001.

4 Twentieth-century social theory and the nonhuman world

Key issues

- Classical sociology and the environment.
- Freud, human nature and the war against nature.
- Existentialism and the 'meaningless' Earth.
- The Frankfurt School: critical theory, nature and the problems of modernity.
- Marcuse and the 'liberation of nature'.
- Jürgen Habermas and the problem of nature in modernity.
- The moral status of the nonhuman world.
- Knowledge, nature and society: towards reconciliation?
- Environmental politics and the defence of the 'lifeworld'.
- Anthony Giddens, globalisation and the environment.
- The rise and meaning of environmental politics.
- The urban environment and the 'town' versus the 'countryside'.
- Environmental politics and conservatism.
- Risk, science and precaution.
- Environment, tradition and identity.

Introduction

Following on from the last chapter, this chapter looks at some of the ways in which social theorising in the twentieth century has dealt with, conceptualised the nonhuman world and the relations between it, and the human social world. Of particular interest are the ways in which the relationship between society and environment is theorised against the backdrop of the continuing legacy of the Enlightenment, and reflections on the development of modern, urban industrialised societies. This theme is particularly evident in critical theory and the work of two leading contemporary social theorists Jürgen Habermas and Anthony Giddens. Another emerging theme within twentieth-century social theory is the connection between social theorising about external nature and internal or human nature.

Classical sociology and the environment

Classical sociology, the 'science of society', has little to say about the environment or the environmental basis of human society. Where the nonhuman world did appear in classical sociology it was usually as one half of the organising dualism – society/nature – and/or as that which humans have historically overcome in their evolution from the Stone Age to the modern industrial age. As Goldblatt puts it:

> The primary ecological issue for classical social theory was not the origins of contemporary environmental degradation, but how premodern societies had been held in check by their natural environments, and how it was that modern societies had come to transcend those limits or had separated themselves in some sense from their 'natural' origins.
>
> (1996: 4)

In this sense, the social theory of such seminal thinkers as Max Weber, Emile Durkheim, Georg Simmel, Vilfredo Pareto and other founding fathers of sociology, while developing increasingly complex and theoretically and empirically informed (if not always persuasive) explanations for social phenomena, had not progressed much beyond the eighteenth-century concern of social theory with exploring and charting the unilinear progress of society from 'rude' or 'primitive' to 'modern' or 'advanced' forms. Key to this approach was that the further a society is removed from nature (as in the transformation of its economy and society from a rural-based agricultural one to one based on urban and industrial manufacturing) and can 'control' the environment, the more advanced, modern and progressive it was.

At the same time, there were disciplinary or knowledge-based reasons we can point to in order to understand the lack of concern within classical social thought with the relationship between society and its environment. Benton, commenting on the early development of sociology, suggests that:

> the conceptual structure or the 'disciplinary matrix' by which sociology came to define itself, especially in relation to potentially competing disciplines such as biology and psychology, effectively excluded or forced to the margins of the discipline such questions about the relations between society and its 'natural' or 'material' substrate.
>
> (1994: 29)

This observation concerning sociology which may be expanded to classical social thought in general, as outlined in the previous chapter, was simply a continuation of the dominant approach to environmental considerations within nineteenth-century social thought.

Freud, human nature and the war against nature

While an unlikely contributor to the engagement of social theory and the environment, Sigmund Freud (1856–1939), the modern founder of psychoanalysis, did have something to say on the issue. In *Civilization and its Discontents*, Freud argues that subduing external nature may be seen as part of the same project of humans imposing their wills on unruly 'human nature'. The positive benefits that could be derived for humanity in uniting in a battle against nature included an unprecedented sense of collective solidarity, and a willed desire to eradicate nature-imposed problems (diseases, pain). In other words, if humanity were to regard nature as its 'common enemy', then it would have a strong reason to unite as a community in solidarity to 'fight' and 'subdue' nature. This aim echoed the early call of Francis Bacon for humanity, via scientific knowledge, to dominate nature. As Freud put it, his ideal for humanity involves 'combining with the rest of the human community and taking up the attack on nature, thus forcing it to obey human will, under the guidance of science' (quoted in Passmore, 1980: 23).

It is important to note that when Freud, like many later social theorists, spoke of 'nature', he meant both 'external nature' (i.e. the natural environment), and 'internal nature' (i.e. 'human nature'). In Freud's case, a central part of his theory concerned how civilisation, advancement and social progress depend not simply on the exploitation of external nature, but also required the 'controlling' and taming of some potentially disruptive aspects of our 'internal nature'.

According to Freud, 'Our civilization is, generally speaking, founded on the suppression of instincts' (1950: 82). According to Freud's view there was 'an inherent conflict between civilization and instinctual pleasure, as though the former was built upon a progressive renunciation of the latter' (Wollheim, 1976: 222). Hence the idea, developed later by critical theorists such as Marcuse (discussed below), that civilisation was 'against nature' in two senses. On the one hand, human civilisation was human, the product of human not nature's agency, and was created in opposition to the natural order (recall the binary opposition of 'culture/nature' in Chapter 2). On the other, the creation of a human social order, a civilised and secure social order, demanded the repression (or at least the sublimation) of human instincts, which were potentially wild, unruly and destructive of that order. However, for other theorists, particularly those associated with a 'deep ecological' perspective, suppression of the 'wildness' within humans was something to be criticised and lamented, and was deeply connected to the loss of 'external' or environmental wilderness. In other words, against Freud, these theorists posit a direct link between the destruction of natural wilderness areas and the progressive taming and domestication of human beings and the loss of human liberty, autonomy and authenticity (Oelschlaeger, 1992).

In Freud's view of the relationship between the social and the natural order, we have not only another underlining of the radical separation of 'nature' and 'society', but also the continuation of a related theme which goes back to Hobbes, for whom life was but a 'motion of limbs' and governed by instinctual desires (e.g. sex, possession, consumption, food) which we struggled to control. Like Hobbes, Freud was concerned that the 'natural impulses' of humans needed to be 'ordered' or 'contained' so that a stable human social order could emerge from natural disorder. Civilisation was not only threatened by 'external nature' but also by 'human nature', the instinctual drives and modes of acting of humans as a species of animal. Hence the need to repress these potentially disruptive instincts within human social relations, and, in arguing for humanity to unite in a 'war against nature', Freud was suggesting that these instincts were best channelled into subduing external nature. As Marcuse puts it, for Freud, 'the diversion of destructiveness from the ego to the external world secured the growth of civilization. However, extroverted destruction remains destruction: its objects are in most cases actually and violently assailed, deprived of their form, and reconstructed only after partial destruction. . . . Nature is literally "violated"' (1955: 78–9).

Thus Freud presented us with a picture of modern society ('modernity') in which the latter is premised on the double repression of 'internal' nature (transforming 'rude human nature' into civilised, ordered codes of conduct and manners) and the domination of external nature. One necessarily entailed the other. In calling for a united humanity to combat and control nature, Freud was simply suggesting a public, collective form of what modern industrial civilisation demanded at the individual, psychological level, an antagonistic and aggressive disposition and attitude towards both internal and external nature. This insight, not limited to Freud's work, set the scene for later social theorising about nature, particularly existentialism and critical theory.

Existentialism and the 'meaningless' earth

While not strictly speaking a form of society theory as opposed to a particular philosophy, existentialism is worthy of a brief discussion because it has an interesting slant on the relationship between humans and nature, and also lays the ground for much of recent social theorising about the environment.

Existentialism, associated with thinkers such as Jean-Paul Sartre, Martin Heidegger and Simone de Beauvoir, is an anti-rationalist philosophy of existence, which flourished from the 1930s to 1960s, though its modern origins go back to the Danish philosopher Søren Kierkegaard (1813–55).

One of the distinctive ideas of existentialism is that unlike pre-modern and Christian thought, the existentialist position holds that humans are simply 'thrown into' a meaningless world. Whereas for both pre-modern, Christian and other forms of social theory, the natural world was basically our 'home', despite periodic problems between it and humans, there is no such comforting view within existentialism. Haught highlights how existential 'homelessness' can both explain the attractiveness of anthropocentrism and also account for environmental problems. For him:

> It appears quite likely that the origins of our environmental crisis lie, in part, in a deeply entrenched suspicion by humans that the cosmos is not really our home. The feeling of cosmic homelessness is, to a great extent, apparently 'religious' in origin. . . . Exaggerating our own importance may be an understandable reaction to the prior conviction that we are exiles from any value-bestowing universe. . . . Anthropocentrism is our way of responding to the feeling of not really belonging to the earth and the cosmos.
>
> (Haught, 1994: 27–8)

Hence existentialism may be viewed as a consequence, how the separation of humans from nature (which as shown in Chapter 2 has a long tradition in Western thought) can easily lead to this existential sense of 'homelessness', isolation and alienation with regard to our relation to the environment.

This human separation and alienation from the natural world is the modern 'human condition'. Modernity and modern society have created a meaningless world which is indifferent to us and our fate. Existentialism is a very anthropocentric philosophical outlook in that its focus is on the meaning of human life, within society, with little reference to the relationship between society and environment. Like Nietzsche, existentialism takes the 'death of God' as a 'given' of the modern secular world in which Christian and other religious views are no longer culturally dominant and philosophically bankrupt in terms of telling us who we are and where we are going.

On the whole existentialism holds that our attitude towards the natural environment has to be an instrumental one, one in which the environment is a passive 'object' rather than an active 'subject'. In many respects, existentialism represents an extreme view of the consequences of the 'disenchantment of nature' in modernity. This is a theme taken up by one of the most influential schools of post-war social theory, the Frankfurt School of critical theory, which we turn to next.

The Frankfurt School: critical theory, nature and the problems of modernity

Critical theory based in the Frankfurt Institute for Social Research (hence its later name, the Frankfurt School) is a body of neo-Marxist social theory. The distinctive features of critical theory were its critical analysis of the Enlightenment and the dominant ways of acting and thinking associated with it, and its linking social theory to social criticism of the prevailing 'modern' social order (both in the liberal-capitalist West and authoritarian-communist East). It included thinkers such as Max Horkheimer (1895–1973), Theodore Adorno (1903–69), Herbert Marcuse (1898–1979), Walter Benjamin (1892–1940), and Jürgen Habermas (1929–) (discussed in more detail below).

Max Horkheimer and Theodore Adorno's famous work, *Dialectic of Enlightenment*, was an examination and explication of the 'dark' side of modernity, the costs and dangers of advanced technological, industrial society, and its dominant modes of thought and behaviours. The danger they point out is quite stark. For them, 'The fully enlightened world radiates disaster triumphant', a sentiment with which many radical greens would agree in terms of the local and global environmental degradation and destruction which modern industrial societies have caused. According to Biro, 'Like both Rousseau and Marx before them, Horkheimer and Adorno question the association of increased mastery over nature . . . with a betterment of the human condition' (Biro, 2005: 123). For Horkheimer and Adorno the technological sophistication, economic organisation and political institutionalisation associated with that mastery has no necessary connection with universalising the benefits of that mastery (here they echo the Marxist critique of the necessary inequalities of capitalism) and that such mastery is itself misplaced and dangerous as the basis for structuring the relationship between society and nature. In addition, and connecting with the point raised above in relation to Frued, Biro points out that 'The lesson of *Dialectic of Enlightenment* is that if we think of nature as "other" to be controlled, the technology we develop will inevitably be turned to control the nature that resides within the self' (Biro, 2005: 148).

As Vogel puts it, 'the Frankfurt school's critique of contemporary society was offered up in a certain sense "in nature's name" – both that of the damaged inner nature of humans stuck in the fatal dialectic of enlightenment and an outer nature robbed of all qualities save those that render it amenable to human use' (1997: 175). As a result of the logic of the Enlightenment, the only value the natural environment can possess is instrumental value; that is, the natural world possesses value insofar as it is useful for human purposes or ends.

The 'disenchantment of nature' (the cultural transformation of nature from a morally significant realm with its own intrinsic value, to being viewed solely as a set of resources for human use and enjoyment, discussed in Chapter 3), as one of the main consequences of modernity, was something not only to be regretted (the Romantic reaction) but was also dangerous for both human society and the nonhuman world. In particular, the increasing rationalisation which was central to the successful technical manipulation of external nature had a tendency to 'spill over' into other spheres of human life in which they were not appropriate and were dangerous. The basic problem was this: the instrumental use of nature developed institutions, modes of thinking and acting which were then 'transferred' illegitimately to human social and personal relations. The domination and exploitation of the natural environment leads to the domination and exploitation of humans – a point echoed by various green thinkers from deep ecologists to eco-anarchists. Enlightenment-derived institutions as modes of rationality suited to human–nature exchanges contained the possibility of being used in human social relations where they were dangerous and harmful. As Horkheimer and Adorno put it, 'Men have become so utterly estranged from one another and from nature that all they know is what they need each other for and the harm they do to each other' (1973: 253). While Adorno in paricular is rather pessimistic about any reconciliation with nature (such a project had to wait for later critical theorists such as Marcuse and Habermas, discussed below), the importance of the early Frankfurt School in relation to social theory and the environment lies in updating and greatly developing the essentially Marxist notion that the root of the link between society and nature lies both within social relations internal to the organisation of society *and at the same time* with how society views, uses and abuses its natural environment/s (Biro, 2005: 159).

Marcuse and the 'liberation of nature'

Herbert Marcuse follows Adorno and Horkheimer's *Dialectic of Enlightenment* in arguing that both inner and outer nature are suppressed and distorted within and by a conflict-ridden, crisis-prone and class-divided modern capitalist social system. As Feenberg points out, '*One-Dimensional Man* (1964) is often compared to *Dialectic of Enlightenment*, it is far less pessimistic' (1996: 49), or as Vogel more aptly puts it, 'Marcuse is just sometimes Adorno in a good mood' (1996: 137)!

In *Eros and Civilisation*, his study of Freud, Marcuse describes the relationship between the 'death instinct' and the exploitation of the external world. For Marcuse:

> the entire progress of civilization is rendered possible only by the transformation and utilization of the death instinct or its derivatives. The diversion

of primary destructiveness from the ego to the external world feeds tech-nological progress. . . . In this transformation, the death instinct is brought into the service of Eros; the aggressive impulses provide energy for the continuous alteration, mastery, and exploitation of nature to the advantage of mankind. *In attacking, splitting, changing, pulverizing things and animals (and, periodically, also men), man extends his dominion over the world and advances to ever richer stages of civilization.*

(1955: 47; emphasis added)

Later, in a critique of the Enlightenment project, Marcuse asserts that 'The ego which undertook the rational transformation of the human and natural environment revealed itself as an essentially aggressive, offensive subject, whose thoughts and actions were designed for mastering objects. It was a subject *against* an object' (1955: 99). In other words, the human subject (the modern self) in order to dominate the external environment to produce the industrial social order had to see that external world as a passive object for human manipulation and control. In so doing, as Marcuse points out, this attitude towards nature (including human nature) revealed an 'aggressive, offensive' human self and by extension, an aggressive, offensive modern social order (capitalism). As he goes on to explain, 'Nature (its own as well as the external world) were "given" to the ego as something that had to be fought, conquered, and even violated' (1955: 99). Human social development and progress under modern industrial capitalism required, as Freud echoing a theme of social theory going back to Hobbes and Rousseau, a warlike attitude towards the natural environment and a will to control the potentially destabilising instincts of internal (human) nature.

However, the 'reconciliation with nature' that Marcuse advocated did not imply going back to some 'pre-modern' or 'organic' relation between humans and nature. On this point Marcuse, following Horkheimer and Adorno, is clear. For him, the '"liberation of nature" cannot mean returning to a pre-technological stage, but advancing to the use of the achievements of technological civiliza-tion for freeing man and nature from the destructive abuse of science and technology in the service of exploitation' (1972: 60). He argues instead for a 'liberating' domination of nature as opposed to the 'repressive' one typical of modern industrial societies, and suggests that 'no free society is imaginable which does not . . . make a concerted effort to reduce consistently the suffering which man imposes on the natural world' (1972: 68). This 'liberating domination' means the 'civilising' or humanising of nature through such liberating physi-cal (and cultural) transformations of nature from 'wilderness' or 'raw nature' into parks, gardens, farmland, landscapes and reservations. Here we find echoes of the Christian 'perfection of nature' idea discussed in Chapter 2. As Feenberg points out this reconfiguration of instrumental rationality:

would lead to a change in the very nature of instrumentality, which would be fundamentally modified by the abolition of class society and its associated performance principle. It would then be possible to create a new science and technology which would be fundamentally different, which would place us in harmony with nature rather than in conflict with it. Nature would be treated as another subject instead of as mere raw materials. Human beings would learn to achieve their aims through realizing nature's inherent potentialities instead of laying it waste in the interest of narrow short-term goals such as power and profit.

(1996: 50)

A central part of reducing this unnecessary suffering for Marcuse requires a transformation in the idea and practice of 'development' in the advanced capitalist nations, basically a shift from a quantitative, consumption-orientated view to a more aesthetic-qualitative one. Anticipating a central green or ecological argument, he suggested that 'the sheer quantity of goods, services, work and recreation in the overdeveloped countries which effectuates this containment [of humanity]. Consequently, qualitative change seems to presuppose a quantitative change in the advanced standard of living, *namely, reduction of overdevelopment*' (1964: 242; emphasis added). And, like the green argument for reducing 'overdevelopment', Marcuse suggested this for both ecological (decreasing environmental degradation) and social/emancipatory reasons (such a transformation would lead to a less oppressive, bureaucratic, work-orientated, consumption-based society).

Towards the end of his life Marcuse saw in the emerging ecological movements in Europe and North America a more promising anti-capitalist social movement than the revolutionary project of Marxist class struggle, which expressed the radical 'emancipatory' goals he endorsed, and the creation of a 'post-capitalist' society. As he put it in a speech published posthumously in 1992 in the journal *Capitalism, Nature, Socialism*:

> The ecology movement reveals itself in the last analysis as a political and psychological movement of liberation. It is political because it confronts the concentrated power of big capital, whose vital interests the movement threatens. It is psychological because . . . the pacification of external nature, the protection of the life-environment, will also pacify nature within men and women A successful environmentalism will, within individuals, subordinate destructive energy to erotic energy.
>
> (1992: 36)

Andrew Feenberg notes in a commentary on this speech:

> The central question of Marcuse's thought appears clearly in this short speech: from what standpoint can society be judged now that it has succeeded in feeding its members? Recognizing the arbitrariness of mere moral outrage,

> Marx measured capitalism by reference to an immanent criterion, the unsatisfied needs of the population. But that approach collapses as soon as capitalism proves itself capable of delivering the goods. Then the (fulfilled) needs of the individuals legitimate the established system. Radicalism means opposition, not just to the failures and deficiencies of that system, *but to its very successes.*
>
> (1992: 38; emphasis added)

In identifying the putative 'successes' of contemporary capitalist-industrial society as a legitimate and central object of critique (alongside highlighting its deficiencies and flaws), Marcuse correctly saw how the ecological and green movements distinguish it from older social movements and thus point to the uniqueness of green social theory which will be discussed later in both the critique of 'economic growth' and material consumption (Chapter 6) and Chapter 11 on 'greening social theory'.

However, since Marcuse, critical theory has not made ecology, the nonhuman world or social–environmental relations the central themes in its critical analysis of modern societies. As Whitebrook has pointed out, 'critical theory, which aspires to provide a comprehensive theory of the crisis of modernity, has little to say about one of its most decisive features, ecology' (1996: 286), although recently there has been some engagement (Brulle, 2002). Partly for an assessment of this claim, we turn next to perhaps the most important theorist of contemporary critical social theory, Jürgen Habermas.

Jürgen Habermas and the problem of nature in modernity

In the main Habermas has sought to show that the only relation we have with the natural environment is an instrumental one, governed by productive, prudential and technical concerns about how best we may exploit it. His concern, relating to what he sees as one of the dangers of modernity, is to prevent human social relations (or what he calls communicative concerns) from being reduced to instrumental norms which are appropriate to the sphere of human techno-logical manipulation of the natural world. That is, he does not want how we treat and view each other to be the same as how we treat and view the nonhuman environment.

Habermas' basic position is that an instrumental, technical or manipulative attitude towards the external world is something that is simply a 'given', a 'brute fact' of the particular character or nature of the human species. For Habermas humans are genetically and evolutionarily disposed towards having this view of the natural environment, or, as Vogel puts it, we have 'hard-wired interests' (1997: 183) in the nature of an instrumental character, and these cannot be changed. The evolutionary character of Habermas' thought, which is in keeping with classical social theory, is central to understanding his position. However, unlike classical social theory, according to Dickens (1992), Habermas is at pains not to be seen to promote a simplistic, unilinear view of social development. However, what Habermas does share with classical social theory is the Enlightenment belief that the progress and development of human society is premised on the exploitation and instrumental use of the natural environment.

Habermas also takes issue with Marcuse's notion of the possibility of a 'new science' and reconfigured instrumental rationality which would overcome the alienation between human society and the natural environment. In his essay 'Technology and Science as "Ideology"', Habermas criticises what he calls the 'secret hopes' of a whole generation of social thinkers, such as Benjamin, Adorno, Bloch and Marcuse, whose implicit ideal was the restoration of the harmony of humanity and nature. According to Feenberg, Habermas 'atacks the very idea of a new science and technology as a romantic myth; the ideal of a technology based on communion with nature applies the model of human communication to a domain where only instrumental relations are possible' (Feenberg, 1996: 48). For Habermas technology is a generic human species-specific project, 'a "project" of the human species *as a whole*', not of some particular historical epoch like capitalist society or of a particular class such as the bourgeoisie (Habermas, 1970: 87). As the quote from Vogel above puts it, technology and its instrumentalising attitude towards nature are 'hard-wired' (and this to all intents and purposes unchangeable) within human nature and is therefore not contingent upon particular social and economic conditions. Thus even a transition to a 'post-capitalist' society such as the one envisaged by Adorno and Marcuse would exhibit an instrumental technology and science, and thus preclude any reconciliation with nature, and indeed more than that; for Habermas any attempt at such a reconciliation is undesirable (even if possible) and dangerous.

The moral status of the nonhuman world

However, Habermas appears uneasy with the fact that his theory 'seems to preclude as irrational a non-objectivistic relation to nature' (1982: 241). That is, it seems that Habermas must dismiss as 'irrational' a non-instrumental view or valuation of the natural environment. As Feenberg puts it:

> Habermas's communication-theoretic conception of the social, it goes along with an unacceptable view of nature as nothing but the reified object of science and technology. In the end this dualism prevents Habermas from dealing

adequately with environmental ethics which crosses the artificial lines he has drawn between nature and society. His narrow positivistic conception of nature leaves no room for a normative relation to the environment. It seems that nature can have only instrumental value.

(1999: 86)

This instrumental-technical attitude to nature is something we must simply accept as the 'price' to be paid for the advantages of the modern social world. Habermas' view of social-environmental affairs is heavily influenced by his theory of human knowledge in which the external environment is for him 'constituted' by the type of knowledge appropriate to its study, namely the natural sciences and the instrumental view and attitude they have towards nature. He does not seem to deny non-instrumental attitudes to nature, but rather maintains that an instrumental attitude is superior in terms of human manipulative control over nature. As he puts it, 'while we can indeed adopt a performative attitude to external nature, enter into communicative relations with it, have aesthetic experience and feelings analogous to morality with respect to it, there is for this domain of reality only one *theoretically fruitful* attitude, namely the objectivating attitude of the natural-scientific, experimenting observer' (Habermas, 1982: 243–4; emphasis in original). At the same time, Habermas is concerned that the alternative to this instrumental-technical attitude to nature (based on its 'disenchantment') is a dangerous strategy of 're-enchanting' nature via an appeal to some mystical or spiritual worldview. For him, 'The phenomena that are exemplary for a moral-practical, a "fraternal", relation to nature are most unclear, if one does not want to have recourse here as well to mystically inspired philosophies of nature, or to taboos (e.g. vegetarian restrictions), to anthropomorphising treatment of house-pets and the like' (1982: 244–5). And the problem with such strategies for re-enchanting nature is that they do not lend themselves to democratic or discursive modes of communication, as well as entailing a breakdown in rational scientific discourse. Mystical attitudes are by definition beyond rational explication and justification, and for Habermas this is precisely what makes them invalid: such forms of thinking and acting constitute potentially anti-democratic forms of authority.

Yet while he rejects attempts to 're-enchant' the environment, he cannot as easily dispel what may only be described as an unease with the implications of his theory for the negative affects of human action on the nonhuman world. Thus, for example, he notes that 'The impulse to provide assistance to wounded and debased creatures, to have solidarity with them, the compassion for their torments, abhorrence of the naked instrumentalisation of nature for purposes that are ours but not its, in short the intuitions which ethics of compassion place with undeniable right in the foreground, cannot be anthropocentrically blended out' (1982: 245).

An almost exact repetition of this line may be observed in a later aspect of his thought, that concerned with 'discourse ethics', where his ambiguity about the claim that our relations with the nonhuman world are normative is evident (if not fully resolved). Habermas asks:

> How does discourse ethics, which is limited to subjects capable of speech and action, respond to the fact that mute creatures are also vulnerable? Compassion for tortured animals and the pain caused by the destruction of biotopes are surely manifestations of moral intuitions that cannot be fully satisfied by the collective narcissism of what in the final analysis is an anthropocentric way of looking at things.
>
> (quoted in Outhwaite, 1996: 200)

Unfortunately, no satisfactory answer from within his own frame of reference is forthcoming, according to green critics such as Eckersley (1990); and on the whole, while the place of nature and social–environmental relations does constitute a gap in Habermas' otherwise impressive and massive social theory, he does not seem unduly bothered by this. However, this may be too sweeping, and we would do well to remind ourselves of the focus on Habermas' emphasis on the priority of social relations. As Dobson states, for Habermas, 'Healing the rift between human beings and the natural world . . . is not a matter of joining what was once put asunder, but of getting the relations between human beings right first' (1993: 198). However, some interpreters of Habermas' thought have been more interested in applying his theory to human–nature relations and it is to certain of these that we turn next.

Knowledge, nature and society: towards reconciliation?

According to Vogel, one of the main roots of Habermas' 'unease' with the place of nature within his theory lies in his theory of knowledge. Habermas maintains a dualism between society and nature and the types of knowledge appropriate to these two spheres. He separates the 'natural' from the 'social' sciences, not least by stating that the former is 'positive' in the sense of yielding objective, value-free knowledge, while the latter is 'normative'; that is, yields knowledge about society and social relations which depend on values and normative principles. According to Vogel, 'The cost of this inoculation of natural science from the normative . . . is precisely the impossibility of conceptualizing nature itself as possessing moral worth, or of recognizing a moral dimension to the way in which we interact with it' (1997: 180). This is simply another way of stressing the 'price' to be paid for the 'disenchantment of nature' as a necessary precondition for increased human technological power and control over nature, and the material and other benefits that follow from this.

Vogel contends that Habermas has an unduly narrow view of scientific knowledge, and that there are other 'theoretically fruitful' (to use Habermas' own terms) ways of studying, conceiving and interacting with the natural environment. For him:

> there is no single 'theoretically fruitful' approach to nature, but rather the question of what is fruitful turns out itself to be answerable in socially and historically varying ways, which means that alternative approaches are certainly imaginable. The view of nature as mere matter for instrumental manipulation criticized by Marcuse and the earlier Frankfurt School is not, as Habermas tried to argue, built into the structure of the species or of 'work', but rather – as they had originally asserted – is associated with a particular social order and a particular historical epoch. A new kind of society, then, might well involve a new science, and with it a new nature as well.
>
> (Vogel, 1997: 187)

In this way Vogel is presenting a form of the social construction of nature in that different social orders may have different forms of science which in turn will reveal or construct new forms of natural environment for study and contemplation. Thus Vogel contends that a 'reconciliation of humans and nature', one of the aims of the early Frankfurt School (and of contemporary deep ecologists and others in the radical ecology movement discussed in Chapter 9), is possible, but only if what he calls the 'sociality of nature' is explicitly acknowledged. For him, the natural is always the social in two senses: 'first, because we perceive and experience it, study and dream about it, in terms that are from the beginning social through and through, *but second also because the objects and landscapes through which we experience it are always themselves – when closely examined – in part the product of earlier social practices*' (1997: 186; emphasis added). While the first can be accepted, in the sense that 'the environment' as a concept is constructed by humans, the second claim that the environment is, in part, always also physically produced by humans is highly contentious.

The reconciliation he outlines is not based on re-enchanting nature, nor premised on recognising the 'Otherness' of nature (that is, how radically different or alien nature is, from a human point of view, discussed later in Chapter 7) or attempting to see nonhuman entities as potential participants in Habermassian 'discourse ethics' as Dryzek (1990), for example, attempts. According to Vogel, 'What is wrong with the way nature appears to us (and our natural science) today is that it seems utterly independent of us and even opposed to us. Its sociality is hidden: we see it as separate from us, as dangerous (and impossibly complex, as always taking its revenge), and fail to see that rather it is always something in which we are deeply and actively enmeshed' (1997: 188). It is precisely because nature appears as 'other' and as 'alien' that we feel separate from it and consequently fear it, and this leads to a desire to control and dominate it. Vogel's strategy is to claim that this seemingly 'alien', strange external nature is in fact always already social in that it is partly the product of (past and present) human transformative practices. As he puts it, 'Only a nature viewed as so separate from us would be something we would feel the (frightened) need to "dominate" or the (equally frightened) need to "preserve". In either case it appears as alien, something to be either overcome or propriated – not simply as part of the (social) world that we both inhabit and continuously transform' (1997: 188). If we see that nature is already social, something which humans have helped produce, this brings home to us that it is not alien, not 'other' and thus lays the basis for harmony between humanity and nature. Vogel's philosophical attempt at reconciliation is based on the (Marxist) idea that we can know that which we make. Hence if 'nature' is, at least in part, 'socially constructed' in the ways Vogel maintains, we can know it, and thus nature is no longer alien or 'other'.

While there is much merit in Vogel's Habermassian 'environmental ethics', in particular his argument which shows the failings of Habermas while also pointing to ways in which Habermas' theory can take on board environmental considerations, there are some problems with it. The largest problem is that Vogel's position cannot extend beyond those parts of the natural environment which are resolutely not, even in part, the product of (past or present) human social practices. Thus while Vogel's argument is generally persuasive in respect to parts of the natural environment such as agricultural landscapes, parks, gardens, and other natural systems which are either wholly or partly the product of human transformation, it is less persuasive when it comes to 'wilderness' areas and other aspects of the natural world, such as global hydrological and carbon cycles, the ozone layer and so on, which are clearly not the product of human social practices. Thus the second of Vogel's claims about the 'social' construction of the environment is, at least for these parts of the natural environment, extremely doubtful. Since we do not make these categories of natural entities, processes and systems, we cannot 'know' them in the way demanded by Vogel's reconciliation thesis. And in this respect, an appeal to the 'otherness' of nature may be unavoidable, though of course this does not mean a sense of 'alienation' from nature (O'Neill, 1993), or necessitate a mystical re-enchantment of nature (Barry, 1993) to overcome that alienation.

However, other critical theorists who follow Habermas have suggested that the most productive line to conjoin his thinking with ecological concerns lies in the area of democratic will-formation and democratic decision-making in relation to the non-human world and different human valuations of nature (a point taken up later in Chapter 6 on the ways in which the non-democratic practices of orthodox economic valuation 'crowd out' non-economic ecological values). According to Brulle:

> We need to develop and institutionalise more adequate procedures to integrate the consideration of ecological values into the decision-making process. To accomplish this task, we do not need to look to metaphysical arguments, but to ourselves and our beliefs, political actions, and social institutions. Constructing an ecologically sustainable society has never been accomplished before, and so we do not know in advance what will or will not work, including ethics, institutions, or individual personality characteristics. Thus any efforts to create this society should proceed through a practice of trial and error in a 'logic of justified hope and controlled experiment' [Habermas, 1971: 283–4]. Critical Theory can provide valuable intellectual resources toward the realisation of this project. It is in the democratic conversation about our fate and the fate of nature that Habermas and green political theory converge.
>
> (2002: 16–17)

Indeed, it is in the area of deliberative democracy (as opposed to integrating the independent ethical status of the nonhuman world) that we find most of the recent scholarship which attempts to bring together Habermas and environmental concerns and/or green thinking (Barry, 1999a; Dryzek,1990; Smith, 2003).

Environmental politics and the defence of the 'lifeworld'

Habermas, like Marcuse before him, sees the rise of environmental politics (as a 'new social movement'), expressing a concern for the care and protection of the natural environment, as significant. Environmental politics are significant according to Habermas because they mean a new politics beyond 'left and right'. As he puts it, 'the[se] new conflicts are not sparked by *problems of distribution*, but concern the *grammar of forms of life*' (1981: 33). By this he means that green politics and concern for the preservation of the natural environment mark a new form of politics, one that is not focused on the distribution of the economic pie, jobs, consumption, income, wealth and so on, which are the main issues in left–right politics. Rather, environmental politics, along with other new social movements, such as gay rights and feminism, seek to challenge and change the existing institutional and moral order, a major part of which involves bringing the question of identity and lifestyle to the centre stage of political struggle. According to Habermas, 'the dissident critics of industrial society start from the premise that the lifeworld is equally threatened by commodification and bureaucratization' and that against these twin imperatives what is needed as an alternative is 'a will formed by radical democratic procedures, which would put participants *themselves* in a position to realize concrete possibilities for a better and less threatened life, on *their own* initiative and in accordance with *their own* needs and insights' (1989a: 63, 69). In the second volume of his *Theory of Communicative Action* he notes that:

What sets off the protest is . . . the tangible destruction of the urban environment; the despoilation of the countryside through housing developments,
industrialization, and pollution; the impairment of health through the ravages
of civilization; pharmaceutical side-effects, and the like – that is, developments
that noticeably affect the organic foundations of the lifeworld and make
us drastically aware of standards of liveability, of inflexible limits to the
deprivation of sensual-aesthetic background needs.

(Habermas, 1989b: 394)

Environmental politics seek, in part, to protect what Habermas calls the
'lifeworld', the world of everyday social interaction and realm of moral action
from the 'systems world', the world of bureaucratic, state administration and the
capitalist market where 'power' and 'money' respectively are the dominant means
of rational organisation. However, whereas for Marcuse environmental politics
is radical and emancipatory, for Habermas it is 'defensive', a reaction against
the technological domination of the natural world in modern advanced societies,
a perception of the environmental movement Habermas shares with Giddens,
discussed below. As Roderick puts it, 'For Habermas, at the present time only the
women's movement belongs to this latter category [of making universalistic
demands] to the extent that it seeks not only a formal equality, but also a fundamental change in the social structure and in real concrete life situation' (1986:
136). Yet to think that the changes which greens and those commited to the
transition to a more sustainable society make are not radical or structural is simply
mistaken and absurd, and represents a significant blind spot within Habermas'
thinking. Habermas continues to be weary and sceptical of environmental politics
and ethics on the grounds that this represents an atavistic conception of nature
with all its reactionary connotations, particularly in relation to facism, Nazism
and ecological themes discussed later in Chapter 6. Another reason for Habermas'
rather dismissive attitude towards new social movements such as the greens lies
in his assessment of their proposals for establishing counter-institutions from
within the lifeworld as unrealistic (Eckersley, 1990: 742).

One of Habermas' main concerns is with preventing what he calls the 'scien-
tisation of politics': that is, ensuring that political discourse is not reduced to a
'technical' one, so that how humans treat each other (through political action and
institutions) does not resemble how humans treat nature. However, for Habermas
environmental problems are essentially 'technical' problems not moral ones,
which goes against how many of those concerned about the natural environment
(and not just greens) feel about environmental degradation and destruction.
As he puts it, 'The intervention of large-scale industry into ecological balances,
the growing scarcity of non-renewable resources, as well as demographic developments present industrially developed societies with major problems; but these
challenges are abstract at first and call for *technical and economic solutions*,

which must in turn be globally planned and implemented by administrative means' (quoted in Outhwaite, 1996: 325). Thus, Habermas may be said to continue the pervasive dualism with Western social theory which insists on maintaining a sharp and morally significant separation between the social and the natural worlds. While not completely the case, it is at the same time not really misleading to see Habermas as being more interested in the human 'lifeworld' than in the 'natural world' or the relations between the two. While he acknowledges the existence and importance of ecological problems he continues to think that they 'can be dealt with satisfactorily within the anthropocentric framework of a discourse ethic' (1982: 268), meaning he does not accept that ecological issues and problems require significant changes to his social theory.

While Habermas himself seems, on balance, to be sceptical of attempts to 'green' his thought, this has not stopped others doing this. Examples include Dryzek (1990) who has argued for an extension of communicative rationality and communicative ethics to the nonhuman world, and that an expanded notion of agency may be extended to the natural world consistent with Habermas' social theory, while Bruelle (2000) has used Habermas to analyse the US environmental movement and point towards ways of democratising and radicalising it, while Torgerson (1999), similar to Bruelle (2002), has extended Habermas' notion of the 'public sphere' to analyse green politics. Other non-green theorists such as White have suggested that the growing environmental crisis could be used 'as a practical catalyst for reflection on how the ways in which we currently assault nature are leading to a more and more frustrating and self-destructive form of life' (White, 1988: 167–8), a suggestion that Habermas as yet does not seem to have taken.

However, Habermas' recent work (Habermas, 2003) on the worrying implications of biotechnology and genetic engineering as applied to humanity and the attempt to shape our own evolution does consttitute something of a shift in his focus, raising the hope of a more engaged relationship with issues of 'external' nature (environment) and 'internal' human nature. In particular, Habermas is concerned that the ability, through biogenetic engineering for humanity to 'screen' out 'undesirable' or 'harmful' traits in humans (a form of 'liberal eugenics which echoes the eugenics argument discussed in the next chapter), threatens to erode the vital distinction between the 'grown' and the artificially 'made' or 'manufactured'.

According to Hall:

> For Habermas, resisting liberal eugenics is a matter of liberal justice: it is the only way to respect the equality and autonomy of future generations. And if we were to engage in the genetic control of future human beings, we would abandon our own self-understanding as morally autonomous individuals bound

by moral respect for others. By trying to control the genetic future, we would erode the foundations of liberal society in the present.

(Hall, 2004: 50–1)

The creation of 'pre-programmed' children for Habermas (and others who also share his concern as to the ethical implications of genetic engineering) would rob them of their autonomy and essential ethical status as free and equal beings, which is why for him the issue of genetic screening and biotechnology is one which relates to the 'ethical self-understanding of the species' of what it means to be 'human'. His basic objection, and one that is as old as Immanuel Kant, is that these technologies open up the possibility (perhaps probability) of 'instrumentalising' human life in the sense that the 'creation' of humans in this way reduces them to a 'means' rather than being 'ends in themselves'.

Anthony Giddens, globalisation and the environment (with Iorwerth Griffiths)

Like Habermas, Anthony Giddens' work is concerned with modernity and its effects, though unlike Habermas he has paid more attention to environmental issues and the place of nature within his social theory. Giddens' engagement with environmental issues begins from an awareness of the lack of attention to ecological issues within sociology. As he puts it, 'Ecological concerns do not brook large in the traditions of thought incorporated into sociology, and it is not surprising that sociologists today find it hard to develop a systematic appraisal of them' (quoted in Cassell, 1993: 287).

Giddens' explicit concern with 'space and time' also means that he is more sensitive than other social theorists to ecological concerns, which clearly have spatial/geographical and temporal dimensions and import for human societies. This geographical dimension of Giddens' work is central to his theory of globalisation. Although the idea has undergone some alteration within Giddens' work, as Goldblatt (1996) points out, globalisation means the linking of the global and the local as a consequence of time–space distantiation. Simply put, globalisation refers to the historical processes which for the past four or five centuries have been connecting and bringing various parts of the world together into one system of cultural, political and, above all, economic relations. While this process has been going on for some time, it has intensified since the Second World War. As a result of various developments politically, culturally, in communications and transport, and in the establishing of a global capitalist market, the creation of an international division of labour, the rise in prominence of transnational corporations, the world today is a smaller place. By this is meant that time and space have been compressed in what Giddens calls the 'late modern age'. As a result of globalisation, distant places on the globe are intertwined and interdependent in that they are tied into a series of shared and common relations, institutions and processes.

Giddens' sensitivity to how central environmental issues are to modern social inquiry may be seen in his analysis of globalisation where he writes that 'the diffusion of industrialism has created "one world" in a more negative and threatening sense . . . a world in which there are actual or potential ecological changes of a harmful sort that affect everyone on the planet' (1990: 76–7). The modern world is a global world in the sense that there is a link between the local and the global as a result of space–time compression, such that changes in one part of the world can have potentially devastating effects on another part. While in Giddens' theory of globalisation these effects range from economic, financial and cultural consequences (not all of which are necessarily bad), one of the most tangible experiences of globalisation are global ecological problems – most saliant amongst which is climate change – and localised ecological problems as a result of the global economic system and the uneven biophysical and socio-economic consequences of 'biosphere–ecosphere' and colonial dynamics as discussed in Chapter 2. In particular, pollution problems, species and habitat loss and climate change transcend territorial boundaries of nation-states and are 'global' in their scope (which is not to say they affect all places and people on the planet equally). Thus, for Giddens and for other thinkers such as Sachs *et al.* (1998), the spread of global ecological problems is a specific consequence of globalisation, the transmission and spread of industrial capitalism to the entire globe, the creation of a global market, and the development of various communication, political and cultural institutions and connections bringing distant societies together within a single globalised world system.

The rise and meaning of environmental politics

According to Goldblatt (1996: 69–71), we can discern at least three different explanations of environmental politics within Giddens. The first is what we can call a 'conservative' view of environmentalism in that for him environmental movements are associated with recovery, the recovering of traditional ways of relating to the environment. Here the urban environment produced environmental movements born of the cultural alienation and spiritual vacuum of the modern urban landscape. A second explanation Giddens has for environmental politics is as a response to perceived ecological threats. On this account, environmental politics emerges as ecological problems and dilemmas become more obvious and discernible to people.

A third explanation is that environmental politics is a 'lifestyle politics' associated with new social movements. Here, Giddens is close to Habermas' view which associates green politics with non-distributional issues. For Giddens, green/ environmental politics is concerned with how one should live and issues of personal identity, rather than the typical issues which dominate 'mainstream' (left–right) politics, such as income levels, employment and economic growth. As Goldblatt puts it, for Giddens, 'environmental politics is not simply the outcome of increasingly perceived environmental risk. It is also fuelled by an increasing demand for the remoralization of abstract systems of social organization that have ceased to be accountable in any meaningful way to those they affect' (1996: 71). Thus Giddens sees the increasing concern with environmental issues and the rise of environmental politics as developments which are explicitly moral, in raising moral questions about the modern social order, its institutions and principles.

The urban environment and the 'town' versus the 'countryside'

According to Giddens:

> A critical theory alert to ecological issues cannot be limited to a concern with the exhaustion of the earth's resources . . . but has to investigate the value of a range of relations to nature that tend to be quashed by industrialism. In coming to terms with these we can hope not so much to 'rescue' nature as to explore possibilities of changing human relationships themselves. An understanding of the role of urbanism is essential to such an exploration. The spread of urbanism of course separates human beings from nature in the superficial sense that they live in built environments. But modern urbanism profoundly affects the character of human day-to-day social life, expressing some of the most important intersections of capitalism and industrialism.
>
> (quoted in Cassell, 1993: 329)

What is interesting about Giddens' writing on environmental issues is the stress he lays on the need for social theory to engage with urban, built environments, not only as they constitute the lived, everyday 'environment' within which individuals are situated, but also because the effects of urbanism have ramifications for how the 'natural environment' is constituted, perceived and acted upon. Here, Giddens follows a tradition within social theory in which urbanisation, cities, buildings and the creation of human-made artificial spaces and places are significant because they represent or express the difference between 'modern' and 'pre-modern' society.

Central to Giddens is the idea of the 'end of nature' which he claims has occurred in two ways. First, there is the spread of built environments meaning that humans are increasingly separated from nature, increasingly living in created locales (Giddens, 1991: 165–6). Second, the created environment thesis of naturally occurring events – such as the seasons, producing food – ceases to exist as more and more of them are influenced by social activities. As he puts it:

> Sequestration from nature in this guise is more subtle, yet more pervasive than in the first sense mentioned. For nature – the alteration of the days and seasons, the impact of climatic conditions – still seems to be 'there': the necessary external environment of human activities, no matter how instrumentally orientated they might be. Yet this feeling is specious. Becoming socialised, nature is drawn into the colonisation of the future and into the partly unpredictable arenas of risk created by modern institutions in all areas subject to their sway.
>
> (Giddens, 1991: 166)

This does not mean that the created environment is a humanly controlled environment. Giddens is aware that while modern societies have attempted to control nature, in doing so they have created new problems. The end of nature happened when we began to worry about what we have done to nature rather than what it can do to us. This difference between 'natural nature' and created environment is another reason why Giddens argues that modernity involves a qualitative break from the past. Furthermore it fits with his thesis of the change in risk from 'external' to 'manufactured' risk, similar to that advocated by Ulrich Beck (discussed later in Chapter 9). External risk fits with natural nature; that is, nature as separate from and external to social practices. On the other hand, manufactured risk fits with a created and humanly manipulated environment, where environmental problems are increasingly ones created by intentional human social (especially economic and technological) practices.

It may be argued that Giddens overstates the case as regards the end of nature and that we live in a created environment (Griffiths, 2006). Others such as Benton have criticised Giddens along the lines that the concept of the created environment is confused in that, at one moment, Giddens says humans have had a greater impact on nature than ever before, yet, at another moment, nature no longer exists (Benton, 1999: 58–62). The reason for this, according to Benton, is Giddens' confused use of the word 'nature'. Benton accuses Giddens of using the word 'nature' in the same way as deep ecology; that is, to signify a pristine Eden which never existed. The reality is that humans have always transformed their environments.

As Griffiths puts it:

> Giddens points out that humans increasingly live in urban environments and processes such as climate change or ozone depletion have been created by humans. However the further claim that this means we live in a created

environment, one in which natural processes no longer exist is an unwarranted leap. The inability of Giddens' concept of 'created environment' to distinguishing between Los Angeles and an English village, or between the West Midlands and the Scottish Highlands . . . puts a question mark over the usefulness of the concept. Benton's point – that humans have always transformed their environment – shows how the identification of pre-modern societies with nature and modern societies with a created environment, in the way that Giddens does, is wholly incorrect. . . . Nature and natural processes are only seemingly absent from urban environments due to continual management. . . . Nature and natural processes are always 'under the surface' of urban environments and not absent from them. . . . Although climate change or ozone depletion may have been created by social activities they are still natural processes in that the effects of social activities are triggering natural processes. Therefore, in the first sense of the created environment humans are indeed separated from nature but that does not mean nature and natural processes cease to exist. In the second sense the effects of social activities do influence natural processes but these are best seen as human-triggered effects.

(Griffiths, 2006: 116–17)

According to Goldblatt, 'urbanism', for Giddens, is the mediator of a new modern relationship to the natural world: 'modern urbanism is the point at which the culturally transformed experience of the natural world is most acutely felt . . . capitalist urbanism is the physical site of the wholesale transformation of the natural into wholly manufactured space' (Goldblatt, 1996: 56). Thus the spread of urban life in modern societies (and modernising societies) marks a profound cultural shift in how the natural environment is experienced, viewed and valued.

Building on Giddens, we may say that one of the consequences of urban modernity is the loss of a sense of being within the 'natural order'. Thus while a modern as much as a pre-modern society is fundamentally dependent upon the natural world (a point stressed by green social theory, discussed in Chapter 9), living in an urban environment distances people from this reality: the reality of the natural world and the material reality of society's dependence upon it. As Goodin puts it, in discussing the normative basis of green political theory, people wish to belong to a whole that is larger than themselves (1992: 38), and for most of human history up until the modern age, nature provided just such a stable, permanent order. However, living in an urban environment, removed from the material or ecological reality of human dependence upon the environment, there is a marked tendency within modern urban societies to assume that natural limits, constraints, have been 'transcended'. Urbanisation removes the natural environment from the everyday lives of people, replacing it with an artificial, human-made one.

At the same time that individuals are removed from the reality of modern productive interactions with the natural world, people do not know exactly how the food, energy and other commodities they consume are produced, where they are from, who produced them, under what conditions and so on. In this sense, modern urbanism complements another ecologically harmful feature of contemporary global capitalism: the increase in distance between production and consumption (Barry, 1999a: ch. 6). Thus while globalisation compresses space and time, it can also by the same token increase distances between production and consumption.

Giddens' account of urbanism needs to be extended beyond a concern with 'manufactured space' to include how urbanism can remove the natural environment from centre stage, as the immediate environment of day-to-day life, diminish its role as the permanent backdrop for human action, push it backstage as it were, and create the impression that humans do not depend upon the natural world.

Environmental politics and conservatism

For Giddens, faced with the alienation and insecurity of the modern, mobile world, environmental politics represents an attempt to establish some substantive moral content and normative security in people's lives. The reason for this has partly to do with the paradox of modernity in relation to the environment. This paradox is that the more society and the natural world are 'modernised', that is, brought under the regulation of the bureaucratic nation-state and the market, the greater the perception of and the higher the value placed on those parts of the natural world which have not been modernised or developed. Giddens' point is that, at least in part, the distinctly modern concern with preserving and protecting the environment, as well as having to do with the reality of environmental problems, has a deeply moral element, and it is this deeply moral element which makes green politics 'beyond left and right' (Giddens, 1994).

On the one hand, the increase in severity and public awareness of environmental problems are motivated largely by collective and individual interests in survival; that is, they are protective and reactive measures against (human-caused) environmental dangers. A central consequence of this is a desire or need to maintain the natural environment in some state at which environmental problems for humans are solved or prevented. On the other hand, the moral content of environmental politics has to do with defending a particular *meaning* of the natural environment. While he does not develop his discussion of the environment in this way, it is compatible with Giddens' social theory to suggest that the particular meaning of the environment in question for environmental politics has something to do with the environment as a separate, independent, stable 'natural order' within which humans can find security and a meaningful order.

It is also related to particular understandings of collective identity, and tradition according to Giddens (1994: 206). While this particular meaning of the natural environment may be detected in many aspects of the environment, it is perhaps most obvious in the dominant meanings attached to the 'countryside' as opposed to the 'town' (Williams, 1988). As Rennie-Short points out, 'In the contrast with city, court and market, the countryside is seen as the last remnant of a golden age. The countryside is the nostalgic past, providing a glimpse of a simpler, purer age. . . . The countryside has become the refuge from modernity' (1991: 31, 34). Hence the connection between environmental protection and tradition for Giddens. Indeed, in discussing Robert Goodin's (1992) argument for the value of nature resting its capacity to be a context larger than ourselves, Giddens notes: 'To say that we need something "larger" than ourselves or more enduring than ourselves to give our lives purpose and meaning may be true, but this is plainly not equivalent to a definition of the "natural". It fits "tradition", in fact, better than it does "nature"' (1994: 206).

Echoing Habermas, Giddens suggests that late modernity brings with it a new form of politics, what he calls 'life politics', which is distinguished from eman-cipatory politics associated with early modernity. Emancipatory politics is all about breaking free from the constraints of tradition and concerns establishing greater freedoms (1991: 210–12). Thus the agenda of emancipatory politics concerns justice, the right to participate in politics, the extension of the vote, legal standing, greater equality and the removal of oppressive social forms. In short, emancipatory politics is all about the promotion of individual or collective autonomy; for emancipatory politics there is always an asymmetry, an 'other'.

Although emancipatory politics remains as part of the political agenda of late modernity the new agenda of life politics emerges. Life politics is unique to late modernity and requires a certain level of emancipation to have been achieved. Emancipatory politics is a politics of life chances, while life politics refers to the organisation of choice. In short, life politics confronts the question of how we should live without the guidance of tradition. Life politics confronts ethical and existential questions suppressed by modernity and its abstract systems, and one of the key questions here is our relationship with nature (Giddens, 1991: 223–5). Interestingly he suggests that 'As the proponents of "deep ecology" assert, a movement away from economic accumulation might involve substituting personal growth – the cultivation of the potentialities for self-expression and creativity – for unfettered economic growth processes' (1991: 223). In this way, Giddens, unlike Habermas, does seem to consider questions of rethinking the 'good life' in the light of ecological concerns.

As Griffiths puts it:

> The colonisation of nature by abstract systems brings our relationship to nature into focus. Therefore life politics is a politics of movement towards rather than away as it is concerned with the development of new morally justifiable ways of life and social forms that will promote self-actualisation. Options are now open, for example abortion, we can now have sex without having children; therefore the right to have an abortion becomes a political issue, does the unborn child have any rights? Ecological politics are, for Giddens, also an example of life politics as they are a return of the sequestered and involve individual lifestyle choices that have an impact on the environment – to drive or take public transport, to recycle, to have children or not, to buy locally produced goods and so on. Ecological politics is an individual and a collective problem.
>
> (Griffiths, 2006: 122)

Giddens characterises the environmental movement as akin to 'philosophic conservatism' seeing both as a politics of loss or lament – eloquently expressed in Bill McKibben's *End of Nature* (1989). This conservative, tragic interpretation of the green movement is also shared by Torgerson (1999) and Samuels, who goes so far as to describe it as 'depressive' and 'authoritarian' (1993: 211). However, the alignment of environmentalism and philosophic conservatism is not wholly pejorative in Giddens' recent thinking, and this connection between conservatism and environmental politics and issues will be discussed in more detail in Chapter 5. Stressing the commonality between green and conservative thought, Giddens states that:

> The conservatives place emphasis on the family, on bonds between generations, and on preserving nature. The greens operate with the same themes, and precisely here one can build a common bridge which can form the basis for a part of the themes which must dominate the radical political agenda of the future. *The destruction of nature and the destruction of solidarity must be halted. We must be concerned with solidarity and the preservation of nature, but not in a traditional way or by traditional means.*
>
> (Giddens, 2000a: 154; emphasis added)

A key aspect of conservative thought is to lament the ending of traditions. Similarly, Giddens sees environmental movements as lamenting the end of nature and trying to defend nature in the natural way; that is, defending nature because it has always existed without realising that we now live in a created environment, and that nature depends on us rather than the other way around. Giddens sees the environmental movement in large part as a reaction to the invasion of modernity's abstract systems into nature, thus ending 'natural nature' and replacing it with a created environment. He remarks that green political theory 'depends for its proposals on calling for a reversion to "nature". Yet nature no longer exists! We cannot defend nature in the natural way any more than we can defend tradition in the traditional way – yet each quite often need defending' (Giddens, 1994: 11). Giddens contends that the environmental movement is founded on a paradox 'that nature has been embraced only at the point of its disappearance' (1994: 206).

The characterisation of the whole environmental movement as a reactionary politics wanting to defend a nature that no longer exists is a huge over-simplification and is blatantly incorrect. The reason why Giddens thinks this way is because of his association of modernity with a created environment; environ-mental movements are therefore associated with a nature that no longer exists. To see the environmental movement as a manifestation of life politics as Giddens does is very useful but *not* as a reaction to the ending of nature. As nature still exists and human societies still depend on and are vulnerable to it, the environmental movement is not a lament for the ending of nature and a desire to turn back the clock to some pre-modern Arcadian idyll of a 'giving environment' as outlined in Chapter 1. An environmental life politics does not attempt to answer the question of how we should live after the end of nature; rather it tries to answer the question how we should live when we are a part of nature and dependent on it for our ultimate survival but are nevertheless apart from it (Barry, 1995c; Griffiths, 2006).

Risk, science and precaution

There are many positive aspects to the ways in which Giddens deals with eco-logical issues. He mentions that the changed nature of risk means that we now have a responsibility for future generations and this must infuse decision-making, thus integrating international justice into this social theory. As risks are now potentially of high consequence and global, future generations must be incor-porated into decision-making to make sure that decisions are made in such a way as to prevent displacement of problems into the future. Science and technological development may have created many of the risks we now face but science remains a key partner in managing and reintroducing nature and that, as nature cannot be disentangled from the effects of social systems any ideas regarding basing policy on a 'nature knows best' attitude is deeply flawed and regressive, not pro-gressive in his view (Giddens, 1994: 211–12). What is also to be recommended are his comments concerning a need for the democratisation of science and tech-nology. His argument is that as they have an increasing effect on our lives and do not speak with one voice they must be subject to greater democratisation and regulation. This will be returned to below. He also mentions the potential of environmental policies in generating solidarity as most ecological benefits are benefits to all social classes (Giddens, 2000b: 109). He proposes a progressive tax on consumption and a system of eco-taxation based on incentives for good behaviour (Griffiths, 2006).

However, for Griffiths (2006), some of Giddens' other remarks on ecological issues are less satisfactory. For example, Giddens has argued that third-way environmental policy should be informed by the precautionary principle, on the grounds that '[S]ince we cannot be totally sure whether or not global warming is occurring, it's probably best on a policy level to proceed in an "as if" manner. As some of the consequences of global warming could be calamitous, it's sensible for nations and the larger world community to take precautionary measures' (1996: 367).

However, he changes his view on this to broadly arguing in favour of the precautionary principle but adding *only where feasible*. His argument is that ecological risk cannot be normalised or managed by applying the precautionary principle 'because in many situations we no longer have the option of "staying close to nature", or because the balance of benefits and dangers from scientific advance are imponderable. We may need quite often to be bold rather than cautious in supporting scientific and technological innovation' (Giddens, 2000b: 61). He also says that the precautionary principle's limiting of innovation rather than embracing it is not applicable to all situations. Giddens uses the example of biotechnology and genetical engineering saying, that although the potential problems with such technologies are huge and imponderable, the unsustainable nature of intensive agricultural practices suggests that we have to give serious consideration to GM. To make science and technology more responsible, innovators should be obliged by law to be liable for the consequences; this will act as a break on irresponsibility. However, somewhat contradicting himself, he states that the precautionary principle may be used to reintroduce responsibility and gives the example that firms should think through the whole product cycle, including disposal after expiry, before introducing new products.

As regards the precautionary principle, Giddens is mistaken in two ways. First, the precautionary principle is not about 'staying close to nature' but about being cautious exactly because of manufactured risk and its consequences in the absence of full scientific evidence and proof. Science has caused such risks and therefore it needs to be slightly more humble in its aims and ambitions – here Giddens' thinking seems to echo Marcuse's hope for a 'new science'. Neither is the precautionary principle anti-innovation; innovations are fine so long as they apply to small-scale scenarios, especially at first, where, should they not work as planned, the potential damage is limited. Second, the responsible risk-taking which Giddens advocates would indeed suggest a precautionary approach as would his remarks regarding the incorporation of future generations into decision-making due to changes in risk. If we accept that science has generated the vast majority of manufactured ecological risks which are, in Giddens' own words, global high-consequence risks – the most pressing and visible of which is global climate change – then this would suggest that science should be rather more circumspect in its future aims so that it does not put itself in a situation where it could create such large-scale risks in the future. Thus, as Griffiths (2006) argues, if Giddens is to be consistent in incorporating the environment and ecological issues into his thinking, this would lead to an 'ecologised' or 'greened' form of social theory more in keeping with the promise of critical theory than Habermas seems to offer.

Environment, tradition and identity

The relationship Giddens talks about between particular forms of collective identity, tradition and meanings attached to the 'natural environment' may be seen by looking at the case of the 'countryside' and how it has been used in recent debates about competing conceptions of 'English' national identity. It must be remembered that when speaking of the countryside, one is not, strictly speaking, referring to a purely 'natural' (in the sense of 'nonhuman' or 'non-transformed') environment. The countryside is a humanised environment in that it is not a naturally occurring ecosystem. Hedgerows, ploughed fields, drystone walls, all of which are central to the idea of countryside, are the product of present and past human transformation of the natural environment. In many ways the countryside is a 'text' which contains the 'trace' of previous human transformative activity. Preserving the countryside can thus be motivated by a desire to maintain the continuity with previous generations, to preserve a particular way of life, and a particular sense of collective identity. For example, there is an argument to suggest that it is a particular conception of 'Englishness' which can, in large part, explain recent movements to 'protect the countryside'. These movements, ranging from the Countryside Alliance to the National Trust and Council for the Protection of Rural England, are motivated not simply by a concern to protect the countryside per se, but to protect a particular type of countryside and land-management system which is a constitutive aspect of a particular understanding of Englishness and 'tradition'. This idea of Englishness, in part, has to do with such claims as England as a 'green and pleasant' land, a land of sturdy yeomen farmers, village greens, a hierarchical but benign class system in which the aristocracy are the main landowners who 'take care' of their land-renting peasants, a society where fox-hunting is not simply a way of getting rid of a pest, but also a constitutive part of the country way of life and so on.

Many of these arguments are explicit in the aims of the Countryside Alliance which saw in the defeated 1998 Parliamentary Bill to outlaw fox-hunting an attack on the country way of life by urban, liberal 'townies' who know nothing about the realities of rural life. Thus in this conflict between 'defenders' of fox-hunting and those opposed to it we find a contemporary example of a long-standing relationship between 'country' and 'town'. Thus, while ostensibly about the practical and moral issues surrounding a particular practice (fox-hunting), it is clear that there is more to this conflict than this, which is not to downgrade the moral commitments and concerns for the welfare of foxes which opponents have. In short, we can say that what is at issue is also a conflict between two different and potentially incompatible views or meanings of the 'countryside'. The important point to note is that it is not the physical reality of 'natural environment' or the 'countryside' as such that is the object of analysis, but the 'meaning' of the countryside (which of course will affect how it is physically affected by human action).

The Countryside Alliance is defending a view of the country which stresses the idea of the rural way of life, and thus emphasises how the countryside is not simply a 'rustic landscape' but a *working environment*, and a system of land management which requires fox-hunting. They accuse opponents of fox-hunting of regarding the countryside as some pastoral landscape, a place of hedgerows, butterflies, copses, forests, babbling brooks and so on, in which the economic reality of country life is missing. From their perspective the urban, metropolitan opposition to fox-hunting on the grounds of its 'uncivilised', 'barbaric' character is in fact the town misperceiving the countryside as the 'pastoral' in Rennie-Short's (1991) terms, and thus 'forcing' the rural way of life into something which it is not.

At the same time, there is a conflict over the relation between 'Englishness' and the countryside in this debate. Defenders of fox-hunting can justify it on the grounds that there is something quintessentially 'English' about fox-hunting. Oscar Wilde's famous definition of an Englishman fox-hunting as 'The unspeakable in pursuit of the inedible' stands with conservative views in which a true Englishman must 'ride to hounds' as examples of this connection between particular environmental practices (and thus particular meanings of the environment) and collective identity. Again Rennie-Short eloquently highlights this identity-forming and identity-affirming function of the countryside: 'In most countries the countryside has become the embodiment of the nation, idealized as the ideal middle landscape between the rough wilderness of nature and the smooth artificiality of the town, a combination of nature and culture which best represents the nation-state' (1991: 35).

Another example of this connection between the countryside and collective identity is the concern expressed over 'foreign' or 'non-indigenous' species of flora and fauna displacing indigenous species. The most striking example here are the debates over grey and red squirrels, and the worrying fact that the red squirrel is being driven from its native habitat by the 'foreign' grey squirrel (and thus its numbers are falling). In this debate over the need to protect the native species from the foreign one, more than simple ecological or animal welfare concerns are at stake, just as was argued to be the case in relation to fox-hunting.

At the same time, there are other movements such as The Land Is Ours and to a lesser extent the Ramblers' Association, which articulate another alternative understanding of the countryside, this time with an emphasis on the right of every British citizen to have access to the land, both in terms of walking and enjoying the countryside (the Ramblers' Association), and the more radical aim of returning ownership and control rights of the land to the people. In such movements, particularly The Land Is Ours, one can see strong connections to past social struggles and environmental practices. For example, the aims of The Land Is Ours movement echo the demands of earlier radical English movements such as the Levellers and the Diggers to defend the 'commons' and the rights of access of the commoners. However, its core democratic aim of redressing the situation today in which 80 per cent of the land in England is owned and managed by 10 per cent of the population also makes it distinctly 'modern' in aspiration. Although beyond the scope of our current discussion, one could suggest that in such contemporary movements there is an attempt to create a newer, less exclusive and monolithic sense of Englishness in relation to the countryside and particular social-environmental practices such as fox-hunting, which marks a decisive break with older conceptions of English national identity that were in large part based on a particular conception of the English countryside, the country way of life, and England as a 'green and pleasant land'.

Unfortunately, these issues are largely missing from Giddens' work on the place of the environment within social theory – largely, I suspect, because he is insufficiently sensitive to the important differences and conflicts between competing meanings of different parts of the environment. Thus while, as Goldblatt notes, for Giddens, 'the conjunction of capitalism and industrialism is responsible for modern environmental degradation . . . whatever the precise causal origins of environmental degradation, the modern world heralds a more wholesale transformation of nature than human societies have been capable of before' (1996: 16–17), Giddens's theory is in many important respects limited. Together with a lack of attention to the full range of issues within the urban/rural or town/country divide, there are other limits to Giddens' social theory.

Relating to the distinction between 'transformed' and 'natural' environments, the idea of 'wilderness' is problematic for Giddens, just as it was for both Habermas and Vogel discussed earlier. An example of this is Giddens' statement that 'In the industrialised sectors of the globe – and increasingly, elsewhere – human beings live in a created environment. . . . *Not just the built environment of urban areas but most other landscapes as well become subject to human co-ordination and control*' (1990: 60; emphasis added). We can note that 'landscapes' are not co-extensive with the natural environment. In this sense Giddens has shifted, and in the process narrowed, the focus of the debate from an analysis of the full range of issues involved in relating the environment to social theory. He focuses on 'landscapes', which for him include the urban landscape, or what has been termed 'blandscapes' by writers such as Porteous (1997), on account of the homogeneity, uniformity and lack of aesthetic content that is typical of most urban, built environments. Giddens misses the essentially ecological character of the environment and its importance for human societies, and substitutes it with an aesthetic-moral concern with the landscape/blandscape of the modern social world. This is not, let me stress, to deny the importance of this aesthetic-moral concern with producing pleasing, beautiful and enjoyable landscapes, but this concern must be placed within its ecological context, and the relationship between society and *environment* and not just society and *landscape*. Here, however, Giddens may be working with a similar notion to Marcuse's 'liberatory domination of nature', discussed above. For Giddens there is no necessary connection between human mastery and exploitation of nature and the destruction of the natural environment. As he puts it, 'Mastery over nature . . . can quite often mean caring for nature as much as treating it in a purely instrumental or indifferent fashion' (1994: 209).

Second, it is surely overstating the case to suggest that 'most other landscapes' which human beings experience or come into contact with are (and ought to become?) transformed by human practices. While this may be true of the highly populated societies of Western Europe, whose indigenous, local environments have been intensively and extensively transformed by previous human activity for hundreds of years (and thus cannot be considered as 'wilderness' and is closer to a 'garden' in terms of the discussion earlier in Chapters 2 and 3), this is not the case with other societies around the world, where wilderness does exist. To the extent that Giddens relies on this idea of the environment as one transformed by 'human co-ordination and control' as the primary orientation around which to integrate the environment into his social theory, to that extent it is limited to the 'humanised environment' rather than to the nonhumanised natural environment. This is confirmed by his statement that 'All ecological debates today, therefore, are about managed nature' (1994: 211). This focus on 'managed nature' becomes even more problematic if we include natural environmental processes such as hydrological cycles, carbon and nitrogen fixing, and ecosystem functions as part of the natural environment we are interested in. These processes are not, and could not be, subject to human manipulation and control.

However, despite these problems, Giddens has developed a coherent and challenging analysis of the place of the environment (both natural and urban) within social theory. He presents us with an account of the environment in modern society and social theory in which the manufactured or managed character of the natural and urban environments, within the context of globalisation, are given central stage. Without looking at urbanisation, any analysis of the emergence of green politics and a moral concern for the preservation of the natural world is deficient for Giddens. Green issues and politics, and the place of the environment in social theory, are distinctly 'modern' phenomena, and the need to be sensitive to the urban experience is the mark of how modern these concerns are. They arise, in large part, as a result of what he calls a paradox of the modern world, which is 'that nature has been embraced only at the point of its disappearance' (1994: 206).

Conclusion

A central theme of twentieth-century social theory's engagement with the environment has focused on the costs (social and psychological as well as environmental) which have arisen as a result of society's technical mastery of the natural environment. It has also looked at the ways in which the natural environment has been transformed into 'humanised environments' such as the urban and built environments. At the same time, it explored the implications of how such 'natural' environments as the 'countryside' are a result of present and past collective human transformation. The 'environment' for twentieth-century social theory is thus not confined to the 'natural' environment. Nor is 'nature' confined to 'nonhuman nature', for, as thinkers from Freud to the Frankfurt School have suggested, one cannot talk of the latter without also talking about 'human nature'. Finally, twentieth-century social theory has also raised questions as to the status, value and direction of modernity and the Enlightenment. The price of modern progress, namely the 'disenchantment of nature' and an almost exclusive instrumental cultural and economic valuation of the natural environment, is increasingly regarded as being in need of re-examination in discussions about what Habermas has called 'the unfinished project of modernity'. Thus the discussion of the environment, our relations to it, its meanings and status, are and have been a central part of debates about and within modernity and the legacy of the Enlightenment (see Hayward, 1995).

What marks both Habermas and Giddens and to a lesser extent the Frankfurt School is their stress on the structural, institutional causes of environmental problems. While there is some slippage and ambiguity within Giddens' work, as Goldblatt (1996) argues, between citing industrialism or capitalism as the cause of these problems, Habermas at least is clear in identifying the economic and political structures of the globalising capitalist world order as the ultimate source of the ecologically unsustainable character of modern (and modernising) societies. However, as Griffiths (2006) points out, Giddens has a lot more to say on ecological issues and environmental/green politics and policies than Habermas and offers a more fruitful basis for the integration of environmental issues on their own terms.

Summary points

- In social theory from the nineteenth to the twentieth centuries, an appeal to the 'naturalness' of principles or view of social relations has often been seen as indicating that such relations are 'given'; that is, they cannot be changed by human will.
- Classical sociology had relatively little to say about the natural environment beyond seeing 'environment' as the opposite of 'culture', which was the main focus of social theory.
- That one cannot discuss nonhuman nature without reference to human nature has been a concern of twentieth-century social theory from Freud to the Frankfurt School.
- Critical theory's assessment of the Enlightenment and modern, industrial societies was the first attempt to systematically analyse the natural environment and its relation to human social practices as part of its critique of the modern social order. The basic view of critical theory is that the domination and exploitation of the natural environment leads to the domination and exploitation of humans.
- Habermas' social theory holds that an instrumental valuation and relationship to the natural world is unavoidable, but he does see the rise of green politics and a concern for environmental protection and preservation as positive developments.
- The work of Anthony Giddens focuses on the urban experience of modernity as central in explaining the rise of environmental concern, which he says can be linked to a defence of tradition and particular forms of collective and individual identity. Threats to the natural environment are modern risks which arise particularly as a result of globalisation.
- For recent social theorists, Habermas and Giddens in particular, while their theories do shed light on the interaction between human societies and their nonhuman environments, there is an unresolved question regarding the proper place of 'wilderness' areas or of natural processes such as global hydrological cycles, nitrogen cycles, ozone production within an environmentally aware social theory.

Further reading

The Frankfurt School/critical theory

For an overview of the overlaps and tensions between critical theory and environmental issues see Stephen Vogel, *Against Nature: The Concept of Nature in Critical Theory*, New York: State University of New York Press, 1996; Andy Dobson, 'Critical Theory and Green Politics' in Andrew Dobson and Paul Lucardie (eds), *The Politics of Nature*, London: Routledge, 1993; and Matthew Gandy, 'Ecology, Modernity, and the Intellectual Legacy of the Frankfurt School', *Philosophy and Geography*, 1: 1, 1996.

Habermas

Habermas has directly discussed the place of the nonhuman world in his (1982) 'A Reply to My Critics', in J. Thompson and D. Held (eds), *Habermas: Critical Debates*, London: Macmillan, and his assessment of ecological politics in his (1981) article, 'New Social Movements', *Telos*, 49. For an early assessment of Habermas' social theory and nature, see J. Whitebook (1979), 'The Problem of Nature in Habermas', *Telos*, 40: 41–69, reprinted with a new introduction in D. Macauley (ed.), *Minding Nature: The Philosophers of Ecology*, New York and London: Guilford Press, 1966. His more recent work on biotechnology and genetic engineering may be found in J. Habermas, *The Future of Human Nature*, London: Polity Press, 2003.

Other commentaries and critical assessments of Habermas include: S. Vogel, 'Habermas and the Ethics of Nature', in R. Gottlieb (ed.), *The Ecological Community: Environmental Challenges for Philosophy, Politics and Morality*, London: Routledge, 1997; S. Vogel *Against Nature: The Concept of Nature in Critical Theory*, New York: State University of New York Press, 1996, also contains a discussion of Habermas; C.F. Alford, *Science and the Revenge of Nature*, Tampa: University of Florida Press, 1985; P. Dickens, *Society and Nature: Towards a Green Social Theory*, Hemel Hempstead: Harvester Wheatsheaf, 1992; R. Eckersley (1990), 'Habermas and Green Political Theory: Two Roads Diverging', *Theory and Society*, 19: 6; R. Bruelle (2002), 'Habermas and Green Political Thought: Two Roads Converging', *Environmental Politics*, 11:4; J. Dryzek, (1990), 'Green Reason: Communicative Ethics for the Biosphere', *Environmental Ethics*, 12: 3.

Giddens

While one can find references to nature and the environment throughout Giddens' work, his most sustained treatment of the topic may be found in the following books: *The Consequences of Modernity*, Cambridge: Polity Press, 1990; *Modernity and Self-identity*,

Cambridge: Polity Press, 1991; *Beyond Left and Right*, Cambridge: Polity Press, 1994; and *The Third Way and its Critics*, Cambridge: Polity Press, 2000b.

David Goldblatt's *Social Theory and the Environment*, Cambridge: Polity Press, 1996, devotes two chapters to Giddens (chs 1 and 2) and one to Habermas (ch. 4), while the edited collection by M. O'Brien, S. Penna and C. Hay (eds), *Theorising Modernity: Reflexivity, Environment and Identity in Giddens' Social Theory*, London: Longman, 1999, contains some good critical essays by Ted Benton and Peter Dickens.

Iorwerth Griffiths' unpublished Ph.D. thesis *Social Theory and Sustainability: Deep Ecology, Eco-Marxism, Anthony Giddens and a New Progressive Policy Framework for Sustainable Development*, Queens University of Belfast, Northern Ireland, 2006, contains the most comprehensive analysis of the relationship between Giddens and environmental issues. It is to be published as a book. Available from the author.

5 Right-wing reactions to the environment and environmental politics

Key issues

- Political ideologies and the environment.
- Right-wing perspectives.
- Conservatism and conservation.
- Eco-fascism and Eco-authoritarianism.
- Limits to growth.
- 'Crisis what crisis?' The right-wing cornucopian and contrarian response.
- Free market environmentalism.
- The controversy over Lomborg's *The Skeptical Environmentalist*.
- Other right-wing responses: the Gaian post-humanism of John Gray.

Introduction

That there cannot be any 'ideologically' or value-free interpretations of the relationship between environment and society has been raised throughout this book. As has been argued from Chapter 1 on, the connection between social theory and the environment is not one of 'facts' about the natural and social worlds which we read or interpret objectively, but rather the connection is through a particular ideological or value-based lens such that facts and values cannot be separated. As discussed in Chapter 3, while anarchists such as Kropotkin looked into nature and saw a realm of equality and harmony which he then used to justify his egalitarian and libertarian anarchist vision of society, others such as the 'social Darwinists' saw nature as 'red in tooth and claw', a competitive, hierarchical and individualist struggle of the strongest, and used this reading of nature to justify unfettered non-egalitarian capitalism. Both of these ideological readings of nature differ in what they 'read into' nature, but share the commitment to legitimising social relations on the basis of natural relations, what was criticised in Chapter 1 as the 'reading-off' hypothesis. Both offer classic left- and right-wing perspectives on the environment respectively.

In this chapter and the next we move on to the examination of more contemporary right-wing and left-wing reactions/perspectives on the environment and the politics of the environment, and look at the variety of ways in which these two broad families of ideology have used (and misused) the environment in the construction of their different ideological perspectives, their construction of various prefix 'sub-ideologies' such as 'eco-socialism', 'eco-Marxism' and 'eco-fascism'; and their often diametrically opposed explanations and solutions to the environmental crisis and reactions to green or environmental political, cultural, economic and policy reponses to that crisis.

Political ideologies and the environment

A simple, but effective, view of political ideologies is to see them as value-based worldviews which inform and help us interpret the political worlds we inhabit. That is, they may be viewed as value-based 'maps' of the political territory. They are the 'isms' which confront one another daily as they vie for our allegiance, and support typically in the form of different political parties or social movements professing to articulate a particular ideological position. Thus 'socialist' parties and movements compete with other political groupings which represent 'conservatism', 'ecologism', 'liberalism' or 'feminism', 'anarchism', 'neo-fascism' or 'religious fundamentalism'.

While each of these (and other) ideologies differ profoundly, they all share at least three features which characterise any ideology. First, each ideology has an analysis and critique of the current social order, pointing out, based on its own fundamental ethical and political principles, what is wrong with the current way society, the economy, the state and so on is set up, or ways in which social organisation can be improved. Second, ideologies will have an alternative vision or perspective on the way society ought to be. That is, each ideology will offer a different picture or aspiration as to the way the political system, the economy, relations between human beings (e.g. men and women) ought to be organised. Third, each ideology will have a 'theory of transition' or 'agency'; that is, how we get from the current structure and organisation of society (which each ideology criticises), to the alternative vision of how society ought to be (based on the ideology's basic principles). For some ideologies, such as Marxism, the theory of agency comes down to a theory of revolution, when one class (the proletariat or working class) will rise up and overthrow the ruling class, or bourgeoisie, and take over the state, abolish private property and institute the equal distribution of socially produced wealth.

Each ideology will also have particular views about the environment and the relationship between human society and the non-human world. However, it is true to say that most ideologies have, up until recently, said relatively little about the natural world and whether left- or right-wing, dominant ideologies have tended to adopt an instrumental view of the natural world, which came to prominence with Enlightenment thinking as discussed in Chapter 3. That is, both left-wing ideologies such as socialism and Marxism, and right-wing ideologies such as conservatism, and other ideologies such as liberalism, which can straddle both left- and right-wing positions, have regarded the nonhuman world as a set of resources for humans to use with no independent moral status. As a consequence, dominant ideologies may be viewed as anthropocentric; that is, concerned solely with the status and relationships between human beings and not, in the main, with the issue of the thinking about a proper ethical relationship between humanity and the environment. This concern for the ethical status of the nonhuman world and making the human–environment relationship a central feature of ethical and political concern is one of the features of green ideology or ecologism, and which serves to sharply distinguish it from the dominant left- and right-wing ideologies. As Kenny has put it, 'The dominant cultural and intellectual traditions in the West have been subjected to a powerful critique by greens who find in its leading cultural and philosophical traditions the justification of the domineering and exploitative attitude towards the natural world' (2003: 154). However, as will be discussed in more detail below, there are some ideological traditions, such as anarchism, or certain liberal theories (such as Mill's defence of the 'stationary state' discussed in Chapter 3), and feminism which do have more to say about the environment. Equally, the emergence of the ecological crisis and green ideology has also meant that new ideological formations have developed as a reaction, such as 'eco-socialism' and 'eco-feminism'. The important point here is that each ideology will have a position or take on the environment, even if a particular ideology has relatively little explictly to say about the environment or the relationship between the environment and the society. In other words, if one had a checklist of issues or criteria by which to assess or categorise ideologies, some of these would relate to the ideology's take on the environment.

Box 5.1 offers a useful checklist by which to record and summarise the main aspects of any ideology. One can add or subtract from the particular items, but the point I wish to emphasise is that one can, and should, include environmental items in one's analysis of each ideology.

For example, one interpretation of conservatism would contend that generally it has an instrumental view of the environment, is anthropocentric, downplays the severity of the ecological crisis and sees state interference, regulation and bureaucracy as the main causes of ecological problems; regards technology and the free

---

Box 5.1 Social theory checklist

Social theory

1   Main ethical and political principles
2   Main theorists and schools of thought
3   Main features of its critical analysis of contemporary society
4   Main features of alternative vision of society
5   Theory of agency
6   Perspective of nature
7   Perspective on social–environment relationship
8   View of the ecological crisis
9   Causes of and solutions to the ecological crisis
10 View of animals
11 View of property
12 Theory of the economy
13 Attitude to the state
14 View of human nature
15 Attitude to the nation

Source: Author

---

market as the main solutions to dealing with the crisis; has a negative view of human nature; and adopts a positive view of the state and stresses the centrality of national identity and loyalty/patriotism in its vision of the 'good society'. One should do this for each ideology to see how they differ and also share or agree on particular issues.

Right-wing perspectives

Right-wing perspectives on the environment and ecological politics can historically be traced back to conservative reactions to the Enlightenment, in relation to the industrial and democratic revolutions which, as outlined in Chapter 2, I take to summarise the main dimensions of the Enlightenment. Right-wing perspectives range from early conservative reactions to the Enlightenment to eco-fascist positions and extreme market ideologies such as 'free market environmentalism' and contemporary 'neo-conservative' views represented by such groups as the Wise Use Movement in America and the growing number of corporate-funded right-wing think-tanks promoting neo-liberal ideas and policies.

Conservatism and conservation

On the face of it, as Dobson (2000: 172–3) suggests, conservatism as a political ideology (or 'disposition' since some hold that conservatism is not an 'ideology') offers a convivial home for the integration of the nonhuman world into social theory. In particular the ineliminable constraints of external nature in *delimiting* the options available to humanity is something that fits with the 'realism' of conservative thinking, and this recognition of external constraints may be termed the 'ecological viewpoint' (Wells, 1982: 3). This viewpoint may be perceived as a conservative one in that 'the basic political question – "what is to be done?" – depends on an account of what can be done?' (Wells, 1982: 15). On this point, conservatives would agree whole-heartedly with green/ecological thinkers such as Jared Diamond and Clive Pointing who have documented the fall of previous human societies and civilisations from Easter Island to ancient Greece due to their failure to regulate their ecological and resource conditions in a sustainable manner. As Diamond comments about the ecological degradation of the once 'fertile crescent' of Southwest Asia (covering the modern land mass of Israel, Jordon, Syria and parts of Iraq and Turkey), 'Thus, Fertile Crescent and eastern Mediterranean societies had the misfortune to arise in an ecologically fragile environment. *They committed ecological suicide by destroying their resource base*' (1997: 411; emphasis added). Ponting's eloquent description and analysis of the collapse of the civilisation on Easter Island in the South Pacific due to the inability of the island's ecosystem to support the society is another reminder of the dangers of forgetting the reality and limits that nature imposes on humanity. As he puts it:

> The history of Easter Island is not of lost civilisations and esoteric knowledge. Rather it is a striking example of the dependence of human societies on their environment and of the consequences of irreversibly damaging that environment. It is the story of a people who, starting from an extremely limited resource base, constructed one of the most advanced societies in the world for the technology they had available. However, the demands placed on the environment of the island by this development were immense. When it could no longer withstand the pressure, the society that had been painfully built up over the previous thousand years fell with it.
>
> (Ponting, 1991: 2)

Both the 'realism' of being guided by what is possible (rather than what is desirable) and the need for a backward-looking respect for and learning from the past are quientessentially conservative traits. For Easter Island read contemporary Western civilisation. The basic conservative position is thus one which reminds us that there is nothing invincible or permanent about our present condition and that human society and human plans are always vulnerable to the vagaries of the external world or prey to the 'all too human' failings of greed, short-sightedness, arrogance and hubris. Thus the constraints of internal human nature are central for conservatives, who unlike 'progressives' such as socialists, greens and feminists, hold a largely negative view of human nature and a scepticism as to the universal and general improvement or perfectability of the human condition (Dobson, 2000: 177), most forcefully expressed by conservative thinkers such as Edmund Burke and Thomas Malthus.

Burke is interesting in that he was one of the first 'modern' thinkers to consider obligations to future generations, itself a traditional conservative trait, and one discussed in more detail below in reference to the landed aristocracy. Burke's view is that society is 'a partnership in all science; a partnership in all art; a partnership in every virtue, and in all perfection. As the ends of such a partnership cannot be obtained in many generations, it becomes a partnership not only between those who are living, but between those who are living, those who are dead, and those who are to be born' (Burke, 1790/1969: 194). Thus the conservative virtue of tradition, or the 'democracy of the dead' as the English theorist and novelist G.K. Chesterton termed tradition (Chesterton, 1959: 48), may be seen as a form of stewardship which looks to the future as well as honouring the past, though as Dobson reminds us, Burke (and conservatives more generally) seems more interested in the past than in the future (Dobson, 2000: 176). Burke's view is that the current generation is a 'steward' of the environment and society's institutions and achievements in government, art, literature and so on for the as yet unborn generations to follow, which they have inherited from the previous generation, themselves viewed as 'stewards'.

In relation to the industrial revolution, right-wing and conservative thinking was initially hostile to the major changes brought about by the profound alteration in the economy and the way societies produced, consumed and distributed goods and services. On the whole, right-wing ideologues represented the landed aristocracy and ruling classes whose dominant position was at times threatened by the transformation of the economy to an urban-based, technologically enhanced manufacturing one founded on the factory system. The economic basis of their political power lay in their position as the dominant property owners of agricultural land and a mode of rural-based agricultural production in which the majority in society were peasants and small farmers and craft-workers who were dependent upon this landed ruling class for their survival. For example, as discussed in Chapter 3, Malthus' arguments may be seen (particularly from a left-wing perspective) as defending the landed aristocracy in Britain from rising demands from the 'lower orders', peasants and the emerging urban working class, for regulation on food production and prices and curbing the power of this landed aristocratic class.

As befits its name, conservatism is instinctively cautious and sceptical of major transformations in society, politics and the economy. Therefore the revolutionary and rapid changes that industrialised production and technology heralded were suspect from a conservative point of view. Change, if any, from a conservative position should be slow, small, piecemeal, based on practical experience and not abstract thinking, and easily reversible. Above all, conservatism holds that there should be a privileging of tradition, and an active disposition to preserve and maintain tradition, especially deference to established authority, such as the landed aristocracy, the instituions of the state, law and order, loyalty to the nation and monarch, and the authority of established religion. Therefore, conservatism is marked by a general disposition towards favouring the status quo and society as it is over any ideological, abstract and 'utopian' visions of alternatives to the way society is currently ordered. In short, because its animating principle is to conserve, it viewed the changes wrought by the industrial revolution with attitudes ranging from scepticism to outright rejection, since these changes, if not properly managed, threatened the settled social order and class organisation of society into those who were the 'natural' rulers of society (the monarchy, established Church and landed aristocracy) and the ruled (the rest of society). While generally conservativism did come around to accepting the new economic realities of the industrial system and in particular came to vigorously defend the benefits of a free market organisation of the economy and the new economic order of industrial capitalism, this was not universally shared by all conservatives. According to Eccleshall:

> The objection of Tory [early British conservatives] paternalists to the Modern Age was that the people were being severed from traditional sources of protection and discipline. Their argument was that some landlords, as well as the captains of a rapidly developing manufacturing industry, had been misled by the 'theoretic folly' of the new science of political economy into supposing that, for the sake of material advance, the emotional ties of benevolent hierarchy had to be displaced by the harsh, impersonal relations of the capitalist market.
>
> (Eccleshall, 2003: 60)

Such thinking also overlapped with other conservative views which reacted negatively to the urbanisation of society, the despoilation of the natural environment and rural landscape as a result of new agricultural practices and technology, which increased productivity but destroyed settled land-use patterns (as well as making farm workers unemployed and who had to move to the cities to find work), and generally evinced a 'Romantic' reaction against industrialisation, which was discussed in Chapter 3. This conservative Romantic rejection or scepticism about industrialism may be seen in such conservative thinkers as Thomas Carlyle who bemoaned the destruction of the English countryside, 'this green and pleasant land', by industrialisation and who, like other Romantically inclined conservatives, wished for a return to the pre-industrial rural and almost medieval class-based, traditional societies in which the wisdom of the landed aristocracy ensured the preservation of the landscape and its rural population, and in which there was a place for everyone and everyone knew their place.

Such Romantic, conservative views essentially sought to conserve particular landscapes (and the associated undemocratic and unequal social and political relations between landowners and peasants needed to maintain that landscape), great estates and wooded areas which were needed for valued aristocratic practices such as hunting, fishing, animal and plant breeding, the keeping and displaying of exotic plants from the colonies, to landscape painting. Conservatives such as Carlyle pointed out that it is only due to the security of inherited property that trees and forests are planted and that the current landowner will never see grow to full maturity. Thus may be found an anticipation of later debates about 'sustainable development' and 'intergenerational obligations' which dominate current thinking about society and environment. The conservative defence of aristocratic landownership and privilege was in this Romantic view based on the claim that only the landowners, who it must be remembered owned most of the territory within European countries and indeed elsewhere in the form of plantations in other parts of the world, could conserve the environment and landscape. This was part of the obligation of the aristocratic class to preserve and hold in trust the land for future generations, through the medium of exclusive ownership of the land and passing the land on, via inheritance, to their sons and daughters in perpetuity.

However, although such conservative reactions to industrialisation can be found from the late eighteenth to the early nineteenth centuries, the general trend in conservative thinking is towards accepting both the reality of the 'new economic system' (i.e. industrial capitalism), and also embracing it as a key part of the 'good society'. That is, conservativism sees that its future lies in articulating the interests of the rising manufacturing class and those landowners who were able to use their inherited wealth to themselves become factory owners and major investors and property-owners in the new economic order. As Eccleshall puts it, 'For other conservatives, however, the future lay not in a revived partnership between some of the propertied classes and the masses, but rather in a new alliance of aristocracy and emerging bourgeoisie strong enough to withstand democratic and egalitarian demands from below' (Eccleshall, 2003: 61).

In relation to the democratic revolution, the initial conservative position is much less varied – conservatives on the whole were completely against the extension of political power to those outside the landowning class, the monarchy and the established Church. They looked in horror at the French Revolution and its slogan of 'liberty, equality and solidarity', which they saw as a direct threat to the 'old regime' or *'ancien régime'* and class organisation of society. Conservative thinkers from Edmund Burke to Thomas Malthus railed against the folly of the French Revolution and its Enlightenment thinkers who believed in the perfectability of humanity based on the transfer and application of new scientific and rational knowledge about the environment to human nature and social organisation. Since conservatives believe in socio-economic inequality, and indeed for most of the history of conservativism conservatives have thought inequality between humans as both 'natural' and desirable, democratic ideas of equality before the law, or one person one vote, were anathema and to be resisted. In particular, conservatives completely ridiculed the idea of gender equality, of the basic equality of men and women. From a conservative point of view, while it was folly and counter-productive to posit equality between men who had property, wealth and experience and governing, and men who had no property, no wealth and who were excluded from governing, it was utter madness to think that men and women were equal. Although discussed in more detail in Chapter 6, it is worth illustrating the strength of the conservative backlash against the Enlightenment-inspired idea of the fundamental democratic equality between men and women. Thinkers such as Mary Wollstonecraft in the late eighteenth century, one of the earliest feminist writers, as well as male democratic thinkers such as Tom Paine and early anarchists such as William Godwin rejected any notion of a 'natural' inferiority or inequality between men and women. Against this conservative notion of a natural inequality between men and women, such that men are naturally superior intellectually, morally and politically to women and therefore ruling out women (no matter if they came from the aristocracy) as equal citizens in the governing of society, the democratic impulse regarded such a position as socially constructed, not naturally given. Therefore, from a democratic perspective equality between men and women is possible; what needs to happen is a change in the way society maintains the artificial (not natural) inequality between men and women. Wollstonecraft wrote *A Vindication of the Rights of Woman* in 1792, inspired by the French Revolution of 1789, which was one of the first systematic attempts to articulate the claim of equality between men and women. Wollstonecraft argued that women should be treated and regarded as the same as men since 'mind has no sex' (i.e. women were as capable of reason as were men), and that therefore the current inequality in power, political and economic rights between men and women was unjustified and based on prejudice, not reason. However, an indication of the ferocity of the conservative reaction to such radical ideas may be gauged from the title of a conservative tract written by Thomas Taylor in response to Wollstonecraft and Paine: *A Vindication of the Rights of Brutes*. In other words, the conservative position was that to suggest extending rights and equality to women was as ill judged and objectionable as to extend rights and equality to animals. Echoing a 'slippery slope' position, Taylor's conservative backlash was basically ridiculing Wollstonecraft's democratic and feminist position, saying that giving women rights was as ridiculous and laughable as giving non-human animals rights.

Democratic demands such as in the American revolution's slogan of 'no taxation without representation' were direct criticisms of the conservative view that government should be reserved for those who are naturally society's 'betters' and 'leaders' – namely the landed gentry, the monarchy, the military and the Church. Politics, the art of governing and the institutions of the state was not for the 'masses' who were, in the eyes of conservatives, unfit to govern themselves never mind society as a whole. Hence paternalism, the view that superior classes know what is in the best interest of 'lower' classes (part of the 'noblesse oblige' honour code of the superior classes), is an essential part of conservative thinking. Yet the democratic imperative proved impossible to hold back. Despite the best efforts of conservatives, from the French and American revolutions sprang movements for democratic reform across Europe and the New World. Throughout the nineteenth century the 'democratic imperative' resulted in the slow, uneven but gradual extensions of democratic equality, equal citizenship and voting rights to more and more people.

However, as already indicated in Chapter 3, along with Malthus, other reactionary and conservative responses to the environment and thinking about the relation- ship between environment and society included 'social Darwinism', and also the racist categorisation of humanity, in the wake of the colonial expansion of Europe throughout the nineteenth century, into distinctive 'races' with deter- minate characteristics which were themselves often judged to be influenced or caused by the environmental context, climate and so on of the different human races. Therein lie the origins of right-wing racist ideologies of 'lazy', 'indolent' and 'non-industrious' non-white races in Africa and the Middle East explained by the fact that the hot climate and ready availability of food and a generally 'giving environment' (to use the term introduced in Chapter 1) meant such races had no reason to follow the 'hard-working', 'inventive' and entrepreneurial' white Northern Europeans (Aryans) who lived in a less giving and harsher environment which consequently explained (and justified) the superiority of the white European race over others.

Eugenics was extremely popular within right-wing (and some left-wing) thinking throughout the late nineteenth and early twentieth centuries, but is today a firmly right-wing phenomenon, and was and is a central component of fascism. It holds that humanity should breed selectively and that there should be no 'mixing of the races' (of inferior non-whites with whites) or 'mixing of the classes' (of lower classes with undesirable traits with other classes with superior traits). Thus particular races and classes are to be 'improved' through deliberate breeding policies, just like animals and plants that are bred selectively to improve particular traits. Above all, eugenics strove for the production of 'pure breeds' and the avoidance of 'mongrel' or 'mixed' offspring. Eugenics in this way purported to be a scientific application of Darwinian and biological thinking to humanity so that humanity could direct its own evolution. Eugenics policies included compulsory sterilisation of certain classes of people (those born with mental or physcial disabilities, such as Down's syndrome, or 'undesirable traits', such as homosexuals), or races (non-white) or ethnic groups (gypsies, Jews, Slavs). Countries which instituted programmes of compulsory sterilisation and other eugenics policies include the United States, Sweden and Australia, as well as the better known example of Nazi Germany in the 1930s and 1940s. An extreme extension of this thinking led to forced euthanasia and mass killing of groups, as in the Nazi 'final solution' and the systematic and industrialised murder of over six million Jews, gays, gypsies and others in concentration camps during the Second World War we now know as the Holocaust. This leads us to a discussion of the more extreme right-wing appropriations of ecological thinking in the form of eco-authoritarianism and eco-fascism which we turn to next.

Eco-fascism and eco-authoritarianism

It was not just eugenic thinking that enabled the articulation of a fascist understanding of the significance of the environment and 'reading off' from natural relations. Other ideas such as organic farming, biodiversity conservation, forest and wilderness protection, and the virtues of a regular experience of nature were also used in the inter-war period by extreme right-wing thinking (Bramwell, 1989; Stephens, 1996). From the 'high-Tory' embracing of the centrality of living according to ecological principles – which also included echoes of the Aristocratic conservative attitude to the land, to the more violent and morally repugnant racist naturalism of the Nazis which divided peoples into 'human' and 'subhuman' categories – right-wing thought did have an ecological dimension, though it is important to stress that this was never a major or constitutive element.

According to Bramwell, while the modern integration of the environment into social theory may be viewed as predominantly a left-wing progressive form, the roots of green and ecological social theory lie to the Right. For her, 'today's Greens, in Britain, Europe and North America, have emerged from a politically radicalized ecologism, based on the shift from mechanistic to vitalist thought in the late nineteenth century' (Bramwell, 1989: xi). This shift from mechanistic to vitalist thinking may be seen as a continuation of the Romantic reaction to the

Enlightenment in general and the industrial revolution in particular, but for Bramwell this shift was a reaction against the soullessness, meaninglessness (prefiguring existentialism) and 'anomie' (Durkheim) of modern 'mass society' and 'mass democracy' which laid the foundations for the rise of National Socialism in Germany. For her, 'German ecologism well predated National Socialism. It formed part of a generic cultural phenomenon that was in part diverted into the Third Reich as an underlying theme. It re-emerged, well after the Second World War, in more obviously left-oriented groups' (1989: 5). Bramwell's biased scholarship is geared towards 'guilt by association' in linking contemporary green ideas, parties and movements with Nazism. For example, she argues that there is an intrinsic link between green ideas about organic farming, its critique of modern chemicalised industrial 'agri-business' and the importance of the rural way of life and Nazi ideas. She argues:

> Between the end of the First World War and the Nazi takeover, the idea that the peasantry had a special 'mission' was widespread. A reaction against the use of artificial fertilizers also occurred. Rudolf Steiner, founder of Anthroposophy, became its leader, before his death in 1925, and inspired the founding of a new school of farming known as 'bio-dynamic agriculture'.
>
> (1989: 200)

This not only attempts to portray contemporary greens, and others concerned with the environment as Nazis but also Rudolf Steiner and his anthroposophical thinking (which is largely represented today in the form of Steiner Waldorf schools) as fascists. The clear ideological bias of Bramwell's book – namely to set out to discredit the progressive, left-wing character of ecological and green politics and to reclaim 'ecology' for the political far right – prefigures in many respects the later ideological scholarship and controversy around the publication of Lomborg's *Skeptical Environmentalist*, discussed below.

Other attempts to establish the reactionary and far-right/fascist origins of ecological thinking include Gasman's (1971) study which claimed that the founder of the modern science of ecology, Ernest Haeckel, laid the foundations for German National Socialism and fascism. According to Gasman, 'Haeckel contributed to that special variety of German thought which served as the seed bed for National Socialism. He became one of Germany's major ideologists for racism, nationalism and imperialism' (1971: xvii). He goes on to point out that 'Near the end of his life he joined the Thule Society, a secret, radically right-wing organization which played a key role in the establishment of the Nazi movement' (Gasman, 1971: 30), all of which is aimed at establishing that the concern for the natural world was intrinsically linked to reactionary and fascist ideas. The *ad hominen* argument here is similar to the faulty (but pervasive) logic which holds that because Hitler was a vegetarian, all vegetarians are fascists!

More recent manifestations of a right-wing environmentalism may be traced to the 1960s and the twin discourses of the 'population bomb' and related ideas around 'limits to growth'. The argument that the major cause of environmental stress and what we would now term 'unsustainable development' was human overpopulation was (and still is) an extremely contentious issue. Authors such as Paul Ehrlich who wrote the bestselling book entitled *Population Bomb* and movements such as the US-based Zero Population Growth (ZPG) claimed that the world was heading for an ecological catastrophy as a result of the pressure on scare resources of human overpopulation. Echoing of course Malthusian thinking of the previous century, writers such as Ehrlich and others such as Garret Hardin promoted an anti-population growth position that was very popular in the Western world, particularly among political and economic elites.

Hardin made his name from his seminar essay on 'The Tragedy of the Commons' (Hardin, 1968) which analysed the disastrous consequences of unregulated human use of scarce resources. The 'tragedy of the commons' posits what would be in the individual rational interest of herders in putting extra animals on common land or 'commons'. A 'commons regime' is one in which the 'commons' is not owned or controlled by any one individual – that is, it is not governed by 'private property' relations. We can think of many 'commons' – for example, public space is 'owned' by no one individual, but is usually under the control of some public body – the state or municipal government. In this case the state owns the space which means all citizens who are members of the state own the space or commons, leading to the conclusion that we all 'own' the space, but 'no one' owns it. Hardin's logic may explain why our public spaces – which are owned by every-one and no one – are often rubbish-strewn and defaced. It is in the individual rational interest of those who drop litter to do so, since the cost of so doing is borne by everyone else. Typically, environmental problems are held to result from the 'tragedy of the commons', the overuse of a resource which no one owns (the seas) or everyone owns (state-regulated resources). As well as authoritarian solutions, another of the solutions proposed by Hardin was the necessity to 'privatise' commonly owned resources – which will be discussed in more detail in the next section on free market environmentalism.

Hardin's conclusion was that while each extra animal decreases the ability of the commons to sustain the animal herd, the cost of this is borne by all the herders while the benefit of the extra animal accruees to the owner. Thus, each herder thinks the same and increases the number of animals on the scarce commons, thus leading to its eventual collapse from over-grazing – hence the 'tragedy of the commons'. As Hardin put it, 'equal rights brings tragedy to all'; that is, the collective effort of individually rational actions leads to an outcome no one desires or wants. This analysis offered by Hardin has led to a whole school of 'rational choice' analysis. Hardin's analysis obviously had direct application to human–environment relations in that he argued that unchecked and unregulated human population increase and human use of scarce natural resources would lead to disaster. His solution was 'mutual coercion, mutually agreed upon' – that is, what was needed was a Hobbesian-type contract between all the herders to limit the number of animals on the common land. In terms of human population control, the Chinese 'one-child' policy and the Indian population control intiatives in the 1970s, which both imposed a non-negotiable limit on the number of children couples could have or enforced reproductive control by the state, are classic 'real-world' examples of policies based on the 'tragedy of the commons' logic.

It is Hardin's solution – the authoritarian use of state power to control individual actions which makes his analysis right-wing in that what he proposes is a modern updating and application of Hobbesian and Malthusian thinking. It is also important to point out that Hardin has been criticised for misinterpreting the 'tragedy of the commons' scenario and that a less authoritarian solution is therefore a more correct outcome of the analysis. Central to the criticism is the claim that a 'commons regime', such as those that historically existed or continue to exist in many parts of the world, are not 'unregulated' and that there exist both systems of informal and formal rules and procedures which regulate the use of the commons. This fundamental criticism of Hardin and the thinking based upon it claims that what Hardin describes and analyses is not a commons regime but an 'open access' one. That is, an example of a 'free-for-all' use of a scarce resource. Defenders of the commons and commons regimes have pointed out that they do not lead to the overuse and/or ecological collapse. As many ecologists, non-Western and green thinkers have pointed out, commons regimes do regulate access to and use of commons resources, but do so without recourse to either the state (via the imposition of authority) or the market (via privatisation) (Goldsmith et al., 1992; Wall, 1994b). Thus commons regimes do not necessarily lead to resource over-exploitation such that authoritative state action or privatisation are the only or the most appropriate solutions.

Hardin built upon his 'tragedy of the commons' analysis into the 1970s and used the same logic to criticise US (and other Western) programmes of foreign aid to the developing world. Echoing Malthus, Ehrlich and other advocates for a reduction in the human population, Hardin concluded in an infamous essay entitled 'Lifeboat Ethics' that it was counter-productive and irrational to give overpopulated countries food and support, since this only prolonged the necessary reduction needed in human numbers to prevent global ecological catastrophe, as outlined by the 'Limits to Growth' analysis. The analysis offered by Hardin and others was interesting in that they only concerned themselves with the population and environmental impact of non-Western countries on the whole. What is telling about this is that the environmental impact of human populations on the environment and natural resources is a function of the overall size and rate of increase of the population multiplied by the intensity of that population's use of the environment – i.e. per capita environmental impact. In other words, to determine any given human population's impact on the environment you need to know the use of materials, energy, food and so on per person, given that different patterns of impact are possible. For example, the lifestyle of a typical individual from the developed world requires vastly greater amounts of energy and resources than a typical individual from the developing world, such that more non-Western individuals can be supported by the same resource base and environmental impact as Western populations. Since the greatest users of the world's resouces and emitters of the world's pollution are the populations of the developed world, we need to ask why Hardin, Ehrlich and others did not look closer to home to reduce the human population and lessen the human impact on the environment. If one American or Western European has the same environmental impact as, say, a hundred Africans, why focus on African population growth and not American? As Robert Paehlke has perceptively put it:

> Why does Hardin not suggest a more sensible scheme, one advocating the eviction of North Americans from the lifeboat? It would cost far fewer persons and would achieve for the species considerably more time to bring about population stabilization that would avoid starvation in the so-called third world. If energy is the most limiting of resources, the heaviest consumers of energy, not the lightest consumers of food, are the greatest threat to long-term human survival.
>
> (Paehlke, 1989: 65–66)

Thus, in particular for those on the Left, the focus on population growth was seen as a quasi-racist and neo-colonial attempt to blame those who were suffering most from a globally unsustainable and unjust economic system for problems which were the result of that system which unfairly benefited a minority of the world's people. Another expression of this racist population focus may also be found in the writings of the influential American deep ecological thinker Edward Abbey, who railed against US immigration policy which was leading to the 'over-running' and 'swamping' of the US with Mexican immigrants. In a chapter entitled 'Immigration and Liberal Taboos', Abbey held that:

> ever continuing industrial and population growth is not the true road to human happiness, that simple gross quantitative increase of this kind creates only more pain, dislocation, confusion, and misery. In which case it might be wise for us as American citizens to consider calling a halt to the mass influx of even more millions of hungry, ignorant, unskilled, and culturally

morally-generically impoverished people. At least until we have brought our own affairs into order. Especially when these uninvited millions bring with them an alien mode of life which – let us be honest about this – is not appealing to the majority of Americans. Why not? Because we prefer democratic government, for one thing; because we still hope for an open, spacious, uncrowded, and beautiful – yes, beautiful! – society, for another. The alternative, in the squalor, cruelty, and corruption of Latin America, is plain for all to see.

Yes, I know, if the American Indians had enforced such a policy none of us pale-faced honkies would be here. But the Indians were foolish, and divided, and failed to keep our WASP ancestors out. They've regretted it ever since.

To everything there is a season, to every wave a limit, to every range an optimum capacity. The United States has been fully settled, and more than full, for at least a century. We have nothing to gain, and everything to lose, by allowing the old boat to be swamped. How many of us, truthfully, would prefer to be submerged in the Caribbean-Latin version of civilization? (Howls of 'Racism! Elitism! Xenophobia!' from the Marx brothers and the documented liberals.) Harsh words: but somebody has to say them. We cannot play 'let's pretend' much longer, not in the present world.

Therefore – let us close our national borders to any further mass immigration, legal or illegal, from any source, as does every other nation on earth. The means are available, it's a simple technical-military problem. Even our Pentagon should be able to handle it.

(Abbey, 1988: 42)

These anti-immigration concerns in particular are the hallmark of neo-fascist movements and parties such as the British National Party in the UK, the Front National in France and right-wing militias in America. Therefore, the articulation of such positions by ecological groups and writers makes them, on this analysis offered here, right-wing, and thus a form of eco-authoritianism or eco-fascism. However, there are others who claim that this interpretation by Abbey is incorrect and that he was not a racist or endorsed eco-authoritarian positions (Cahalan, 2001).

An extreme form of this thinking is contained in the following infamous example of Hardin's 'Lifeboat Ethics' thinking. Writing under the pseudonym 'Miss Ann Thropy' at the time of the Ethopian famine, a member of the radical US environmental group Earth First! welcomed AIDS as a necessary corrective to curb growing human population numbers:

If radical environmentalists were to invent a disease to bring human population back to ecological sanity, it would probably be something like AIDS. So as hysteria sweeps over the governments of the world, let me offer an ecological perspective on the disease . . . I take it as axiomatic that the only real hope for the continuation of diverse ecosystems on this planet is an enormous decline in human population. Conservation, social justice, appropriate technology, etc., are great to discuss and even laudable, but they simply don't address the problem. . . .

Barring a cure, the possible benefits of this [AIDS] to the environment are staggering. If, like the Black Death in Europe, AIDS affected one-third of the world's population, it would cause an immediate respite for endangered wildlife on every continent. More significantly, just as the Plague contributed to the demise of feudalism, AIDS has the potential to end industrialism, which is the main force behind the environmental crisis. None of this is intended to disregard or discount the suffering of AIDS victims. But one way or another there will be victims of overpopulation – through war, famine, humiliating poverty. As radical environmentalists, we can see AIDS not as a problem, but a necessary solution (one you probably don't want to try for yourself). *To paraphrase Voltaire: if the AIDS epidemic didn't exist, radical environmentalists would have to invent one.*

(Miss Ann Thropy, 1986; emphasis added)

Limits to growth

Other putative ecological writers in this vein include Robert Heilbroner and William Ophuls, both of whom based their analysis of the growing crisis between humanity and the global environment on the 'Limits to Growth' arguments of the late 1960s and 1970s. The publication of *Limits to Growth* (Meadows *et al.*, 1972) was a watershed in the development of the relationship between environment and social theory in terms of left-wing and right-wing reactions, and also for ecological thinking and the emerging environmental movement in particular. Using early computer modelling software, systems thinking and the new field of futures research, the 'Limits to growth' project explored the human impact on the global environment according to five factors – population growth, energy use, productive agricultural land, water and pollution. It was the first attempt to map and predict to extrapolate what would happen to the environment and resources if human population and resource use continued to increase. Its conclusions were dramatic and stark – if the use of fossil fuels kept increasing as well as the human population, pollution levels would also increase, productive land use would decrease and the exhaustion of non-renewable energy would all result in the collapse of human populations, leading to conflict, famines, resource wars and the ending of economic growth and prosperity as we know it in the developed world. The implications of the report were that we needed to reduce human numbers, begin the process of transforming our economies to use renewable energy and prepare for something resembling the 'stationary state economy' described by J.S. Mill in Chapter 3.

There were (at least) two different right-wing and authoritarian reactions to the analysis and conclusion of 'Limits to Growth'. On the one hand there were those like Robert Heilbroner and William Ophuls who broadly accepted the 'Limits to Growth' position and for whom its analysis meant the ending of liberal democracy and Western civilisation as we know it in the developed world, a 'survivalist' discourse according to Dryzek (2005: 27–51). On the other hand, there were those who rejected the whole notion of there being ecological (or any other) limits to endless economic growth and material prosperity – who may be termed contrarians, cornucopians or Prometheans (Dryzek, 2005: 52–73).

A good starting point to explore the relationship between liberal democracy and Western civilisation and the environment is Tocqueville's suggestion that:

> General prosperity is favourable to the stability of all governments, but more particularly of a democratic one, which depends upon the will of the majority, and especially upon the will of that portion of the community which is most exposed to want. When the people rule, they must be rendered happy or they will overturn the state: and misery stimulates them to those excesses to which ambition rouses kings.
>
> (Tocqueville, 1956: 129–30)

This assumption of the positive correlation between material affluence and the stability of a democratic political order is one which what I am here calling eco-authoritarian authors such as Hardin, Ophuls and Helbroner accepted. The eco-authoritarian implication of the link between scarcity and political arrangements has been forcefully made by Ophuls. He begins from the assumption that:

> The institution of government whether it takes the form of primitive taboo or parliamentary democracy . . . has its origins in the necessity to distribute scarce resources in an orderly fashion. It follows that assumptions about scarcity are absolutely central to any economic or political doctrine and that the relative scarcity or abundance of goods has a substantial and direct impact on the character of political, social and economic institutions.
>
> (Ophuls, 1977: 8)

Calling the affluence experienced by Western societies over the past two hundred years or so 'abnormal', a material condition which has grounded individual liberty, representative democracy, social stability and progress (1977: 12), Ophuls concludes that with the advent of the ecological crisis, interpreted as a return to scarcity (following 'Limits to Growth' (LTG) thesis), 'the golden age of individualism, liberty and democracy is all but over. In many important respects we shall be obliged to return to something resembling the pre-modern closed polity' (1977: 145). He argues that what we need to replace (liberal) democratic institutions is a (benign) technocratic and quasi-theocratic dictatorship. Now while it is perhaps true that democracy does require some degree of material affluence, it is a completely different issue to argue that a diminution in material prosperity heralds the end of democracy as we know it such that the only political alternative is an eco-authoritarian dictatorship.

The resolution of the ecological crisis according to Ophuls requires that 'the steady-state society will not only be more authoritarian and less democratic than the industrial societies of today . . . but it will also in all likelihood be much more oligarchic as well, with only those possessing the ecological and other competencies necessary to make prudent decisions allowed full participation in the political process' (1977: 163). Ophuls' argument for a 'priesthood of responsible technologists' (1977: 159) to take charge of society has echoes of Plato's anti-democratic position that only those with competent knowledge are 'fit' to govern while the rest are only fit to be governed, and also echoes conservative fears of and scepticism about popular democracy.

The other right-wing response to the LTG is a rejection of its central arguments, what is often termed as cornucopian, Promethean (Dryzek and Schlosberg, 2004; Dryzek, 2005: 51) or contrarian response. The cornucopian response was an initial reaction to the argument that there were limits to economic growth which basically rejected the notion that there were any natural limits in terms of either resource (input) or pollution (output) limits. Mostly coming from economists, cornucopian reactions included authors such as Julian Simon and Herman Kahn. Contrarian reactions rejected the notion of an ecological crisis and included authors such as Lomborg (discussed in more detail in Box 5.2).

'Crisis what crisis?': The right-wing cornucopian and contrarian response

The other main right-wing reaction to the LTG thesis was to simply reject it, undermine its scientific foundations and to do so loudly and often. This contrarian position posited that there were no limits to human material progress, that the threats to the ecoystem, natural resources and human well-being and health were, at best, exaggerated, and, at worst, deliberate lies and hype on behalf of the green movement. Early proponents of this position included Julian Simon, who (with futurist Herman Kahn) edited *The Resourceful Earth* as a reply to the Carter adminstration's *Limits to Growth*-inspired Global 2000 Report, rejected the latter's

Box 5.2 The Lomborg controversy

In 2001, Cambridge University Press published a book by Danish statistican Bjørn Lomborg entitled *The Skeptical Environmentalist: Measuring the Real State of the World*. There was an extremely well-organised media campaign around the author's ideas and he wrote in a variety of newspapers from the *Guardian* in the UK to the *New York Times* and *Washington Post* in the US as well as television interviews and debates. The main thrust of the book was to 'debunk' the 'myths' of environmental 'doom-sayers' using 'sound science', economics and statistics to demonstrate that on the whole, things are getting better, not worse. However, unlike other contrarians, Lomborg accepts the reality of climate change, but his view is that we should be spending resources on providing safe water and sanitation for the developing world, rather than trying to deal with a reality we cannot change. The book contains little new by way of contrarian thinking about the environment but has gathered together many of its basic and organising ideas and perspectives. Clearly inspired by another earlier contrarian Julian Simon, Lomborg may be said to be one of the best-known contrarian right-wing thinkers, the new 'public face of Prometheanism' according to Dryzek (2005: 55).

Lomborg's position may be gleaned from an article he wrote for the *Wall Street Journal*:

> I am Danish, liberal, vegetarian, a former member of Greenpeace; and I used to believe in the litany of our ever-deteriorating environment. You know, the doomsday message repeated by the media, as when Time magazine tells us that 'everyone knows the planet is in bad shape'. We're defiling our Earth, we're told. Our resources are running out. Our air and water are more and more polluted. The planet's species are becoming extinct, we're paving over nature, decimating the biosphere. The problem is that this litany doesn't seem to be backed up by facts. When I set out to check it against the data from reliable sources – the U.N., the World Bank, the OECD, etc. – a different picture emerged. We're not running out of energy or natural resources. There is ever more food, and fewer people are starving. In 1900, the average life expectancy was 30 years; today it is 67. We have reduced poverty more in the past 50 years than we did in the preceding 500. Air pollution in the industrialized world has declined – in London the air has never been cleaner since medieval times.
>
> (Lomborg, 2003)

The book had been published in Denmark in 1998 and led to some of Lomborg's colleagues writing rebuttals to the main points raised. This set the pattern for the predicted clamour in the media upon the publication of its English edition, mainly between those who enthusiastically supported him as providing objective and scientific proof of the fallacy of environmental arguments and those who either demonstrated the faulty scientific thinking contained in the book and/or highlighted its ideological 'hidden agenda'.

Of particular note is that the controversy was such that Lomborg's book was discussed by the Danish Committee on Scientific Dishonesty, which in January 2003 ruled that, 'Objectively speaking, the publication of the work under consideration is deemed to fall within the concept of scientific dishonesty. In view of the subjective requirements made in terms of intent or gross negligence, however, Bjørn Lomborg's publication cannot fall within the bounds of this characterization. Conversely, the publication is deemed clearly contrary to the standards of good scientific practice' (DCSD, 2003). This ruling was overturned in December 2003 by the Danish Ministry of Science, which noted that the DCSD judgment was not backed up by documentation, and was 'completely void of argumentation' for the claims of dishonesty and lack of good scientific practice (Environmental Assessment Institute, 2003). However according to Dryzek's reading of the verdict:

> In December 2003 the Danish Ministry of Technology overturned the committee's negative verdict, by which time Lomborg had been appointed as Director of Denmark's Environmental Assessment Institute by a right-wing government under a prime minister to whom Lomborg had access. This episode reveals much about the peculiar politics of science in Denmark than it does about the veractity or falsity of Lomborg's analysis.
>
> (2005: 56)

Lomborg certainly made a name for himself, and has since gone on to write other books and research which offer a contrarian position (Lomborg, 2003). He has been championed by right-wing publications such as *The Economist* magazine, as well as right-wing think-tanks and media outlets, while he has been criticised in scientific publications such as *Scientific American* – which ran an eleven-page critique of his book without offering him the right to reply – and *Nature*.

In 2004 he launched the *Copenhagen Consensus*, which focuses on using expert economic thinking to prioritise ten of the world's most pressing problems from climate change to communicable diseases. Of particular note about the whole controversy is the way in which ideology, values and political claims are mixed with scientific claims, thus undermining (or at least muddying) the latter's unified and undivided sense of offering 'objective' and 'value-free' knowledge.

central thesis of rising prices for scarce resources, declining supplies of vital resources and worsening ecological conditions. For Simon:

> The standard of living has risen along with the size of the world's population since the beginning of recorded time. And with increases in income and population have come less severe shortages, lower costs, and an increased availability of resources, including a cleaner environment and greater access to natural recreation areas. And there is no convincing economic reason why these trends toward a better life, and toward lower prices for raw materials (including food and energy), should not continue indefinitely.

Contrary to common rhetoric, there are no meaningful limits to the continuation of this process. . . . There is no physical or economic reason why human resourcefulness and enterprise cannot forever continue to respond to impending shortages and existing problems with new expedients that, after an adjustment period, leave us better off than before the problem arose. Adding more people will cause us more such problems, but at the same time there will be more people to solve these problems and leave us with the bonus of lower costs and less scarcity in the long run. . . .

Following on this is the fact that economic history has not gone as Malthusian reasoning suggests. The prices of all goods, and of the services they provide, have fallen in the long run, by all reasonable measures . . . there is no meaningful physical limit – even the commonly mentioned weight of the earth – to our capacity to keep growing forever. There is only one important resource which has shown a trend of increasing scarcity rather than increasing abundance. That resource is the most important of all – human beings.

(Simon, 1996)

More modern representatives of this strain of thinking include writers such as Greg Easterbrook (1995), Wilfred Beckerman (1974, 1995, 2002), Matt Ridley (1995) and Bjorn Lomborg (2001); right-wing and free market think-tanks such as the Adam Smith Institute in the UK, the Freedom Foundation and the Competitive Enterprise Institute in the US and others like it in other (mostly industrialised) countries; right-wing media outlets, such as the *Daily and Sunday Telegraph* in the UK or *Fox News Network* in the US. They share a common purpose in questioning or undermining the scientific basis of global ecological crises, such as climate change sceptics and promoting a right-wing critique of the green movement, what has become known as the 'Green backlash' (Rowell, 1996; Beder, 1997, 2001a; Paterson, 1999) and seek to promote and popularise a free market vision of how to deal with environmental problems through deregulation, rolling back the state and encouraging private sector innovation and ownership of environmental goods and services. As Beder puts it, 'Anti-environmentalism has been a response to the rise of environmental consciousness first in the late 1960s and early 1970s and then again in the late 1980s and early 1990s. It is a backlash against the success of environmentalists in raising public concern and pressuring governments to protect the environment' (2001b: 19). Such right-wing reactions are also united in branding environomentalists as 'socialists' or 'eco-fascists' who are the new 'enemy of freedom' in a post-Cold War era. As Rowell puts it, 'The tide is turning against the environmental movement worldwide. Environmental activists are increasingly being scapegoated by the triple engines of the political Right, corporations and the state. The backlash has one simple aim: to nullify environmentalists and environmentalism' (1996: 4). For example, the US-based Competitiveness Enterprise Institute has organised an annual 'Resourceful Earth Day' to revile what it sees as the negative and misleading messages of the celebration of 'Earth Day' on 22 April each year.

A good example of this type of right-wing reaction to the ecological crisis and the green movement is Matt Ridley's *Down to Earth: A Contrarian View of Environmental Problems* published by the Institute of Economic Affairs in London in 1995. On the back cover of this publication Ridley claims that 'World population growth is decelerating; food, oil and copper are all cheaper and more abundant than ever before', 'The ozone layer is getting thicker, not thinner, over temperate latitudes', and 'Environmental lobbying organisations are spending more money on lawyers and marketing men to grow their own budgets and less on naturalists and volunteers'. The book rejects that biodiversity loss and species extinction are as bad as the greens make out, that state ownership and regulation are the real enemies of conservation and green groups are utopian Marxists who wish to destroy capitalism and the Western way of life who base their arguments not on 'sound science' but on false ideology. More recent publications from the same institute have questioned the reality and severity of global climate change (Bradley, 2004) and promoted the benefits of and need for genetically modified foods (Bate, 2003). These positions are standard and may be found in most right-wing anti-environmentalist perspectives. The 'contrarian' position is basically to challenge or deny the scientific and evidence bases for environmental arguments, such as that there are limits to growth or that human-caused climate change is happening. This has the not inconsiderable political effect in the general population (and political elites and policy-makers) of sowing seeds of doubt about the reality, existence or severity of environmental problems (Ehrlich and Ehrlich, 1996). According to Buell (2003) this constitutes a vital part of the American Right's 'politics of denial'. As he puts it:

> Something happened to strip environmental crisis of what seemed in the 1970s to be its self-evident inevitability. Something happened to allow environmentalism's antagonists to stigmatize its erstwhile stewards as unstable alarmists and bad-faith prophets – and to call their warnings at best hysterical, at worst crafted lies. Indeed something happened to allow some even to question (without appearing ridiculous) the apparently commonsensical assumption that environmentalists were the environment's best stewards. The most important explanation for these events isn't hard to find. In reaction to the decade of crisis, a strong and enormously successful anti-environmental disinformation industry sprang up. It was so successful that it helped midwife a new phase in the history of US environmental politics. . . . Despite scientific evidence and even, in a number of cases, virtual scientific consensus to the contrary, issue after issue was contested. The ozone hole was denied and trivialized, food and population crises were debunked, and global warming was hotly denied, doubted and dismissed as unproven.
>
> (Buell, 2003: 3–4)

Some of the more extreme (if that were possible!) right-wing contrarian reactions are found in America in organisations such as the ultra-libertarian Ayn Rand Institute. A typical polemic is the following piece from 2002 which links direct action environmental activism to the 'war on terror' and goes so far as to call environmentalism a form of 'terrorism'. The author, one Onkar Ghate, holds that 'Even more ominous than the growing environmental terrorism is the fact that – unlike in the case of Islamic terrorism – no one has yet risen to defend the irreplaceable values under attack by the environmentalists. Their targets are not, fundamentally, a particular ski resort or logging truck or research project, but what these represent: human technology, human progress, human life' (2002). Another associate of the same right-wing institute goes further in linking environmental activism in the US to the 9/11 attacks: 'Our inaction in the face of Islamic terrorists prior to Sept. 11 helped to embolden them; our inaction in the face of eco-terrorism is doing the same. We dare not wait for eco-terrorists, motivated by their own nihilistic ideology, to mount their own Sept. 11. They must be stopped by the force of government, now' (Journo, 2002).

Other examples in this vein include the founder of the Wise Use Movement, Ron Arnold, who has used the case of the lone individual, like the Unabomber who was responsible for a number of bombings in America motivated by environmental concerns, to smear the entire environmental movement (Arnold, 1997). Such is the nature of polemic and ideologically charged 'backlash' against the environmental position. Although less offensive, and also notable for initially being based within the academy as opposed to the popular press and partisan media outlets (although this is where the debate also went), the controversy over the publication of Lomborg's *Skeptical Environmentalist* is instructive (see Box 5.2). Reflecting on the US context where this 'green backlash' is most acute, Brian Tokar notes:

> The [anti-environmental] movement is closely allied with Republicans in the US Congress, several of whom were elected in 1994 – when Congress became Republican-dominated for the first time in 40 years – on an overtly anti-environmental platform. As a result, renewal of many of the landmark environmental laws passed in the 1970s such as the Clean Air and Clean Water Acts, the Endangered Species Act and Superfund [are] being held up by unexpected obstacles, delays and consistent efforts to weaken them.
>
> (Tokar, 1995: 152)

In the US, groups such as the Wise Use Movement and allied organisations such as the Centre for the Defense of Free Enterprise have taken these contrarian and cornucopian positions and put them into action in terms of well-funded (and popular) campaigns to privatise federal lands (a modern form of 'enclosure' echoing the discussion in Chapter 2), to reduce state regulation (specifically environmental and labour standards and laws) of extractive industries such

as mining and forestry, and also lowering the 'regulatory burden' within industrialised agriculture and food processing. It is closely related to the Bush
administration's refusal to acknowledge the reality and severity of climate change
and to sign up to the Kyoto Protocol. In this, Bush Jr. is following the same anti-
environmental position as his father. The comment from Burke below is equally
if not more applicable now to Bush Jr. as to Bush Snr.:

> In the last days of the 1992 presidential campaign, George Bush denounced
> 'environmental extremists' who sought to lock up natural resources and destroy
> the American way of life. At the heart of this imagined green conspiracy was
> the 'Ozone Man', Senator Al Gore Jr., author of *Earth in the Balance*. Bush's
> attack on environmentalism failed to save his candidacy, but it was a high
> water mark for the political influence of the 'Wise Use' movement, a network
> of loosely allied right-wing grassroots and corporate interest groups dedicated
> to attacking the environmental movement and promoting unfettered resource
> exploitation.
>
> (Burke, 1993)

Free market environmentalism

While also a key feature of right-wing thinking, the market-based approach
of free market environmentalism (FME) is also right-wing in being the other main
interpretation of Hardin's tragedy of the commons – namely the privatising of
the commons. For FME, environmental problems are largely due to open access
and the inability to exclude others from using and abusing the resource. Hence
FME proposes private property rights as a solution to over-exploitation and degradation of the 'ecological commons'. Thus, free market environmentalism is
a modern form of the 'enclosure of the commons' as discussed in Chapter 3.

FME holds that to conserve resoures we need to privatise them since it will be in
the interests of owners not to deplete them, unlike the overuse and abuse of commonly owned resouces. FME holds that by aggregating individual preferences
for environmental resources, as 'revealed' by supply and demand intersecting at
the equilibrium price, the 'efficient' economy–ecology metabolism will be determined. Unfortunately, an economically 'efficient' metabolism may not be an
ecologically sustainable one (Daly, 1987). Pure economic rationality is assumed
to lead to an ecologically rational outcome. However, for many environmental
public goods, for which no market can exist, such as biodiversity protection and
prevention of global warming, it is clear that a free market environmentalist
approach is fundamentally flawed. That is, there are environmental problems
whose nature is such that they cannot be disaggregated into component parts which
can be solved by exchange on the open market between property holders.

A real-world example of free market environmentalism is the Wise Use Movement in America which has lobbied Congress to privatise federal land (Dryzek, 2005: 65; Harvey, 1996: 383–386). The aim of the Wise Use Movement is to transform environmental goods from public goods in which their 'use' was limited to access and recreation, into privately owned goods where use would be much more extensive, intrusive and ecologically damaging. 'Use' for the Wise Use Movement implies forms of economic development, such as mining, logging, building, hunting and generally the exploitation of (public) environmental resources for (private) economic gain. From this free market environmentalist point of view, objections to such 'abuses' can only take the form of other private agents, such as environmental protection organisations, purchasing these lands themselves and thus preserving them. This has the effect of making environmental protection a function of wealth, where only those with sufficient purchasing power can buy it. For Eckersley, the fact that free market environmentalists are not interested in the *distribution* of private environmental property rights, such that, for example, there is equal distribution for both rich and poor, 'must be seen as a thinly disguised endorsement of the existing distribution of property rights and income. . . . Indeed, the long-term consequence of zealously pursuing the privatisation of environmental resources . . . is likely to be the intensification of the already wide gap between the propertied and the propertyless, and the rich and the poor, both within and between nations' (Eckersley, 1993: 15).

However, there may be some positive aspects that may be taken from free market environmentalism. One is the notion of stewardship and care implicit in the emphasis on private property and ownership. As Anderson and Leal (1991: 3) point out, the idea of having property, a claim of ownership, to the land or to some environmental resource involves a commitment to look after it. However, as pointed out above with reference to the 'tragedy of the commons', the steward-ship ideal implicit in the idea of property is not confined to *private* ownership. *Common* ownership can also deliver the virtues of stewardship and careful husbanding. As Goldsmith *et al.* (1992) argue, there is plenty of empirical evidence from around the world that commons regimes can deliver sustainable levels of resource exploitation. As suggested above, the 'tragedy of the commons' is as a result of 'open access' to an environmental resource; a commons regime is not a 'free-for-all' (Goldsmith *et al.*, 1992: 127). Indeed, according to this line of argument, so-called environmental tragedies of the commons are in fact the result of tragedies of enclosure (Bromley, 1991). It is only if one accepts the initial argument that environmental problems are caused by common ownership that private property in environmental resources can be regarded as the only solution. Commons regimes in which the commons are the property of all, or all members have an equal right of access, is a different form of collective ownership than state ownership, which is the real target of free market environmentalism. Under central state ownership, it is less likely that individuals will feel they, conjointly with others, actually own the resource and thus have responsibility for it. Under these conditions it is truer to say that the resource is owned by nobody. *Collective* ownership and regulation of environmental resources (which is not the same as centralised state ownership and control) does not stand condemned on ecological grounds. For example, one could envisage a hybrid 'commons-type regime', where environmental resources were managed by a combination of local state co-ordination together with local community, including business, participation.

One type of private ownership of environmental resources which may be seen as positive from a green point of view is the private ownership of the land as part of farming viewed as a social practice. According to Thompson, 'Stewardship does not arise as a constraint on the farmer's ownership and dominion of the land, but as a character trait, a virtue, that all farmers would hope to realise in service to the self-interests created by ownership of the land' (1995: 74). However, in this instance the argument for private ownership of the land is not for the same reasons as put forward by free market environmentalism. In the case of agricultural stewardship, private ownership is not justified on the grounds of economic productivity or private profit alone, nor is its content determined by market exchange. Indeed, according to Thompson (1995), the family-owned farm, properly speaking, restrains the productivist imperative which would transform agri*culture* into 'agri-*business*'. That is, private ownership (or secure tenure) of the land, within the context of farming as a way of life and not simply as an industry, may be justified from a green point of view. Private ownership within the context of a socially embedded cultural practice is not the same as private ownership within the context of a market system.

If property rights and markets themselves are politically created and maintained, they can – and from a green perspective ought to – be politically and morally constrained. Market-based solutions to social-environmental problems, such as that proposed by free market environmentalism, do not represent 'depoliticised' solutions, any more than purely technological solutions are non-political. Market-based approaches are as inherently political and just as normative as any non-market alternatives.

In the end one must conclude that free market environmentalism is transparently ideological, the environmentalist dimension of the right-wing libertarian political project. It is a reaction to the 'statist' implications of neo-classical environmental economics and any other solution to environmental problems in which the state has a central role or in which the market does not. This view is extended to the 'green movement' as a whole, which is perceived as a bulwark against the privatisation of environmental resources, another form of social resistance to the extension of the discipline and advantages of the free market. For most free market environmentalists, greens are simply another pressure group using the political process to undermine the inexorable advance of free market principles. Seen within the historical context of the collapse of communism and the crisis within the Left, environmentalism is portrayed as an alternative legitimation for socialist collectivism and the illegitimate supression of individual liberty (Anderson and Leal, 1991; North, 1995; Ridley, 1995). Indeed, if the arguments concerning the centrality of democratic planning and regulation to collective ecological management of the previous chapter are right, then it may be that ecological issues can serve to relegitimise aspects of socialist politics, which will be discussed in Chapter 6. However, first, we will conclude this chapter by outlining a revealing case study of other right-wing reactions to the environmental crisis, the controversy around and reactions to Bjorn Lomborg's book, *The Skeptical Environmentalist*.

Other right-wing responses: the Gaian post-humanism of John Gray

With the exception of political and social theorists such as Brian Barry, Ulrich Beck and Anthony Giddens (and to a lesser extent Charles Taylor and David Miller), John Gray is one of the few major modern political thinkers within what one would call the 'mainstream' of contemporary political thinking to have discussed green or ecological issues and ideas. Equally, he is one of the most intriguing and idiosyncratic. While the overwhelming bulk of contemporary political theory has little if anything to say about ecological concerns, Gray has been a notable exception in his interest in and integration of ecological themes and issues into his thinking. An important landmark in Gray's journey from supporter of the free market nostrums of the 1980s to a green, scientifically based post-humanism in his latest works may be traced to a chapter entitled 'An Agenda for Green Conservatism' in his book, *Beyond the New Right* (Gray, 1993), which offers his first conservative exploration into distinctly 'green' political territory. From this exploratory article onwards, significant aspects of Gray's thought and the themes he has tackled in his 'post-liberalism' has seemed to move in a 'green' direction. From his trenchant critique of the Enlightenment and the self-defeating Prometheanism of a technologically focused conception of progress, his rejection of globalising capitalism as a dystopian fantasy, to his wholesale embracing of James Lovelock's 'Gaia hypothesis', to his defence of better treatment of animals and championing of non-Western and mystical traditions of thinking and the integration of these normative concerns with a hard-headed 'realpolitik' analysis of geopolitical conflict in an age of ecological scarcity are all themes that, on first gloss, make him a 'green' thinker or someone who takes seriously the import of the environment and human relations/dependence on it for social theory and practice.

However, the level of engagement with Green thought is, as suggested below, rather cursory and superficial, given that it is confined to some marginal thinkers and ideas within Green theory – a pattern that has continued to mark the 'greening' of Gray's thought in an idiosyncratic manner. Thus Gray's take on green ideas is one that is, to say the least, unique to Gray and may be construed as a right-wing interpretation of the ecological crisis and green thinking. In particular, Gray's rejection of the non-ecological progressive agenda of green thinking relating to political and economic issues serves to characterise his thinking as right-wing.

Gray's frequent positive endorsement of James Lovelock's 'Gaia hypothesis' may be understood as underwriting his post- or anti-humanist or 'Earth-centred' thinking. The Gaia hypothesis suggests that the Earth is a living organism, with its own in-built cybernetic feedback mechanisms. It is often used within green thinking to buttress arguments for saving the planet, but the real upshot of the Gaia hypothesis is that the planet does not need saving, as Lovelock himself has forcefully suggested (Lovelock, 1988: 212). The reality of the Gaia hypothesis is that if humans continue to inflict ecological damage on global and local scales then the Earth will no longer function as our species' life support system. In other words, the planet will continue and prosper without humans. This is the upshot of Gray's 'post-humanist', Gaia-inspired vision – a vision of the planet that is, quite literally, 'post-human'. The main reason for this, as Lovelock himself indicates and as some Greens have recognised (Dobson, 1995), is that the Gaia hypothesis is essentially anti- or non-anthropocentric/humanist, in the sense that its focus is not humanity or its interests or its long-term survival, but the planet and its living and non-living entities and processes considered as an interrelated whole. For example, as Lovelock rightly points out, 'The very concept of pollution is anthropocentric and it may even be irrelevant in the Gaian context' (Lovelock, 1988: 110). It is clear that Gray's use of the Gaia hypothesis moves him in this 'post-humanist' direction, since as he puts it: 'It may indeed be that the Gaian vision, being free from anthropocentrism which privileges humans in the universe . . . is the most appropriate antidote to this [sentimental-humanist] malady of the spirit that parades as enlightenment' (Gray, 1993: 177).

As even a cursory knowledge of the evolution of human society and the histories of human societies indicates, relations between humans and parts of their environment, particularly domesticated animals, have never been entirely devoid of moral content. Indeed, in keeping with Gray's traditional conservative stress on history and practices (following Edmund Burke) – that is, focusing on what human beings *did* and *do*, rather than hypothesising abstractly about what they may or should do – his position is that we should learn from the past and not rush into new and uncharted practices in how we relate to and use nature.

According to Gray:

> Humanism is a doctrine of salvation – the belief that humankind can take charge of its destiny. Among Greens, this has become the idea of humanity become the wise steward of the planet's resources. But for anyone whose hopes are not centred on their own species the notion that human action can save themselves or the planet must be absurd.
>
> (Gray, 2002: 16–17)

Gray's rejection of any notion of 'stewardship' as 'absurd' and the almost palable resignation to the inevitability of human extinction all betray a distinctly pessimistic conservative mindset.

However, from his 1993 essay Gray cannot be called a cornucopian/Promethean in that he accepts the reality of definite biophysical limits to economic growth, accusing those who promote the idea of infinite and exponential economic growth of peddling a 'wholly unrealisable fantasy' (1993: 142), thus putting him at odds with Promethean thinkers like Julian Simon and Lomborg. Rather, he is closer to Ophuls in acknowledging that 'the political legitimacy of Western capitalist market institutions depends upon incessant economic growth; it is endangered whenever growth falters' (1993: 152). It is clear in his *volte face* on the neo-liberalism he promoted in the 1980s that he now sees free market economics as a form of insanity, without foundations or proven contribution to human well-being, but buttressed by the brute political power of the 'minority world' – led by the United States – and backed by a willingness to use military strength if 'free trade' fails to procure the necessary resources. Yet he also rejects green and left-wing calls for a new politics and economics of quality of life, socio-economic equality and distributive justice.

In a phrase chillingly echoing the main thrust of the logic of the 'Miss Ann Thropy' piece cited above, Gray holds that 'It is not of becoming the planet's wise stewards that Earth-lovers dream, but of a time when humans have ceased to matter' (Gray, 2002: 17). Gray's almost exclusive focus on population control (especially since it seems to be focused only on the developing world) places his 'green thinking' firmly within an eco-conservative framework, and one that has echoes of the debate over the 'population bomb' associated with writers such as Paul Ehrlich and Garret Hardin in the 1960s and 1970s. One of the reasons why most Greens reject the exclusive focus on population as either the main or only reason for the ecological crisis is that there is more than a whiff of racism and xenophobia often underpinning arguments for population reduction – as in the thoughts of Edward Abbey above. It is easier to demand population reduction in the teeming hoards of Africa and Asia in the name of lessening the human impact on the global environment, than to challenge the unjust and profligate use of finite resources by a comfortable minority; easier to demand 'they' reduce their population than 'we' curb our consumer lifestyles.

For Gray:

> A high-technology Green utopia, in which a few humans live happily in balance with the rest of life, is scientifically feasible; but it is humanly unimaginable. If anything like it ever comes about, it will not be through the will of homo sapiens. So longer (*sic*) as population grows, progress will consist in labouring to keep up with it. There is only one way that humanity can limit its labours, and that is by limiting its numbers. . . . Zero population growth could be enforced only by a global authority with draconian powers and unwavering determination. There has never been such a power, and there never will be.
>
> (Gray, 2002: 184–5)

In this stress on population as the only way to lessen ecological degradation and the explicit link to authoritarianism, Gray's thinking mirrors that of other eco-authoritarians, who like Gray have a clear scientific analysis of the causes of the ecological crisis and come to an equally dogmatic and eminently logical conclusion. Gray's authority, charged with zero population growth using draconian powers, has been imagined and defended by one of the most well-known eco-authoritarian writers, Garret Hardin.

Gray perhaps can be better described as an *ecological* rather than a *green* thinker – accepting the ecological 'human condition', namely our utter dependence upon nature and the futility of attempts to 'master' and 'dominate' it through technology, our continuity with nonhuman animals and so on, but rejecting all the 'humanist' and progressive baggage of social, global and environmental justice, socio-economic equality as a precondition for a politics of quality of life for all, the extension of democracy and citizen activism and so on with which most 'green' and left-wing social theory is concerned.

Conclusion

Right-wing reaction to the environment has a long history. Its ideological appropriations of and engagements with ideas of 'nature' and the 'environment' range from the 'aristocratic' Romantic critique of the industrial revolution in the early nineteenth century; to the prima facie 'fit' between conservative traits of 'realism' and concern with human vulnerability in terms of how the external environment (and internal human nature) constrain and de-limit what human societies can do; to the more recent polemical 'anti-environmentalist' tirades of right-wing contrarian and free market thinkers and think-tanks, and the articulation of an 'ecological' conservatism in the thought of thinkers such as John Gray. Anticipating Gray, Paehlke in 1989 explored the strategic or pragmatic intersection of environmentalist and neo-conservative ideology, pointing out the commonalities between them, namely distrust of large, centralised bureaucracies, a commitment to decentralisation, and a 'restoration of morality and traditional values, although they do not agree on what those values are' (Paehlke, 1989: 217). Today the discourse of 'security' – long a conservative/right-wing value – has taken on new environmental overtones in the proliferation of high-level policy discussion and popular discourse about 'environmental security' (Barnett, 2001), and increasing 'energy security' in the light of 'Peak Oil', especially in foreign policy and international relations debates (Homer-Dixon, 2001). The relationship between right-wing thinking and the environment is thus a rich and varied pattern and legacy of social theorising, involving a selective use and abuse of normative arguments and scientific evidence, ideological 'spin' and contested readings of 'nature' and the import of the environment for social theory (both in terms of 'importance' and 'importation' of these readings into thinking about how human society ought to be organised).

Summary points

- Aspects of conservative thinking, namely its realism and acceptance of limits (environmental and those arising from 'human nature') to human action, indicate its acknowledgement of the importance of the environment and enables a connection to be made between conservative thought and environmental themes and issues.
- Two broad conservative reactions to the industrial revolution may be discerned: a 'landed aristocratic' one which had affinities with the 'Romantic reaction' discussed in Chapter 3 and which sought to resist capitalist industrial organisation of the economy, and an opposite and much stronger one which accepted the 'new industrial order' and sought to maintain traditional landed class privilege through an alliance with the rising manufacturing class. In relation to the democratic revolution, early conservative reaction was universally against the granting of political rights and voting to the 'masses'.
- Thomas Malthus, together with Edmund Burke, represent the two traditional conservative thinkers who took the natural environment seriously in their social theories. Burke in particular was one of the first 'modern' thinkers to

discuss and stress the importance of what we would now call 'intergenerational justice' and the normative significance of obligations both to past and future generations.

- The publication of Charles Darwin's theory of evolution led to 'Social Darwinism', which was a right-wing appropriation of Darwinian thinking applied to human society and which was used to justify and legitimate the inequalities of capitalist society and the reality and virtues of 'the competition of the fittest' and a *laissez-fair* free market economic system.

- Authoritarian and fascist interpretations of ecological and green concerns have been made in terms of linking the fascist ideology of German National Socialism to its origins in ecological themes.

- Eco-authoritarian reactions to the landmark and hugely influential *Limits to Growth* report and general thesis held that the era of liberal democracy, unchecked population growth, economic growth and consumerism was now at an end and that the only way (for some, read Western) human societies could survive the ecological crisis was by authoritarian political means, echoing the idea of Malthus and Hobbes.

- Garret Hardin is perhaps the most well known of the eco-authoritarian writers who argued that it was counter-productive for the developed world to give aid to the developing world, and, following Malthusian thinking, it would be better to let the developing world deal with its own problems.

- The focus on population growth (in the developing world) within certain strands of right-wing reactions to the ecological crisis sometimes led in racist and morally repugnant directions, as in injunctions against immigration and against support for dealing with the AIDS crisis in the 1980s.

- A cornucopian/Promethean right-wing reaction to the 'Limits to Growth' thesis was to say that there was no 'ecological crisis' and that there was no resource scarcity threat to continuing economic growth.

- Part of this reaction was a 'contrarian' response which sought to deny the scientific evidence for the worsening ecological crisis, from biodiversity loss to global climate change. Both Promethean and contrarian right-wing responses may be seen as key aspects of the ideological 'green backlash' against environmentalism and environmental regulation of economic activity.

- Free market environmentalism holds that the cause of the environmental crisis is the lack of clear enforceable property rights, and that the solution lies in the privatisation of common or state-owned resources, and for these resources to be traded and used in the free market.

- The controversy over the publication of Bjorn Lomborg's *Skeptical Environmentalist* in 2001 is an excellent case study in the ideological battle of ideas between pro- and anti-environmental camps, in which science and statistical data are not 'objective' or 'value free' but weapons in a battle to establish the 'truth' about the 'real state of the environment'.

• John Gray represents an interesting and idiosyncratic conservative take on environmental issues and may be best described as an 'ecological' rather than a 'green' thinker.

Further reading

On traditional conservatism and the environment, see J. Gray (1993), 'An Agenda for Green Conservatism', in his *Beyond the New Right: Markets, Government and the Common Environment*, London: Routledge; A. Dobson (2000), *Green Political Thought* (3rd edn), London: Routledge, pp.172–8.

On eco-authoritarianism, see W. Ophuls (1977), *Ecology and the Politics of Scarcity*, San Francisco, CA: W.H. Freeman; R. Heilbroner (1974), *An Inquiry into the Human Prospect*, New York: W.W. Norton; G. Hardin (1977), *The Limits to Altruism*, Bloomington: Indiana University Press; R. Barnett (1980), *The Lean Years: Politics in the Age of Scarcity*, New York: Simon & Schuster; R. Paehlke (1989), *Environmentalism and the Future of Progressive Politics*, New Haven, CT, and London: Yale University Press, chs 3, 4 and 8; D. Wells and D. Lynch (2000), *The Political Ecologist*, Aldershot: Ashgate; J. Barry (1999), *Rethinking Green Politics: Nature, Virtue and Progress*, London: Sage, ch. 8; J. Dryzek (2005), *The Politics of the Earth* (2nd edn), Oxford: Oxford University Press, ch. 2.

On fascism/eco-fascism see A. Bramwell (1989), *Ecology in the 20th Century: A History*, New Haven, CT: Yale University Press; D. Gasman (1971), *The Scientific Origins of National Socialism: Social Darwinism in Ernst Haeckel and the German Monist League*, New York: Transaction Publishers; P. Stephens (2001), 'Blood, not Soil: Anna Bramwell and the Myth of "Hitler's Green Party"', *Organization and Environment*, 14: 2; I. Coates (1993), 'A Cuckoo in the Nest: The National Front and Green Ideology', in J. Holder *et al.* (eds), *Perspectives on the Environment: Interdisciplinary Reseach in Action*, Aldershot: Avebury.

On Promethean/cornucopian right-wing responses, see W. Beckerman (1974), *In Defence of Economic Growth*, London: Cape; W. Beckerman (1995), *Small is Stupid: Blowing the Whistle on the Greens*, London: Duckworth; W. Beckerman (2002), *A Poverty of Reason: Sustainable Development and Economic Growth*, Oakland, CA: The Independent Institute; J. Simon and H. Kahn (1984), *The Resourceful Earth: A Response to Global 2000*, Oxford: Blackwell; J. Simon (1996), *The Ultimate Resource 2*, Princeton, NJ: Princeton University Press; J. Dryzek (2005), *The Politics of the Earth* (2nd edn), Oxford: Oxford University Press, ch. 3.

On the 'contrarian' right-wing position, see M. Ridley (1995), *Down to Earth: A Contrarian View of Environmental Problems*, London: Institute of Economic Affairs, in association with the *Sunday Telegraph*; F. Buell (2003), *From Apocalypse to Way of Life: Environmental Crisis in the American Century*, New York: Routledge; G. Easterbrook (1995), *A Moment on the Earth: The Coming Age of Environmental Optimism*, New York: Penguin; C. Mooney (2005), *The Republican War on Science*, New York: Basic Books; P. Ehrlich and A. Ehrlich (1996), *Betrayal of Science and Reason: How Anti-environmental Rhetoric Threatens our Future*, Washington, DC: Island Press; R. Bailey (1993), *Ecoscam: The False Prophets of Ecological Apocalypse, Global Warming and other Eco Myths*, New York: St Martin's Press.

On the right-wing 'green backlash', see S. Beder (1997), *Global Spin: The Corporate Assault on Environmentalism*, Devon: Green Books; S. Beder (2001a), 'Neoliberal Think-tanks and Free Market Environmentalism', *Environmental Politics*, 10:2; S. Beder (2001b), 'Anti-environmentalism' and 'Green Backlash', in J. Barry and E.G. Frankland (eds), *International Encyclopedia of Environmental Politics*, London: Routledge; M. Paterson (1999), 'Understanding the Green Backlash', *Environmental Politics*, 8:2.

On the Lomborg controversy, see B. Lomborg (2001), *The Skeptical Environmentalist: The True State of the World*, Cambridge: Cambridge University Press; B. Lomborg (2003), 'Smearing a Skeptic: Something is Rotten in the State of Denmark', *The Wall Street Journal*, 13 January; Lomborg's official website at www.lomborg.com and critics of Lomborg at www.anti-lomborg.com; J. Dryzek (2005), *The Politics of the Earth: Environmental Discourses*, (2nd edn), Oxford: Oxford University Press, ch. 3; Special Edition of *Grist Magazine* (2001), 'Something is Rotten in the State of Denmark: A Skeptical look at The Skeptical Environmentalist' at http://www.grist.org/advice/books/2001/12/12/of/ (accessed 15 May 2006).

On John Gray, see J. Gray (1993), *Beyond the New Right: Markets, Government and the Common Environment*, London: Routledge; J. Gray (1997), *Endgames: Questions In Late Modern Thought*, Cambridge: Polity Press; J. Gray (1998), *False Dawn: The Delusions of Global Capitalism*, London: Granta; J. Gray (2002), *Straw Dogs: Thoughts on Humans and Other Animals*, London: Granta; J. Gray (2004), *Heresies: Against Progress and Other Illusions*, London: Granta; J. Barry (2006), 'Straw Dogs, Blind Horses and Post-Humanism: The Greening of Gray?', *Critical Review of Social and Political Philosophy*, 'Special Issue: The Political Theory of John Gray', 9:2; J. Gray (2006), 'Reply to Critics', *Critical Review of Social and Political Philosophy*, 'Special Issue: The Political Theory of John Gray', 9:2.

6 Left-wing reactions to the environment and environmental politics

Key issues

- Eco-Marxism and eco-socialism.
- Marxism, socialism and the environment.
- The ecological crisis as the 'second contradiction of capitalism'.
- Logics of displacement within the ecological restructuring of global capitalism.
- The environmental justice movement.
- The ecological restructuring of socialism.
- Towards eco-Marxism and eco-socialism?
- Eco-anarchism and social ecology.
- The anti-globalisation/global justice movement.

Introduction

Left-wing approaches to the environment, ecological questions and green issues are as complex as 'left-wing' thought itself. While there are a wide variety of schools of thought which one can label as 'left-wing', ranging from revolutionary Marxism-Leninism, Maoism, Trotskyism, communism, to trades union activism to reformist social democracy, to anarcho-syndicalism, anarchism and much more, this chapter will confine itself to two main historical trends in left-wing social theory and action, namely Marxism and anarchism.

Eco-Marxism and eco-socialism

It is curious to look back on the previous two centuries as in some ways a better time, when things were simpler, less threatening and the ever-onward march of the industrial revolution and technologically based progress was both taken for granted and taken for granted to be unquestionably good. Indeed, the Marxist and socialist critique was essentially that capitalism was a fetter holding back the inevitable tide of progress, denying its fruits to be enjoyed equally by all. How different things seem now after almost two centuries of industrialisation. Western societies are characterised as 'risk societies' (Beck, 1992a) – discussed in more detail in Chapter 9 – increasingly sensitive to a widespread sense of powerlessness in the face of economic and ecological forces outside of their control. A large part of this has to do with the growing sense that nature is taking its revenge, as the ecological life support systems of the planet are degraded and destroyed, from the devastation of the Asian Tsunami of 2004 which killed over a quarter of a million people, to Hurricane Katrina which wreaked havoc on New Orleans in August 2005. At the same time, globally, after four decades of officially sanctioned 'development' policies, the vast majority of the world's human population go with basic needs unfulfilled, and perhaps for the first time in human history, there is a widespread feeling that things will not be better in the future. At the dawn of a new millennium, for a growing number, capitalist industrial progress simply ain't what it used to be.

Marxism, socialism and the environment

Despite the perception that ecological issues are new, there is a history of the relationship between Marxism and ecology. It is fair to say that historically classical Marxism, being a product of its time, did not address the range and significance of ecological issues that have come to play such an important part of late twentieth-century political and ethical discourse (Dickens, 1997). Indeed, insofar as ecology stresses natural or absolute limits to economic development, early Marxist and socialist theory was anti-ecological, since talk of 'natural limits' was interpreted as an ideologically conservative attempt to mask socially unjust limits on progress as 'natural' and therefore 'unalterable' and beyond human capacity to change or improve. Thus, ecological arguments were viewed as 'reactionary' not progressive by Marxists. As discussed in Chapter 3, it is in the Marxist attack on Malthus' theory of population and the latter's argument for the need for subsistence wages that we can trace the predominant reaction of left-wing social theorising to ecological issues up until recent years.

Marx's attack on Malthus' ideas set the tone, and often the parameters within which the interaction between Marxism, and its various offshoots, and ecology took place. In this encounter are all the main ingredients which mark, and continue to mark, the relationship between ecology and Marxism. First, there is the Marxist perception of ecology as anti-enlightenment in general and anti-industrial in particular. Second, and following on from the latter, is the equation of anti-industrial with anti-working class, such that even to this day the first reaction of some Marxists and socialists to political ecology and concerns of the environment is to see it as an intrinsically bourgeois ideology with middle-, not working-class, interests (Weston, 1986; Eckersley, 1989; Norton, 2003), as articulated in a classic article by Hans Magnus Enzensberger (1974). A common left-wing reply to the *Limits to Growth* report and analysis was to portray it as an attempt to protect the lifestyles of the rich. Pavitt, for example, stated that 'the movement hostile to economic growth can be seen as supporting the interests of the materially well-off, who feel that life is less pleasant for them when an ever larger number of people begin to approach the same living standards as their own, and, in particular, when they start using the same scarce infrastructure' (Pavitt, 1974: 154).

Third, we have the importance of science and technology on both sides. On the one hand we have Marx completely optimistic in the ability of technology, once free of capitalist relations, to transcend so-called 'natural limits'. On the other we have Malthus' claim that his theory was fully supported by scientific and statistical data, which led to the opposite conclusion from that of Marx and indeed was at odds with the dominant belief in progress that characterised the early development of industrialisation under capitalism. In this way ecology from a classical Marxist perspective was another fetter holding back the onward and inexorable rise of the revolutionary proletariat and the creation of a communist, post-capitalist society. In many ways ecology was worse than bourgeois political economy, because unlike the latter, ecology was seen to be anti-industrial and anti-modern, held to desire a return to a pre-modern, agrarian, social order. As Marx and Engels put it in the *Communist Manifesto*, 'The bourgeoisie has subjected the country to the rule of the towns. It has created enormous cities, has greatly increased the urban population as compared with the rural, and has thus *rescued a considerable part of the population from the idiocy of rural life*' (Marx and Engels, 1978: 477; emphasis added). Where ecological thought was expressed as a romantic defence of the natural world against industrialisation, as in Wordsworth, Carlyle, J.S. Mill or Emerson or Thoreau in America – as discussed in Chapter 3 – this merely confirmed its regressive, conservative, elitist and anti-democratic character for Marxists and socialists.

A dominant Marxist response to the rise in ecological concerns in the early 1970s typified by the (in)famous *Limits to Growth* report (Meadows *et al.*, 1973) was largely negative. For many Marxists, socialists and social democrats this report, and the nascent environmental movement which concurred with it, were simply ecological and more modern versions of Malthusianism, or 'Malthus with a computer' (Freeman, 1973). And the reaction by left-wing movements, parties and thinkers to ecology was more or less within the paradigm set by Marx's critique of Malthus. Ecological and green talk of 'post-industrialism', and the necessity for a 'steady-state-economy' (Daly, 1973), replacing the pre-modern complexion of nineteenth-century political ecology, was nevertheless still taken to express the anti-working-class, anti-socialist, anti-progressive character of ecological politics. In opposition to the ecological argument for a less expansionist, simpler lifestyle, Marxists and socialists still held the domination and control of nature as a precondition for the creation of a free and equal society. A typical example of this was Markovic's view that 'Man must master the forces of nature in order to develop freely all of his creative powers. For this reason, Marx is aware of the historical significance of industrialisation, private property, and reification (which are necessary consequences of an intense struggle with the natural environment). He understands that *there is no other road to universal human emancipation*' (Markovic, 1974: 149; emphasis added). A more recent reiteration of this point is eco-Marxist thinker John Bellemy-Foster who holds that the common ecological/green critique of the 'Prometheanism' (the technological domination and control of nature) of Marxism should be rejected as groundless:

> This charge of 'Prometheanism', it is important to understand, carries implicitly within it certain anti-modernist (postmodernist or pre-modernist) assumptions that have become sacrosanct within much of green theory. True environmentalism, it would seem, demands nothing less than the rejection of modernity itself.
>
> (Bellemy-Foster, 2000: 135)

Again, just as in the debate between Marxism and Malthus in the nineteenth century, talk of limiting human technological control over nature either because of natural constraints as revealed by ecological science, or circumscribing human control and use of nature on normative grounds to do with the intrinsic value of the nonhuman world, are regressive, reactionary and need to be rebutted, from the Marxist theory of historical materialism and socialist political project of achieving a post-capitalist society.

However, there were also discernible signs of a more positive Marxist engagement with ecological issues. The origins of this dialogue are to be found in the early Frankfurt School and especially in Horkheimer and Adorno's *Dialectic of the Enlightenment*, the New Left, and its humanist interpretation of Marxism, particularly in writers such as Marcuse – discussed in Chapter 3 – and Erich Fromm (1976), all of which constituted some significant revisions of 'scientific' and classical Marxism. In this New Left tradition we find authors such as Gorz (1982, 1994) and Marcuse (1992) seeking to incorporate ecological concerns. In a nutshell, the Frankfurt School and the New Left held that the domination of nature (the means) undermined Marxism's emancipatory ends.

Thus we come to the basic dichotomy within Marxist-based left-wing responses to ecology. On the one hand there are those for whom its central political message is regressive and what is of value within ecology can be easily incorporated within Marxism. Another example of this negative Marxist reception to ecology is Costello, who claims that 'The "green awareness" of the last decade was, in fact, a product of conservative and anti-industrial ideology radicalized due to the absence of a left political alternative' (1991: 8–9). On the other hand, in terms of Marxist theory there has been a greater willingness to examine the ecological critique of industrialism and to see if Marxism could learn anything from ecology. Here writers such as Benton (1989, 1993), Hayward (1992, 1995), Soper (1991), Mellor (1992a, 1995) and Dickens (1996, 1997, 2002) have accepted core aspects of the ecological critique (though by no means all of it) and on this basis reconstructed a more ecologically sensitive Marxist/socialist political theory, while others such as O'Connor (1995), Bellemy-Foster (2000, 2002) and Hughes (2000) have begun to flesh out an ecological expansion and reinterpretation of Marxist political economy and the creation of a coherent eco-Marxist and eco-socialist political project. This ecological reconstruction of left-wing political economy and politics will be developed within and from the context of the eco-Marxist analysis of the ecological crisis as the 'second contradiction of capitalism'.

The ecological crisis as the 'second contradiction of capitalism'

The dominant Marxist analysis of the ecological crisis begins from a search for its underlying economic causes within capitalism as expressed in class conflict and the management or masking of that underlying conflict between workers and owners of capital. From a Marxist perspective the ecological crisis may be analysed either as an economic crisis *within* capitalism or the more complex process whereby the ecological crisis becomes a crisis *of* capitalism (Hay, 1994). As a crisis within capitalism, the ecological crisis reveals itself as an increase in the costs of production, lower profits due to ecological sinks filling up, polluter-pays legislation, more expensive raw materials as resources run out and so on. In response to this capitalist industrialism attempts to restructure itself both ideologically and economically, are discussed below. However, in its attempts to incorporate ecological externalities, pollution, loss of biodiversity, global climate change, all of which it has produced by 'displacing' rather than solving these problems (Dryzek, 1987), this restructuring process comes up against both absolute ecological limits and increasing social and ethical resistance. In other words, the ecological crisis *within* capitalism cannot be contained by the logic of displacement and this results in an ecological crisis *of* capitalism; that is, threatening the entire system and thus (potentially) opening up the possibility of a 'post-capitalist' social system.

Whereas the first contradiction of capitalism is premised on the contradiction between the forces and relations of production, the second contradiction of capitalism has to do with the disjuncture between the capitalist mode of production (both forces and relations) and what James O'Connor (1991) has called the 'conditions of production'. Following Marx, he holds that there are three such conditions:

1   'personal conditions' (i.e. human labour power);
2   'communal general conditions' (i.e. urban space, communications and infrastructure);
3   'external conditions' (i.e. nature or environment).

From this perspective the ecological crisis may be regarded as a crisis of the 'external conditions' of capitalist production. In short, capitalism destroys the very natural basis upon which it exists. Since this external condition is not 'produced' by capitalism (or by any form of human social agency for that matter), the capitalist state secures and regulates capital's access to them, since if left to themselves individual capitalist firms and corporations would undermine and destroy the very natural conditions (or 'natural capital', as environmental economics discussed in a later chapter terms these inputs) upon which capitalist production depends. Thus the capitalist state ensures the long-term availability of these essential productive conditions to capital and acts in their long-term interests, whereas they themselves are very short-term focused on profit-making.

The first contradiction referred to the inability of capitalism to sustain itself internally due to the contradiction between socialised production (production produced by the many) and individual appropriation (profits from production accruing to a few), and may be seen as a crisis engendered by capitalism being parasitic upon the *non-capitalist social world*. The second contradiction on the other hand is caused by capitalism being parasitic and dependent upon the *nonhuman world*. It is this viewpoint which distinguishes the Marxist analysis from those green or ecological analyses which locate the crisis in the anthropocentrism (human-centredness) or particular worldview or 'paradigm' that characterises Western societies, as discussed in Chapter 10. From the Marxist position the ecological crisis is part of wider economic and political contradictions of capitalism.

Marxist political economy would *not* say that the ecological crisis arises from capitalism running up against natural or absolute limits. The ecological crisis within capitalism arises in the form of *higher costs* of production as the conditions of production are degraded, made scarce and thus more costly, as a result of the actions of individual capitals. In a sense, this systemic irrationality of capitalism is similar to the 'tragedy of the commons' understood as an 'open access regime' with no regulation, where individual capitals systematically destroy the 'capitalist commons' that are its 'conditions of production'. As O'Connor puts it, "'Limits to growth" thus do not appear, in the first instance, as *absolute shortages* of labor power, raw materials, clean water and air, urban space, and the like, but rather as *high-cost* labor power, resources, infrastructure and space' (1995: 163; emphasis added). When the conditions of production become scarce and thus costly, the system responds to the short-sightedness and collective ecological irrationality of the rational actions of individual capitals seeking to maximise profits, by the state (and supra-state agencies) regulating access to the conditions of production.

The economic system causes ecological damage which as a 'market failure' becomes the responsibility of the state, and the rest of society has to 'pay' for this capitalist ecologically irrational behaviour. Witness the growth in environmental legislation, for example, the creation of state agencies to regulate water, air, soil, forests, urban space, analogous to the expansion of the welfare state to regulate personal conditions of production, such as health, education and housing (Meadowcroft, 2005). In managing the ecological crisis the capitalist state takes it upon itself to administer the collective and long-term interest of capital in cheap and available conditions of production. State regulation of environmental conditions of production acts to displace potentially system-threatening economic and ecological-economic crises into the political realm (Hay, 1994: 219). Ideologically, the way the state does this is via the extension of capitalist economic rationality to encompass ecological goods and services, as may be seen in the recent development of 'environmental economics' discussed in Chapter 7. The state's 'crisis management' function is extended to cover ecologically based economic crises.

Logics of displacement within the ecological restructuring of global capitalism

Capitalist responses to pollution problems, for example, do not seek solutions, since solutions would require the restructuring of the economy and the transformation of capitalism. Rather, 'displacement strategies' are deployed; from one media to another (water pollution becomes solid waste), from one place to another (the export of toxins from the North to the South), or in time to future generations. So rather than 'problem solution' we have 'problem displacement' (Dryzek, 1987: 10): instead of dealing with the causes, the negative effects are simply 'removed', both from individual subjective experience as well as from the national accounts. Yet herein lies the rub: the strategy of displacement presupposes that there is an 'away' where environmental pollutants can be sent, but within the confines of a small planet there is no 'away' in the long term. At the present time one can say that the ecological restructuring of global capitalism arises from the necessity of addressing the 'ecological crisis' as a series of discrete problems, and thus to some degree 'displaceable', in order to prevent this crisis from inducing a total breakdown in the whole system. The reason for this is quite simple and is related to the globalised options available to capitalism. Just as the imperialist expansion of the core capitalist nations permitted them to overcome realisation crises domestically by finding new markets and sources of cheap raw materials in those parts of the world they colonised from the eighteenth century onward, domestic ecological problems may for a time be 'exported', displaced to another part of the world, or viewed as a 'technical/bureaucratic' problem, or a problem for the future but not for the present.

Just as capitalism is systematically unjust with regard to the distribution of the goods it produces, so from a Marxist perspective it continues this pattern in respect of the distribution of ecological risks or 'bads'. The externalities of capitalist economic growth are distributed such that it is the marginalised in the core capitalist nations in the North, the Southern hemisphere, women and ethnic minorities that suffer the most. This is the basic starting point of the environmental justice movement (see Box 6.1). This is not new. From the earliest evolution of capitalism the rich and powerful have always been able to insulate themselves against the inevitable 'negative externalities' of capitalism, whether those externalities be crime, urban decay or worsening ecological conditions. It is at this point that Marxist political economy demonstrates its effectiveness in diagnosing the causes of the ecological crisis. Not only is it the poor who suffer from the inegalitarian distribution of ecological bads, but poverty and inequality are also causes of global and local ecological degradation. As Weston remarks, 'It is the accumulation of wealth and its concentration into fewer and fewer hands which creates the levels of poverty that shape the lives of so many people on our planet, thus making it a major determinant of the environment which people experience. It is poverty which forces people to place their own short-term interests above the long-term interests of the Earth's ecology' (1986: 4–5). This is particularly true in the South, where the penetration of capitalism has displaced non-capitalist and often sustainable forms of resource use. For example, neo-colonialism and the global dynamics of the world economy force the 'developing nations' into adopting forms of industrialised and monocultural agriculture for the global market which are environmentally destructive, such as large-scale deforestation leading to soil erosion and desertification. These indebted, hard-currency starved nations are also obliged to offer themselves as convivial sites

---

Box 6.1 The environmental justice movement

The environmental justice movement is a broad term covering a variety of different theoretical perspectives and environmental movements. What unites them is a concern with the distribution of environmental 'bads' or injustices such as the siting of toxic dumps, polluting forms of production along class, race or ethnicity lines such that, as empirical studies have shown, the poor and marginalised are the ones most likely to suffer such environmental injustice and environmental racism (Bullard, 2005). As Barry Commoner put it in his seminal book *The Closing Circle* over thirty years previously, long before the rise of the official environmental justice movement, 'One thing that does clearly emerge from nearly all statistical studies of the effects of air pollution on health is that they are most heavily borne by the poor' (1971: 76). It also goes under different names. For example, Martinez-Alier's (2001) highlighting of an 'environmentalism of the poor' concerned with marginalised peoples securing the ecological conditions for their survival, in contrast to the middle-class environmentalism concerned with aesthetic concerns of wilderness or animal protection. As he puts it, 'The unequal incidence of environmental harm gives birth to environmental movements of the poor' (2001: 54).

As well as pointing out the correlation between the lower environmental standards of water or air quality, higher levels of ill-health and marginalised status, others within the environmental justice movement also highlight the issue of lack of voice accorded to marginalised people. Author's such as Schlosberg (1999), drawing on the work of Iris Marion Young, have also pointed out the injustice of marginalised people and groups being denied *recognition* by public authorities and established institutions in the processes through which environmental decisions are made.

Delegates to the First National People of Color Environmental Leadership Summit held on 24–27 October 1991 in Washington, DC drafted and adopted seventeen principles of environmental justice. Since then, *The Principles* have served as a defining document for the growing grassroots movement for environmental justice:

> We, the people of color, gathered together at this multinational People of Color Environmental Leadership Summit, to begin to build a national and international movement of all peoples of color to fight the destruction and taking of our lands and communities, do hereby re-establish our spiritual interdependence to the sacredness of our Mother Earth; to respect and celebrate each of our cultures, languages and beliefs about the natural world and our roles in healing ourselves; to insure environmental justice; to promote economic alternatives which would contribute to the development of environmentally safe livelihoods; and, to secure our political, economic and cultural liberation that has been denied for over 500 years of colonization and oppression, resulting in the poisoning of our communities and land and the genocide of our peoples, do affirm and adopt these Principles of Environmental Justice:

1) **Environmental Justice** affirms the sacredness of Mother Earth, ecological unity and the interdependence of all species, and the right to be free from ecological destruction.

2) **Environmental Justice** demands that public policy be based on mutual respect and justice for all peoples, free from any form of discrimination or bias.

3) **Environmental Justice** mandates the right to ethical, balanced and responsible uses of land and renewable resources in the interest of a sustainable planet for humans and other living things.

4) **Environmental Justice** calls for universal protection from nuclear testing, extraction, production and disposal of toxic/hazardous wastes and poisons and nuclear testing that threaten the fundamental right to clean air, land, water, and food.

5) **Environmental Justice** affirms the fundamental right to political, economic, cultural and environmental self-determination of all peoples.

6) **Environmental Justice** demands the cessation of the production of all toxins, hazardous wastes, and radioactive materials, and that all past and current producers be held strictly accountable to the people for detoxification and the containment at the point of production.

7) **Environmental Justice** demands the right to participate as equal partners at every level of decision-making, including needs assessment, planning, implementation, enforcement and evaluation.

8) **Environmental Justice** affirms the right of all workers to a safe and healthy work environment without being forced to choose between an unsafe livelihood and unemployment. It also affirms the right of those who work at home to be free from environmental hazards.

9) **Environmental Justice** protects the right of victims of environmental injustice to receive full compensation and reparations for damages as well as quality health care.

10) **Environmental Justice** considers governmental acts of environmental injustice a violation of international law, the Universal Declaration on Human Rights, and the United Nations Convention on Genocide.

11) **Environmental Justice** must recognize a special legal and natural relationship of Native Peoples to the U.S. government through treaties, agreements, compacts, and covenants affirming sovereignty and self-determination.

12) **Environmental Justice** affirms the need for urban and rural ecological policies to clean up and rebuild our cities and rural areas in balance with nature, honoring the cultural integrity of all our communities, and provided fair access for all to the full range of resources.

13) **Environmental Justice** calls for the strict enforcement of principles of informed consent, and a halt to the testing of experimental reproductive and medical procedures and vaccinations on people of color.

14) **Environmental Justice** opposes the destructive operations of multi-national corporations.

15) **Environmental Justice** opposes military occupation, repression and exploitation of lands, peoples and cultures, and other life forms.

continued

16) **Environmental Justice** calls for the education of present and future generations which emphasizes social and environmental issues, based on our experience and an appreciation of our diverse cultural perspectives.

17) **Environmental Justice** requires that we, as individuals, make personal and consumer choices to consume as little of Mother Earth's resources and to produce as little waste as possible; and make the conscious decision to challenge and reprioritize our lifestyles to insure the health of the natural world for present and future generations.

(Environmental Justice/Environmental Racism, 1991)

The environmental justice movement is also marked by being a grassroots movement often in opposition both programmatically/ideologically and organisationally from professional environmental organisations and offering localised and often radical forms of resistance to development decisions by state agencies and corporations which they think will negatively impact on the health, well-being, identity or cohesion of local communities.

for footloose capital, competing with each other to lower pollution and health controls, wages, and taxes in a desperate effort to attract investment. This is the 'race to the bottom' that both socialist and green analyses claim to be the inevitable outcome of the creation of a global and globalising world capitalist economic system (McMurtry, 1999; Perkins, 2000).

It is within the context of ecological restructuring that Marxists would place the recent rhetoric of 'sustainable development' as a way in which global capitalism seeks to alter its conditions of production in order to reduce costs and gain legitimacy as a viable social form. At the global institutional level, especially since the 1992 Rio Earth Summit and the 2002 Johannesburg World Summit on Sustainable Development, there has been much debate about 'sustainable development' as the way to reconcile ecological and economic (read capitalist) imperatives. For many Marxist thinkers the whole concept of sustainable development is suspect. As Benton puts it, 'The emergence of the discourse of "sustainable development" in the mid-1980s has played a very important part in this process of legitimation, and in integrating diverse strands of the popular environmental movement into a policy agenda largely shaped by capital and technocratic interests in the state' (2000: 101).

That the logic of this discourse is a bureaucratic one of the undemocratic management of the global ecological commons is one of the many points of overlap between some green and Marxist critiques. For example, according to Gorz, 'In the context of industrialism and market logic . . . recognition of ecological constraints results in the extension of techno-bureaucratic power. . . . It abolishes the autonomy of the political in favour of the expertocracy, by appointing the state and its experts to assess the content of the general interest and devise ways of *subjecting* individuals to it' (1993: 60; emphasis in original). This suspicion is also shared by those in the green movement for whom the 1992 Rio 'Earth Summit' and subsequent international meetings and treatises about sustainable development marked the rise of a new unelected and undemocratic 'global ecocracy' (Sachs, 1995) removed from the radical and democratic potentials of the creation of sustainable societies. What the rise in global environmental governance portends is nothing less than the creation of global institutions functionally similar to welfare state institutions at the national level. Hence the emphasis on such abstract, quantitative concepts such as 'biodiversity', 'carrying capacity', and above all else right-wing and conservative concerns about 'population control' which peppers the global discourse of this new transnational global class of 'ecocrats'.

That this whole global ecology discourse and practice is in the interests of the core capitalist nations of the affluent North is something which both Marxists and greens agree upon. Where Marxists differ from some ('deep') greens is in refusing to see the global ecological crisis in terms of an undifferentiated 'humanity' making excessive demands on an equally undifferentiated 'nature'. It is not 'humanity' as a whole that is destroying the web of life on Earth, but the capitalist system and those classes within that global system who gain most from it. To simply say that human population growth is the major cause of the ecological destruction of the planet lumps together the marginalised who are forced to precipitate soil erosion by clearing forests to survive, and those who consume for luxury. To adopt this ideological stance is to adopt the perspective of the global ecocracy for whom 'No matter if nature is consumed for luxury or survival. No matter if the powerful or the marginalized tap nature, it all becomes one for the rising tide of ecocrats' (Sachs, 1995: 435). In other words, the issue of social and environmental justice needs to be always considered as the starting point for any analysis of social–environmental interaction.

The ecological restructuring of socialism

The dialogue between socialism and ecological concerns has led some socialist and Marxist thinkers to revise certain core aspects of Marxism, to update socialist theory and practice in light of ecological concerns. This revising of Marxism maintains that the ecological crisis of capitalism arises from the collective inability of capitalism as a global and national socio-economic system to regulate sustainable access to the 'conditions of production'. These productive conditions are:

1 'External conditions': environmental or nonhuman contributions to production.
2 'Personal conditions': human labour power, including reproductive labour.
3 'Communal general conditions': urban space, communication, infrastructure.

The first revision suggested by ecology for Marxism relates to the Marxist self-understanding of its political project as intimately related to the development of the forces of production. What an ecological view requires is a reassessment of the Marxist idea that the post-capitalist social order is premised on material abundance. The central Marxist thrust about the production of material wealth in order to satisfy the needs of all, which is systematically impossible under capitalism, needs to be tempered by devoting more attention to the question of distribution. An ecologically sensitive Marxism is one which accepts the green argument about the existence of ecological limits to growth (Soper, 1996). In another sense this revision of Marxist 'Prometheanism' and productivism suggests that Marxism can be reconciled with the *desirability* and not just the *necessity* of this revision.

This has recently been discussed by eco-socialist writers who have suggested that the 'Promethean myth' of Marxism, understood as the premise that human emancipation is built upon the 'domination of nature' and the production of material abundance, must be radically revised. Writers such as Benton (1993, 1989, 1996) and Hayward (1995) have suggested that the claim that only by transcending natural limits (both external and internal) can humans be eman-cipated needs to be seriously questioned. What needs to be assessed is that the expansion of the forces of production *without reference to the conditions of production* can no longer be taken as conducive to social progress in general and human well-being in particular. Indeed, it may be that social progress and human emancipation requires shifting attention away from liberating the forces of production towards altering their development within the context of securing the conditions of production. Whereas the forces of production are *instrumental* to human well-being, the conditions of production are *constitutive* of human well-being.

A Marxism and socialism that accepts limits turns its attention to issues of social/ distributive and environmental justice, both globally and nationally (Pepper, 1993). For example, an eco-Marxist position would be that the industrialised North is 'overdeveloped' and what is required is an ecological redirection of material development in the North, and ecologically rational forms of economic development in the South (Lee, 1993; Sarkar, 1999). This may be viewed as a shift from (capitalistic) undirected and undifferentiated economic growth towards socialist and democratically planned economic and social development across the globe. One of the implications for Marxist theory of this ecological analysis is the abandoning of productivity as the primary criterion by which to judge economic performance, and the notion of maximum consumption as the criterion of human well-being. Self-limitation, sufficiency and greater stress on reproductive activities and the realisation of the internal goods of work rather than the external rewards of labour point the way towards the vision of a better society. Marxism must evolve into a project of self-limitation, part of which involves the democratic limitation of economic rationality (Gorz, 1993). Both market and administrative logics disempower citizens from collectively and democratically deciding the direction, composition and ends of economic activity. Ecology and economy can only be brought back into synch via a self-conscious, democratic organisation of the economy. However, the main point here is that any choice between indiscriminate economic growth and economic self-limitation is precluded at the present time. Whereas continuous economic growth is a structural requirement of capitalism, it is not a requirement of socialism if Marxist Prometheanism is abandoned.

A second issue that ecology calls into question is the orthodox Marxist understanding of historical development. The claim that each society must pass through the capitalist historical stage in order to develop the preconditions for an emancipated society has increasingly been criticised. If the relationship between forces and relations of production are no longer the primary focus of Marxist historical materialism, it becomes free of the linear and rigid logic implied by classical Marxism's narrative of the progression of societies through changes in the mode of production and a rather simplistic model of modern societies and economies (Goldblatt, 1996: 190). Introducing the conditions of production as a third category of historical materialism which emphasises ecological sustainability as a precondition for a free and equal society may permit Marxism to endorse non-capitalist, non-Western models of social–environmental productive interaction. Thus rather than the linear historical model of classical Marxism, the way is open to seeing *non-capitalist* as opposed to *post-capitalist* forms of socio-economic organisation as possible frameworks for emancipation.

The third major change which ecological considerations bring to Marxism is methodological. As outlined above, the second contradiction of the capitalist thesis highlights how in the ecological stage of capitalism Marxism needs to place the forces and relations of production (i.e. the mode of production) within the context of the conditions of production. For O'Connor (1992) Marxist political economy is necessary to make capitalist relations of production transparent while ecology (as the study of nature's economy) makes the forces of production transparent. The green critique of capitalist forces of production has shown that capitalist forces of production degrade nature. However, analysing the forces without looking at their place in the capitalist mode of production, that is, in connection with capitalist social relations, mistakes the *effects* of ecological destruction for *causes*. Simply laying the blame on industrial forces of production, technology and science offers a shallow and mistaken critical analysis. Many green analyses simply 'naturalise' the forces of production as if there is some inherent anti-ecological dynamic within them, so that irrespective of prevailing social relations industrial productive forces are by their *very nature* (independent of productive relations) ecologically damaging. An understanding of the source and meaning of this anti-ecological assumption must be sought within social relations. This involves greens generating a critique of capitalist relations of production, rather than abstracting their critique of the productive forces and reifying it as a super-ideology of 'industrialism'.

Expanding the analysis to include the conditions of production as a third explanatory category alongside the forces and relations of production suggests a way in which both green and Marxist concerns may be combined. This third category may be understood as 'The natural limits of both human and non-human nature which regulate the metabolism (of humans and nature) from the side of nature' (Hayward, 1995: 120). This metabolic interpretation of economy–ecology exchanges opens the way towards a reconceptualisation of Marxism and the 'greening' of Marxism' or creation of an 'eco-Marxism' and 'eco-socialism' (Benton, 1996; Dickens, 1997; Bellemy-Foster, 2000; Hughes, 2000). On the one hand, recognition of the limits and contributions of 'internal nature', that is, the reproductive conditions of labour, to economic activity is long overdue as feminists and eco-feminists have argued. This will be dealt with in more detail in Chapter 7 on gender and the environment. On the other hand, the introduction of an analytical category of the conditions of production reveals that the human economy is a subsystem of the wider natural economy, and that this is an ineliminable reality that cannot be altered.

Towards eco-Marxism and eco-socialism?

The upshot of the engagement between Marxism and ecology is profound, setting out as it does the necessity for major revisions within Marxist political economy and the Marxist/socialist political project. The ecological critique of capitalism reminds socialists that capitalism is a *mode of production* and not simply a set of property relations. Capitalism is also a particular, historically specific physical *metabolism* between society and nature, and is thus subject to natural laws and limits. Ecology reveals these modes of social–environment interaction as having a definite material content, not just in the usual Marxist sense of material, but in the ecological and biological sense of physical, emotional and psychological material interaction. Thus ecology requires a deeper understanding of *materialism* within Marxism, such that 'historical materialism is the ecology of the human species' (Benton, 1989: 54). But as the eco-feminist critique demonstrates, this ecological interpretation of historical materialism requires a further understanding of capitalism as a *mode of reproduction*.

The import of ecology for Marxist politics is also profound. What the ecological perspective calls for is nothing less than a complete reconstruction of its emancipatory political project (Eckersley, 1992). As noted above, eco-Marxism would have to reconceptualise emancipation not as freedom *from* but freedom *within* our ecological embeddedness (in external nature) and our embodiedness (internal human nature), hinted at in 'New Left' thinkers such as Marcuse and more recent environmental political and social theorists who continue this tradition such as Benton (1989), Dobson (2003), Eckersley (1992, 2005), Kassiola (1990) and Kovel (2002). As Benton puts it:

> Once the underlying theme of progressive mastery of nature through technological innovation as the key presupposition of human emancipation is called into question, then it becomes possible to think in quite different ways about the possibility of qualitatively different ways of scientific and technical innovation, shaped by different sorts of embedding in human social relations and value commitments.
>
> (Benton, 2000: 91)

As regards the moral critique of the capitalist mode of production in terms of its purely instrumental view of nature (internal and external) as simply a storehouse of resources (from raw materials to genetic information) or a waste bin, an ecologically updated Marxism is sufficiently flexible to accommodate the thrust of the green moral position. However, unlike some deep green moral critiques, Marxism sees the moral problem as arising from the 'capitalisation of nature', rather than the 'humanisation of nature'. For many greens the blame for the ecological crisis is to be found within this humanisation of nature project: 'industrialism', modernity's 'disenchantment of nature', all of which are held to be morally underpinned by an anthropocentric (human-centred) moral outlook. They therefore advocate an 'ecocentric' or non-anthropocentric moral position to rectify what they see as the 'arrogance of humanism' (Ehrenfeld, 1980). The aim of this ecocentric politics is the 'emancipation of nature' (Eckersley, 1992). Eco-Marxists on the other hand would see the moral problem as residing within the particular capitalist interpretation and implementation of humanism. Unlike radical green theory, eco-Marxism is firmly anthropocentric, and sees 'arrogance' rather than 'humanism' as the main problem to be addressed. A reconceptualisation of humanism, one that sees human beings and their emancipation as resting within rather than against nature can be made compatible with the basic ethical thrust of ecocentrism. A revisioning of human emancipation which is premised upon the idea that humans are *part of* as well as *apart from* nature, in the sense that humans represent a *differentiation within* rather than a radical or permanent *separation from* the nonhuman world, would have as a practical consequence the flourishing of that nonhuman world (Barry, 1995c). At the same time, as recent eco-socialist writers have shown, a revised Marxism can provide a compelling and practical normative framework within which human moral concern may be extended to the nonhuman world (Kovel, 1988, 2002; Benton, 1993, 2000; O'Neill, 1993; Hayward, 1995). But as indicated above, emancipation would have to break with the assumption of material abundance premised on the domination of nature. Marx's original goal of the 'humanisation of nature' from an ecological perspective is not compatible with the domination of nature, or with the limiting of its moral critique of capitalism to its destructive effects on human beings alone.

At the cusp where the ecological crisis becomes a crisis of capitalism we can place the green critique of capitalism in terms of its unjustified abuse of the natural world, its inability to ensure the global meeting of human needs, and its in-built bias against future generations. It is the political-normative task of new social movements such as the green movement, the peace and the anti-nuclear movements, together with socialists and Marxists and above all women, to render the opaque economic view of the ecological crisis *within* capitalism publicly visible as a crisis *of* capitalism (Hay, 1994). It is at this point that capitalism becomes not just economically unsustainable, but has so disrupted the metabolism between human societies and their environments that it is both ecologically and socially unsustainable. In capitalising more and more parts of nature, including human nature, capitalism runs up against the social resistance of those for whom the environment is not simply a set of resource inputs or a waste sink but constitutive of who and what they are both individually and collectively. That is, it is a matter of identity. In a sense, capitalism threatens the very lifeworld of individuals in its attempt to render all that exists as capital. Thus both science and ethics are necessary in order to elaborate and comprehend when ecological crisis becomes a crisis of rather than within capitalism, and it is only when these two perspectives are brought together politically and internationally that a political project aimed at the transition to a post-capitalist, socially just and ecologically rational, global order can be created. But as indicated above, the first step of an ecologically updated Marxism requires the reconceptualisation of 'progress' and 'emancipation', and an abandoning of the Promethean claim that the 'domination of nature' is a pre-condition for human emancipation.

Eco-anarchism and social ecology

Another prominent left-wing reaction to ecological issues has been anarchist social and political thought. As pointed out in Chapter 3, the origins of this connection may be found in Kropotkin's late nineteenth-century reading of Darwin which posited nature as a realm in which co-operation and solidarity were as, if not more important, than competition (the right-wing interpretation of Darwin). In particular, ecological science is held to undermine hierarchical and competitive interpretations of natural relations. Equally, the basic anarchist rejection of centralised political and economic power, whether in the nation-state or in large corporations, is one that finds more than an echo in many varieties of green and ecological thinking (Dobson, 1995; Barry, 1999a). This suspicion and rejection of centralised power is one of the main principles which distinguish anarchist from socialist and Marxist thinking, as well as anarchism not being as fixated on material abundance (and the domination of nature to achieve that) as a precondition for a post-capitalist, free and equal society.

One of the most prominent schools of eco-anarchist thought is 'social ecology', most closely associated with the American thinker and political activist Murray Bookchin. The basic argument of social ecology is that the domination of nature by human society is a product of a system of domination within society whereby some groups and classes stand in a hierarchical relationship to others. Bookchin's social ecology combines libertarian anarchism, elements of New Left and critical theory (but a resolute critic of Marx, Marxism and Soviet communist ideology); the emancipatory potential of technology in the tradition of Lewis Mumford (Guha, 1996; see also White, 2003); coupled with a consistent radical critique of the human and ecological irrationality of most of the institutions of modern Western industrial society in the same mode as other radical critics such as Ivan Illich (1971, 1973, 1975).

The main principles of social ecology may be summarised as follows:

- The domination of nature by humans has its roots in the historical emergence of patterns of hierarchy and domination within human society.
- A dialectical approach to understanding the relationship between human society and the natural world. Underpinning many of Bookchin's ideas is a reworking of the dialectical thinking combining Hegel's system with ecological thinking to produce what Bookchin calls 'dialectical naturalism'.
- A rejection of 'deep ecology' and eco-centrism and the idea that humans as simply one species among others' but a rejection of an arrogant humanism which sees humanity as separate from and above nature.
- A philosophy of nature in which values and practices such as freedom, subjectivity and co-operation are present in nature and part of its evolutionary

trajectory, such that human freedom is a natural outworking of natural evolution, and where human culture is 'second nature' which is dependent and derived from the nonhuman world or 'first nature'.

- A positive attitude towards technological innovation which is 'human scale', democratic, and enhances human freedom, need-fulfilment and autonomy.
- A rejection of both the centralised nation-state and corporate capitalism and a revolutionary-utopian vision of a decentralised, sustainable, egalitarian participatory democracy – libertarian municipalism – which is characterised by 'post-scarcity', understood not in the classical Marxist sense of material abundance, but as 'a sufficiency of technical development that leaves individuals free to select their needs autonomously and to obtain the means to satisfy them' (Bookchin, 1980: 251).

As Bookchin puts it:

> The truth is that human beings not only belong in nature, they are products of a long, natural evolutionary process. Their seemingly 'unnatural' activities – like the development of technology and science, the formation of mutable social institutions, of highly symbolic forms of communication, of aesthetic sensibilities, the creation of towns and cities – all would be impossible without the large array of physical attributes that have been eons in the making, be they large brains or the bipedal motion that frees their hands for tool making and carrying food. In many respects, human traits are enlargements of nonhuman traits that have been evolving over the ages.
>
> (Bookchin, 1993: 470)

Bookchin termed the social ecology vision of stateless social order 'libertarian municipalism' (1986: 37–44; 1990: 179–85; 1992). This is defined as 'a confederal society based on the co-ordination of municipalities in a bottom-up system of administration as distinguished from the top-down rule of the nation-state' (Bookchin, 1992: 94–5). Libertarian municipalism as an eco-anarchist theory may be argued to represent a novel form of anarchism. Limiting the scope of communities to simply go their own way marks a decisive break with traditional anarchist thought, which took the communal right to self-governance as its principal and highest political norm. The aim of libertarian municipalism is to recapture the classical political values of the polis, and 'authentic' politics, in opposition to the 'inauthentic' (and illegitimate) modern politics as 'statecraft' (Bookchin, 1992).

Although sharply critical of orthodox Marxism, Bookchin is also at one with the Marxist critique of capitalism and the need for the creation of a post-capitalist society. In a passage which highlights the commonalities and differences between social ecology and orthodox socialism/Marxism, Bookchin holds that a 'libertarian municipal' sustainable society would mean that:

Property . . . would be shared and, in the best of circumstances, belong to the community as a whole, not to producers ('workers') or owners ('capitalists'). In an ecological society composed of a 'Commune of communes', property would belong, ultimately, neither to private producers nor to a nation-state. The Soviet Union gave rise to an overbearing bureaucracy; the anarcho-syndicalist vision to competing 'worker-controlled' factories that ultimately had to be knitted together by a labor bureaucracy. From the standpoint of social ecology, property 'interests' would become generalized, not reconstituted in different conflicting or unmanageable forms. They would be *municipalized*, rather than nationalized or privatized. Workers, farmers, professionals, and the like would thus deal with municipalized property as citizens, not as members of a vocational or social group. Leaving aside any discussion of such visions as the rotation of work, the citizen who engages in both industrial and agricultural activity, and the professional who also does manual labor, the communal ideas advanced by social ecology would give rise to individuals for whom the collective interest is inseparable from the personal, the public interest from the private, the political interest from the social.

(Bookchin, 1993: 23)

The emphasis on 'appropriate scale' within social ecology is a principle supported by almost all greens and one which can be 'read off' by certain (anarchistic, co-operative or mutualist) readings of nature and imported into society as a regulating principle. It is usually taken as expressing the need for 'appropriate scale' in political decision-making procedures and especially the sphere of production within which the particular economy–ecology metabolism of the community is located. According to Porritt, 'In terms of restoring power to the community nothing should be done at a higher level than can be done at a lower' (1984: 166). This principle is compatible with state institutions because for some things, particularly international negotiation on global commons issues, the state is the lowest level. The very term 'municipal', with its strongly urban character, resonates and is compatible with the demand to strengthen local and regional tiers of government/governance away and beneath the centre.

The principles of libertarian municipalism seem to accord with T.H. Green's assessment of those sceptical of the state. According to Green, 'The outcry against state interference is often raised by men whose real objection is not to state inter-ference but to centralisation, to the constant aggression of the central executive upon local authorities' (1974: 217). Thus the critique of the state is in large part a critique of centralisation, and conversely eco-anarchism may be translated as a demand for decentralisation, democratisation and devolved decision-making powers (de Geus, 1999). According to Paterson, 'Thus the state is not only unnecessary from a Green point of view, it is positively undesirable. Whether or not we term the result "anarchist" (which often ends up as a simple terminological dispute about what we mean by the "state") the dominant political prescription within Green politics is for a great deal of decentralisation of power to communities much smaller in scale than nation-states' (2001: 62).

An important characteristic of the social ecology position is that its clear eco-anarchist vision of the future 'sustainable society' vividly encapsulated, in shorthand form, the basic principles and values of the (early and/or radical) environmental/green movement; *inter alia*, ecological and social harmony, decentralisation, simple living, quality of life, community, direct democracy and critique of the military-industrial complex. Within the context of the early development of green/ecological social theory, it was simply assumed that an ecological transformation of society required anarchism to be updated for the age of ecological limits and technological advances. Such an eco-anarchist vision may be found beyond Bookchin's 'social ecology' in other writers and schools of thought such as Alan Carter (1993, 1999) who offers a sophisticated eco-anarchist political theory, and those who differ from Bookchin in their interpretation of social ecology (Rudy and Light, 1996; Watson, 1996; Light, 2003; Clark, 2004). Clark, a former close ideological follower of Bookchin, has sharply criticised the inflexibility and increasing dogmatism of Bookchin's thought, claiming that:

> social ecology [is] an evolving dialectical, holistic philosophy, and the increasingly rigid, non-dialectical, dogmatic version of that philosophy promulgated by Bookchin. An authentic social ecology is inspired by a vision of human communities achieving their fulfilment as an integral part of the larger, self-realizing earth community. Eco-communitarian politics, which I would counterpose to Bookchin's libertarian municipalism, is the project of realizing such a vision in social practice. . . . Bookchin has made a notable contribution to this effort in so far as his work has helped inspire many participants in ecological, communitarian, and participatory democratic projects. However, to the extent that he has increasingly reduced ecological politics to his own narrow, sectarian program of Libertarian Municipalism, he has become a divisive, debilitating force in the ecology movement, and an obstacle to the attainment of many of the ideals he has himself proclaimed.
>
> (Clark, 2004)

The dominance of the eco-anarchist vision has to do with the 'either/or' style of thinking adopted by green/environmental social theories and movements, which is common to most radical forms of analyses. One of the earliest and most influential instance of this was Tim O'Riordan's fourfold typology of the institutional choices open to how to deal with the ecological crisis, namely:

1   New global order,
2   Authoritarian commune,

3   Centralised authoritarianism, or
4   The anarchist solution.

(O'Riordan, 1981: 404–7)

In reality the choice comes down to the anarchist solution or the rest, given that it was the only one which, a priori, embodied the green values and principles noted above. In particular, as green political theory developed, it was assumed that the 'sustainable society' was 'anarchistic' (Dobson, 2000: 83–4) and un-ashamedly utopian in intention. This is particularly the case in respect to 'deep ecology'. Arne Naess, the founding father of this branch of ecological thinking, has said that 'supporters of the deep ecology movement seem to move more in the direction of non-violent anarchism than toward communism' (1989: 156), though Bookchin, for example, has been a trenchant critique of deep ecology, at one stage dismissing it as 'eco-la-la'.

This anticipatory-utopian and totalising form of political critique is directly related to the evolution of the green movement, and its roots are manifold. First, the practical requirement as a 'new social movement' for it to maintain its distinct identity, and to prevent existing ideologies from stealing their ideas and proposals, presented a good case for accentuating the radical and the utopian. Second, in common with other new social movements, the green/ecological movement seemed to be particularly obsessed with questions of self-identity, to demonstrate (to themselves as much as to anyone else) their 'newness' and 'authenticity'. Thus the green movement was at pains to portray itself as a completely new type of politics, 'neither left nor right, but in front'. For example, Porritt declared that 'For an ecologist, the debate between the protagonists of capitalism and communism is about as uplifting as the dialogue between Tweedledum and Tweedledee' (1984: 44), a statement noteworthy for lumping these alternatives together as different versions of the super-ideology of 'industrialism'. Green politics was 'post' or 'anti-industrial', which cast greens as the vanguard of the future society (Milbrath, 1984), and green politics as the politics of the twentieth first century (Sessions, 1995). In this way, in being 'beyond the old/Marxist left and neo-conservative right', anarchism seemed be the most convivial 'base' for green thinking within the existing ideological canon.

Third, added to these internal dynamics was the simple fact that, as a new social actor, it had little or no access to the policy-making process, and therefore did not need to outline programmes, budgets or detailed policies in the language of that process. Broad-brush strokes rather than attention to the fine print characterised early green discourse. The overriding imperative was to distance itself theoretically and practically from the reality surrounding it, and in the case of the 'eco-monastic' strand of eco-anarchism, to turn one's back on the existing social order and create 'liberated zones' from the 'industrial mega-machine' (Eckersley, 1992: 163–7; Bahro, 1994).

A good example of how these radical, utopian and anarchistic trends (together with many elements of the Marxist critique) have evolved in a social movement with strong environmental claims is the anti-globalisation/global justice movement (Box 6.2).

Box 6.2 The anti-globalisation/global justice movement

The anti-globalisation/global justice movement is a 'catch-all' term to encapsulate a wide 'movement of movements' that are broadly united in their opposition to the creation of a global capitalist system dominated by corporations and the imposition of neo-liberal trade, economic and social policies by global institutions such as the World Bank, the International Monetary Fund, and above all the World Trade Organization (WTO) and the powerful role of individual countries such as the United States of America and other developed or Northern states who benefit unfairly from the global economic system, and their promotion of the 'Washington Consensus' and the inevitable 'resource wars' such as the illegal invasion and occupation in Iraq by American, British and other states. The global justice movement comprises trades unionists, environmentalists, feminists, indigenous peoples and human rights movements, peace activists, religious groups, radical left, green and anarchist movements that claim, to use one of the movement's slogans, 'another world is possible'.

The Global Justice movement can be seen as a reaction to neo-liberalism according to Doran:

> The *Washington Consensus* is the title of a business roundtable manifesto published in the USA in 1979. The 'Consensus' view is informed by a neo-liberal discipline of balanced budgets, tax cutting, tight money, deregulation, and anti-union laws. With the ascendancy of this prior economics of power domestically there followed, by virtue of USA influence on international institutions such as the IMF, a globalisation of the same logic or rule. Starting from a tiny embryo at the University of Chicago with the philosopher-economist Friedrich von Hayek and his students like Milton Friedman at its nucleus, the neo-liberals and their funders created a huge international network of foundations, institutes, research centers, publications, scholars, writers and public relations hacks, to package and push their ideas and doctrine. The institutionalisation of neo-liberalism as an authoritative and globalised form of political knowledge is more than an achievement at the level of policy; it is an achievement insofar as it has produced a 'politics of truth': leading one British Prime Minister to celebrate her commitment to the project with the words 'There is No Alternative (TINA)'. This, of course, is the antithesis of the claim now championed by the globalisation movement: 'Another world is possible'.
>
> (Doran, 2006: 155)

It is incorrect to say that the 'movement of movements' is anti-globalisation per se, and more accurate to describe it as demanding a more democratic, grassroots, bottom-up and transparent form of globalisation. Although it came to prominence in the media in the anti-WTO protests in Seattle in 1999, its origins lie in a series of anti-IMF demonstrations in developing countries throughout the 1990s. In particular, these mobilisations were organised against the imposition of IMF structural adjustment programmes. Podobnik (2002) states that 56 per cent of the sixteen million individuals recorded as having participated in globalisation protests during the period 1990 to 2002 did so in the developing not developed world. Thus, movements such as the Zapatistas in Mexico or the election of President Hugo Chavez on an unashamedly left-wing and populist platform in Venezuela and other left-wing leaders in other countries of South America could be considered as elements of the global justice movement, or at least as evidence of its success in gaining ground and support. The global justice movement has also been prominent in protesting against the meetings of the G8 – the annual meetings of the eight most powerful states and the war and occupation in Iraq. The following excerpt from the playwright Harold Pinter's Nobel Prize speech captures this element of the global justice movement:

> The invasion of Iraq was a bandit act, an act of blatant state terrorism, demonstrating absolute contempt for the concept of international law. The invasion was an arbitrary military action inspired by a series of lies upon lies and gross manipulation of the media and therefore of the public; an act intended to consolidate American military and economic control of the Middle East masquerading – as a last resort – all other justifications having failed to justify themselves – as liberation. A formidable assertion of military force responsible for the death and mutilation of thousands and thousands of innocent people. We have brought torture, cluster bombs, depleted uranium, innumerable acts of random murder, misery, degradation and death to the Iraqi people and call it 'bringing freedom and democracy to the Middle East'.
>
> (Pinter, 2005)

What is worth noting here is the connection between a war based on Malthusian 'resource scarcity' and the logic of the *Limits to Growth*, and the common left-wing view (encompassing Marxist/socialist and anarchist views) that the war and occupation of Iraq represents a move to protect the 'national interest' of powerful states, such as the United States, and client corporations and the global capitalist system.

The World Social Forum (WSF), and regional forums such as the European Social Forum (ESF), are worth examining in the context of embodying elements of both socialist/Marxist and anarchist thinking and organisation. Established in Porto Alegre, Brazil in 2001 to rival the World Economic Forum held in Davos, Switzerland each January, the WSF has grown into something approaching a global 'movement of movements'. The founding Charter of Principles of the

WSF makes it clear that the WSF is neither an organisation nor a common platform, but:

> The World Social Forum is an open meeting place for reflective thinking, democratic debate of ideas, formulation of proposals, free exchange of experiences and interlinking for effective action, by groups and movements of civil society that are opposed to neo-liberalism and to domination of the world by capital and any form of imperialism, and are committed to building a planetary society directed towards fruitful relationships among Mankind and between it and the Earth.
>
> (World Social Forum, 2001)

These gatherings of civil society and social movement groups represent a new phenomenon in world political terms, the first stirrings of a global civil society and the creation of a global social movement. If the WTO and the 'Washington' consensus represent what the global justice movement is against, it is within the processes of the World Social Forum that its alternatives are being debated and constructed.

Conclusion

This chapter has discussed some of the main left-wing reactions to the environment, ecological and green issues and politics. While differing in some respects, both socialist/Marxist and anarchist reactions share a commitment to critiquing the current capitalist organisation of society and the global economic system, while differing in their proposed institutional solutions. Both see that explanations for environmental degradation are to be found in socio-economic relations and not between 'humanity' and 'nature'. In particular, both point to how inequality and the domination of some groups and classes over others are usually the reason for environmental degradation. This connection between social domination and environmental degradation is demonstrated in the environmental justice and global justice movements, both of which arise from and are animated by analyses which stress the fact that the marginalised and less powerful groups are those who suffer the most from environmental degradation, even as they are often forced to participate in that degradation. The global justice movement is a unique historical phenomenon, the global creation of a 'movement of movements' which is not 'led' by any one party, organisation or individual, and one that seems to represent elements of earlier Marxist and anarchist theories of revolutionary change. However, it remains to be seen if the global justice movement will achieve the transformations it seeks globally, though there are signs of this in the success of anti neo-liberal governments in South America, and the continuing vitality of protest movements in North America, Europe, India and other parts of the developing world.

Summary points

- Marxism and socialism historically have expressed a hostile at worse or ambivalent attitude at best towards environmental issues, seeing them as 'middle class' and reactionary.
- Marxism's engagement with the environment has largely been shaped by Marx's vehement critique of Malthus, leading many left-wing perspectives to label environmental movements and concerns as a form of reactionary, right-wing neo-Malthusianism. This may be seen in left-wing reactions to the *Limits to Growth* report in the early 1970s.
- There has been some recent rethinking of Marxist social theory in response to the ecological crisis and the rise of the green/environmental movement. This ranges from the creation of an 'eco-Marxist' social theory which questions the 'Prometheanism' and productivism of Marxism and the need for left-wing theory and practice to take the natural conditions of production into account in their analyses.
- Eco-Marxism holds that the ecological crisis is a result of capitalist social and economic dynamics, property relationships and systemic injustice and periodic crises. A viable eco-Marxism/eco-socialism holds that the ecological crisis is not just a crisis *within* capitalism, but *of* capitalism.
- Class and class struggle are central to any eco-Marxist analysis of environmental issues, a good 'real-world' example of which is the environmental justice movement.
- The other main left-wing reaction to environmental issues has been anarchist social theory, the most dominant school of which has been 'social ecology' associated with Murray Bookchin.
- Social ecology sees the domination of the natural world and the ecological crisis as stemming from hierarchical and exploitation relations within society, but until Marxism/socialism sees the state as part of the problem not the solution.
- There are other varieties of 'eco-anarchist' theories and movements apart from Bookchin's social ecology which for some has become dogmatic and inflexible.
- The 'global justice movement' may be seen as combining elements of the Marxist and anarchist left-wing reaction to the ecological crisis and green issues, and the World Social Forum may be regarded as one of the first global civil society movements which challenges the institutions and actors who are the objects of the critiques of both eco-Marxism/socialism and eco-anarchism, namely transnational corporations, global institutions such as the World Trade Organization and the promotion of the interests of the most powerful nation-states, particularly the United States of America, through the G8 organisation.

Further reading

On Marxism, socialism and environmental issues, see T. Benton (2000), 'An Ecological Historical Materialism', in F. Gale and M. M'Gonigle (eds), *Nature, Production, Power: Towards an Ecological Political Economy*, Cheltenham: Edward Elgar; J. Bellemy-Foster (2000), *Marx's Ecology: Materialism and Nature*, New York: Monthly Review Press; J. Bellemy-Foster (2002), *Ecology against Capitalism*, New York: Monthly Review Press; J. Barry (1999), 'Marxism and Ecology', in P. Dickens (1996), *Reconstructing Nature: Alienation, Emancipation and the Division of Labour*, London: Routledge; P. Dickens (1997), 'Beyond Sociology: Marxism and the Environment', in M. Redclift and G. Woodgate (eds), *The International Handbook of Environmental Sociology*, Cheltenham: Edward Elgar; P. Dickens (2002), 'A Green Marxism? Labor Processes, Alienation, and the Division of Labor', in R. Dunlap, F. Buttel, P. Dickens and A. Gijswijt (eds), *Sociological Theory and the Environment. Classical Foundations, Contemporary Insights*, New York: Rowman & Littlefield; J. Hughes (2000), *Ecology and Historical Materialism*, Cambridge: Cambridge University Press; J. Kassiola (1990), *The Death of Industrial Civilization: The Limits of Economic Growth and the Repoliticization of Advanced Industrial Society*, Albany, NY: The State University of New York Press; J. Kovel (2002), *The Enemy of Nature: The End of Capitalism or the End of the World?*, London: Zed Books.

For a summary of 'social ecology' as articulated by Murray Bookchin, see M. Bookchin (1993), 'What is Social Ecology?', in M. Zimmerman (ed.), *Environmental Philosophy: From Animal Rights to Radical Ecology*, Englewood Cliffs, NJ: Prentice Hall. Bookchin's other works include: (1971), *Post-scarcity Anarchism*, London: Wildwood House; (1980), *Towards an Ecological Society*, Montreal and Buffalo: Black Rose Books; (1990), *Remaking Society*, Montreal and New York: Black Rose Books; (1991), *The Ecology of Freedom: The Emergence and Dissolution of Hierarchy* (revised edn), Montreal and New York: Black Rose Books; (1992), 'Libertarian Municipalism', *Society and Nature*, 1: 1; (1995), *Re-enchanting Humanity: A Defence of the Human Spirit against Antihumanism, Misanthropy, Mysticism and Primitivism*, London: Cassell; (1998), *Anarchism, Marxism and the Future of the Left*, Edinburgh: AK Press.

For critiques and analyses of Bookchin see the volume edited by Andrew Light (2003), *Social Ecology after Bookchin*, New York: Guilford Press; J. Clark (2004), 'Municipal Dreams : A Social Ecological Critique of Bookchin's Politics', *Research on Anarchism Forum*, http://raforum.apinc.org/article.php3?id_article=1038; D. Watson (1996), *Beyond Bookchin: Preface for a Future Social Ecology*, New York: Autonomedia; A. Rudy and A. Light (1996), 'Social Ecology and Social Labor: A Consideration and Critique of Murray Bookchin', in D. Macauley (ed.), *Minding Nature: The Philosophers of Ecology*, New York: Guilford Press; A. Light (1993), 'Rereading Bookchin and Marcuse as Environmental Materialists', *Capitalism, Nature, Socialism: Journal of Socialist Ecology*,

4:1; D.F. White (2003), 'Hierarchy, Domination, Nature: Considering Bookchin's Critical Social Theory', *Organization and Environment*, 16: 1.

Other variants of eco-anarchism may be found in A. Carter (1999), *A Radical Green Political Theory*, London: Routledge; A. Carter (1993), 'Towards a Green Political Theory', in A. Dobson and P. Lucardie (eds), *The Politics of Nature: Explorations in Green Political Theory*, London: Routledge; M. Paterson (2001), *Understanding Global Environmental Politics: Accumulation, Domination, Resistance*, Basingstoke: Palgrave.

Critiques and analyses of the anarchist position may be found in R. Goodin (1992), *Green Political Theory*, Oxford: Oxford University Press; J. Barry (1999), *Rethinking Green Politics: Nature, Virtue and Progress*, London: Sage; A. Dobson (1995), *Green Political Thought* (3rd edn), London: Routledge; R. Eckerlsey (1992), *Environmentalism and Political Theory*, London: UCL Press.

On the environmental justice movement see D. Schlosberg (1999), *Environmental Justice and the New Pluralism*, Oxford: Oxford University Press; R. Bullard (1990), *Dumping in Dixie: Race, Class, and Environmental Quality*, Boulder, CO: Westview Press; R. Bullard, (ed.) (1993), *Confronting Environmental Racism: Voices from the Grassroots*, Boston, MA: South End Press; D. Schlosberg (1999), 'Networks and Mobile Arrangements: Organisational Innovation in the US Environmental Justice Movement', *Environmental Politics*, 6:1; L. Pulido (1996), *Environmentalism and Social Justice: Two Chicano Struggles in the Southwest*, Tucson, AR: University of Arizona Press; J. Agyeman (2000), *Environmental Justice: From the Margins to the Mainstream*, London: Town and Country Planning Association; B. Boardman, with S. Bullock and D. McLaren (1999), *Equity and the Environment*, London: Catalyst Trust; C. Williams (ed.) (1998), *Environmental Victims*, London: Earthscan.

On the global justice movement see D. Korten (2002), *When Corporations Rule the World* (2nd edn), Bloomfield, CT: Kumarian Press; G. Monbiot (2003), *The Age of Consent: A Manifesto for a New World Order*, London: Flamingo; P. Doran (2006), 'Street Wise Provocations: The "Global Justice" Movement's Take on Sustainable Development', *International Journal of Green Economics*, 1:1/2; F. Gollain (2002), 'Anti-globalisation Movements: Making and Reversing History', *Environmental Politics*, 11:3; D. Wall (2001), 'Green anti-capitalism', *Environmental Politics*, 10:3; D. Hoad (2000), 'The WTO: The Events and Impact of Seattle 1999', *Environmental Politics*, 9:4; B. Gills (ed.) (2000), *Globalization and the Politics of Resistance*, London: Palgrave; V. Bennholdt-Thomsen *et al.* (eds) (2001), *There is an Alternative: Subsistence and Worldwide Resistance to Corporate Globalization*, London: Zed Books; B. Podobnik and T. Reifer (eds) (2004), *Global Social Movements Before and After 9–11*, special issue of *Journal of World Systems Research*, 10:1.

On the World Social Forum see J. Leite (2005), *The World Social Forum: Strategies of Resistance*, Chicago, IL: Haymarket Books; J. Smith (2004), 'The

World Social Forum and the Challenges of Global Democracy', *Global Networks*, 4:4; T. Teivainen (2002), 'The World Social Forum and Global Democratisation: Learning from Porto Alegre', *Third World Quarterly*, 23:4; W. Fisher and T. Ponniah (eds), *Another World is Possible: Popular Alternatives to Globalization at the World Social Forum*, London: Zed Books.

7 Gender, the nonhuman world and social thought

Key issues

- Gendered hierarchies in Western thought and culture.
- Eco-feminist spirituality/essentialist eco-feminism.
- Materialist eco-feminism.
- Eco-feminist political eonomy.
- Resistance eco-feminism.

Introduction

Up until the modern era, the idea of the inequality between men and women and the subservience of women to men as 'natural', something 'given' and beyond human powers to alter was a taken-for-granted perspective. Women were held to be physically (and psychologically) weaker than men and seen as occupying a position somewhere below 'man' but above 'animals' or nature. It is these (and other) historical and conceptual connections between women and nature that makes the adoption of a gendered approach to the discussion of social theory and the environment not just interesting but absolutely essential. As will become clear in later discussion of eco-feminist social theory, social theorising about the environment is not a gender-free zone.

The connection between gender and the environment within social theory is something that has received much attention in the past three decades or so. Tracing this relationship has been a key, if not always a central part, of the feminist movement and feminist social theory since the 1960s. However, the connection between gender, environment and social theory has its origins in the late eighteenth century with Mary Wollstonecraft's book *A Vindication of the Rights of Women* (published in 1792) and reactions to it, as discussed in Chapter 5. In this seminal book, Wollstonecraft argued for the extension of some limited equal rights to (some) women, such as the right to hold property, capital and education. However, her proposal for 'women's rights' provoked a reaction which starkly reveals the dominant view of women at the time. This was the publication of a book in response to Wollstonecraft entitled *A Vindication of the Rights of Brutes*. The basic position of the response was that if women were to have rights, then why should animals not also have them? Here, an explicit connection was made between the status of women and animals, a similarity between women and animals which was to last both historically and conceptually up until recent times and the creation of the feminist movement which challenged such 'sexist' assumptions (although as we shall see this is not the whole story).

It may seem odd to attempt to link ecology and feminism, since as Mellor points out, 'While feminism has historically sought to explain and overcome women's association with the natural, ecology is attempting to re-embed humanity in its natural framework' (1997: 180). However, for Salleh (1997) this identification of women with nature, upon which the whole edifice of hierarchical dualisms within Western culture and thought has been built, should be welcomed by feminists. As she puts it, 'Feminists should not fear the double-edged metaphor of Mother = Nature. This nexus both describes the source of women's power and integrity, and at the same time exposes the complex of pathological practices known as capitalist patriarchy' (1997: 175). It is to these gendered dualisms that we turn next.

Gendered hierarchies in Western thought and culture

Following eco-feminist social theorists such as Plumwood (1993) and Merchant (1990), we can begin to understand the relationship between gender and the environment by first noting a series of gendered dualisms that exist within, and define, Western culture as a patriarchal culture. The origins and effects of patriarchy are central to any feminist analysis, and it is worthwhile to note Marylin French's definition of it as 'an ideology founded on the assumption that man is *distinct from the animal and superior to it*' (quoted in Zimmerman, 1987: 25; emphasis added). According to this line of analysis, Western patriarchal culture is based on a gendered separation of 'culture' from 'nature' such that male attributes and values are associated with culture, while female attributes and values are associated with nature. However, it is not just that there is this dualism within Western culture, it is also the case that both historically and conceptually male attributes have been seen as not only separate from but also 'superior' to female ones. For some eco-feminists, such as Plumwood (1993) and Merchant (1990), the creation and maintenance of these sets of hierarchical gendered dualisms has its roots in Judaeo-Christianity in general, and the 'domination of nature' thesis in particular, discussed in Chapter 2.

The association of women and nature has historically produced the following sets of hierarchies (dualisms or binary oppositions), such that items on the left-hand side are accorded more importance or value than those on the right.

Gendered hierarchies in Western thought and culture
Culture / Nature
Men / Women
Human / Non-human
Reason / Emotion
Mind / Body
Abstract / Concrete
Objective / Subjective
Public / Private
Production / Reproduction
Rationality / Intuition
Competition / Co-operation
Violence / Non-violence

'Human' or the 'really' human came to be associated and identified with those 'male' characteristics and properties on the left-hand side. 'Female' characteristics on the right-hand side have been historically viewed as not representing what is 'truly' or distinctly human about human beings on account of being too closely tied to nature, the body, animality, sensuality, emotions and so on. This is the basic proposition and historical analysis of most feminist critiques of sexism and patriarchy, namely that Western culture (and many non-Western cultures too) privileges certain (i.e. 'male') attributes and properties (reason, abstract thought, mind, culture, production) above others (i.e. 'female') attributes and properties (emotion, concrete thought, the body, nature, reproduction). As Merchant points out, 'Anthropologists have pointed out that nature and women are both perceived to be on a lower level than culture, which has been associated symbolically and historically with men. Because women's physiological functions of reproduction, nurture, and child rearing are viewed as closer to nature, their social role is lower on the cultural scale than that of the male' (1990: 143).

The implications of this are substantial for the examination of the relationship between social theory, social practice and the environment. Since this set of dichotomies/dualisms is at the heart of Western culture, this means that one cannot examine 'nature' or the environment in Western thought without adopting a gender perspective. That is, one needs to look at gender in examining environmental issues generally, since social–environmental interaction, views of the environment and its significance for society are not 'gender-free' zones. For example, the very language used in social theory and in everyday discourse about the environment, ecological relations and the interaction between society and the nonhuman world is saturated with gendered terms. For example, we speak of 'virgin lands', the 'rape of the wild' (Collard, 1988), the 'despoliation of nature', 'Mother Earth' and so on, all of which are clearly gendered terms. Thus, some feminist social theorists have made the gendered (i.e. socially constructed) connection between women and nature explicit, and in so doing began the process of developing an *eco-feminist* perspective. As Karen Warren has noted:

> As I see it the term eco-feminism is a position based on the following claims: (i) there are important connections between the oppression of women and the oppression of nature, (ii) understanding the nature of these connections is necessary to any adequate understanding of the oppression of women and the oppression of nature; (iii) feminist theory and practice must include an ecological perspective; and (iv) solutions to ecological problems must include a feminist perspective.
>
> (Warren, 1987: 4–5)

The joining together of the ecological and feminist movement is for some eco-feminists a condition for success of both struggles. That is, feminists cannot achieve their ends within incorporating ecological concerns, and ecological aims will be frustrated without a feminist dimension. This strong argument for connecting feminism and ecology has been made by Rosemary Reuther. According to her, 'Women must see that there can be no liberation for them and no solution to the ecological crisis within a society whose fundamental model of relationships continues to be one of domination. They must unite the demands of the women's movement with those of the ecological movement to envision a radical reshaping of the basic socio-economic relations and the underlying values of this society' (quoted in Pietilä, 1990: 200–1). The term 'eco-feminism' was first coined in 1974 by a French feminist writer Françoise D'Eubonne who argued that the conjoining of ecological and feminist concerns was necessary for the creation of a radical critique of patriarchal industrial capitalism against the background of a growing ecological crisis.

Perhaps the best way of illustrating the complexities, variety and potential insights to be gained from paying close attention to gender and the environment within social theory is by outlining some of the main lines of thinking within eco-feminism. There are at least three main schools of thought which adopt this explicitly gendered approach to theorising the environment and social–environmental relations. These different approaches to eco-feminist social theory are discussed below.

Eco-feminist spirituality/essentialist eco-feminism

The basic argument of this school of eco-feminist thought is that a necessary condition for solving the ecological crisis and attaining more equality for women is to reverse the gendered dualisms indicated above. For eco-feminists such as Susan Griffin (2000), the Western dualistic mindset is a form of mental illness, 'a form of insanity that lies at the heart of our destruction of the environment' (Glotfelty, 2001: 298). There is a strong sense within this branch of eco-feminism which states that the ecological crisis is not simply due to the anthropocentrism (i.e. human-centredness) of the 'modern' or Enlightenment worldview, but can be traced to the androcentrism (i.e. male-centredness) underpinning it. To cure this cultural mental illness we need to adopt an Earth-focused eco-feminist worldview. As Plant explains:

> the world is rapidly being penetrated, consumed and destroyed by this man's world – spreading across the face of the earth, teasing and tempting the last remnants of loving peoples with its modern glass beads – televisions and tanks. . . . As the Amazon rainforest is bulldozed to provide cheap beef for American hamburgers, the habitat of peoples who once lived and loved with this earth and the fragile womb of planet Earth is dealt yet another killing blow. This 'man's world' is on the very edge of collapse.
>
> (Plant, 1989: 1–2)

And if it is this 'man's world' which is the problem, then the solution is the creation of a 'women's world', since women are 'closer to nature' and thus better suited than men as guardians and protectors of the natural environment. This is the 'essentialist' element: *women are essentially closer to nature because of their particular biological natures and abilities/capacities.* Like nature women reproduce, are life-givers and nurturers, they are in touch with 'natural cycles' such as menstruation, which have affinities with natural cycles such as those of the seasons. It is important to note here that what is being criticised (and this is true of all schools of eco-feminist thought) is not 'men' per se as individuals or as a group. Rather the main point of criticism is 'male' forms of thinking, institutions and practices which have led both to the degradation of the natural world, and the oppression of women and the denigration of female values and attributes. Again as Plant puts it:

> Making the connection between feminism and ecology enables us to step outside of the dualistic, separated world into which we were all born. From this vantage point, this new perspective, we begin to see how our relations with each other are reflected in our relations with the natural world. The rape of the earth, in all its many guises, becomes a metaphor for the rape of woman, in all its many guises. In layer after layer, a truly sick society is revealed, a

society of alienated relationships all linked to a rationalization that separates 'man' from nature.

(Plant, 1989: 5)

A central part of this school of eco-feminism is the practical importance of 'earth-bonding' spirituality and rituals, and the need for men to discover or become more in tune with their 'female side', that is, to become more associated with what have been traditionally regarded as 'female' values and practices such as emotion, caring, intuition (the items on the right-hand side of the list above). Some of these rituals take the form of a return to 'Wicca' or 'witchcraft' (which is not the same as Satanism), as part of a return to older more woman-centred, natural or pagan religions which Judaeo-Christianity had destroyed. In opposition to the Judaeo-Christian God which is male and located in Heaven (i.e. not on Earth) and directed towards humanity, eco-feminist spirituality sees divinity, meaning and spirituality as existing throughout 'Mother Nature' and not as the special preserve of humans alone.

The main points of eco-feminist spirituality may be summarised as follows:

1   There needs to be a reversal of the binary oppositions within Western culture, such that 'female' attributes, concepts and ways of thinking and acting form the basis of a new ecologically sensitive, life-affirming culture.
2   Action and real experience of the natural world is prioritised over abstract theorising about it.
3   Women rather than men ought to represent nature.
4   Personal (inner) change must precede political or social (external) change.

However, like any subdivision of a larger body of social theory, this branch of eco-feminism has been criticised by other schools of eco-feminism, particularly materialist eco-feminism (as well as by other schools of feminist thinking more generally). One of the main criticisms levelled concerns the 'essentialist' view of women. Here, it has been claimed that there is a confusion of sexual characteristics (biology) with social roles (gender). 'Biology is not destiny' as the feminist slogan has it. Thus eco-feminist spirituality is found to be offering a biological determinist view of women, which is dangerous and misleading. For example, one can ask about the status of women who do not fulfil their 'biological duty/function' in having children. According to the logic of essentialism, women are seen as essentially mothers, which seems to imply that women who choose not be mothers are somehow denying their 'essential nature'. The latter is a conservative or traditional view of women which other feminists are struggling to deconstruct. In short, it seems as if essentialist eco-feminism simply wants to 'celebrate' the very attributes and views of women that have (1) been formed under sexist conditions, and (2) been traditionally used to keep women down.

A second problem following on from the first is the confusion between the *feminine* as opposed to the *feminist* character of this form of eco-feminist social theorising. In urging the celebration of qualities such as mothering, caring, feeling, and nurturing, by basing it on an appeal to the essential, biological, psychological 'nature' of women, they advance a revaluation of the 'feminine' as opposed to a feminist politics aiming for equality between women and men. An illustration of this is in the way green politics is often seen to be less sexist and more 'pro-women' than other political ideologies such as liberalism or socialism. However, as Mary Mellor points out, 'Where male green thinkers claim that a commitment to feminism is at the heart of their politics, this often slides into a discussion of feminine values' (Mellor, 1992b: 245). Celebrating the feminine (as if this were a biologically 'given' concept) according to Mellor demonstrates a lack of awareness of the fundamental distinction between biology and gender. As she puts it, 'The feminine is not the missing half of the masculine; the feminine is what men need to create the masculine in patriarchal culture' (Mellor, 1992a: 81). A social theory and politics based on the feminine, as opposed to a feminist perspective, does not challenge sexism and gender inequalities, rather it serves to reinforce them. Thus essentialist eco-feminists may succeed in 'saving the planet' (and men) at the cost of the liberation and equality of women.

A third and final problem lies with a whole series of questions relating to defining 'female' characteristics. How can we determine what are 'female' traits if women's lives have been determined, structured and influenced by patriarchy and sexism? Are submissiveness and unassertiveness 'female' characteristics and therefore something to be celebrated and encouraged? As Mellor puts it, 'Feminists have long argued that until women have control over their own fertility, sexuality and economic circumstances, we will never know what women "really" want or are' (1992b: 237). A more charitable analysis for the negative reception of this school of eco-feminist thought is given by Mellor (2001) who notes that the early articulations of eco-feminist ideas in the 1980s were often poetic and romantic. This meant that 'much early eco-feminism was rejected by mainstream feminism as being reactionary, encouraging a cult of the Goddess or Mother Earth that seemed to push women back into an elemental association with motherhood and Nature that they were trying to escape' (Mellor, 2001: 145). Taking up the challenge that the relationship between women and nature is not 'elemental' but one based on common exploitation has been one of the main aims of materialist eco-feminism which we turn to next.

Materialist eco-feminism

In opposition to many of the tenets of essentialist eco-feminism stands what I have termed 'materialist eco-feminism', which represents a more nuanced and sophisticated approach to the relationship between social theory, gender and the environment, one based on the material realities of economic inequality of women within the sexual division of labour. For Mary Mellor (1992b: 162), one of the main social theorists in this area, 'despite the influence of cultural and spiritual feminism, eco-feminism is necessarily a materialist theory because of its stress on the immanence (embodiedness and embeddedness) of human existence'. Ariel Salleh echoes this point, noting that:

> It is nonsense to assume that women are any closer to nature than men. The point is that women's reproductive labour and such patriarchally assigned work roles as cooking and cleaning bridge men and nature in a very obvious way, and one that is denigrated by patriarchal culture. Mining or engineering work similarly is a transaction with nature. The difference is that this work comes to be mediated by a language of domination that ideologically reinforces masculine identity as powerful, aggressive, and separate over and above nature.
>
> (Salleh, 1992: 208–9)

Here the connection between women and nature is not on the basis of some 'essentialist' or 'biological' grounds; rather it is the fact that women's work (including reproductive work) means that they are closer to nature. Materialist eco-feminism thus has affinities with eco-socialism and eco-Marxism. Whereas the essentialism of eco-feminist spirituality located the connection between women and nature in *sex* (biological characteristics of women), materialist eco-feminism locates the connection in *gender* (social constructions of practices, characteristics and roles based on sex). Women and nature both suffer at the hands of patriarchy and industrial capitalism. That is, *what unites women and nature, is not the biological closeness of women to nature, but the fact that both are exploited and oppressed by male, sexist culture, its institutions, values and practices*. Thus the root of the connection (both historical and conceptual) between women and nature lies in their material exploitation, as a result of the sexist organisation of society and the economy. As Mellor puts it, 'A feminist green politics must begin with women's work of nurturing and caring and the sexual division of labour that largely excludes men from that work' (1992b: 278). In material terms women are assigned to the (disvalued or non-valued) sphere of 'reproduction' (e.g. child-rearing, food preparation, home-making, nurturing), while men are assigned to the (valued) sphere of production (producing commodities for the market, earning income). The disvalued or non-valued status of 'women's reproductive work' may be easily seen in that the work women do in the home – food preparation, cleaning, child-rearing, caring, comforting – is not paid, and since it is not paid it is not publicly visible or valued. From the perspective of orthodox theories of economics, which are themselves gendered (surveyed in Chapter 8), women's reproductive and nurturing work is 'free', just as environmental services are also wrongly seen as 'free'. Buckingham-Hatfield (2000) goes further in highlighting the fact that it is not just that women and nature are 'free' but also exploitable. As she puts it:

> The globalisation of industry both utilises gender and environment in the search for cheap production and market share. Both environment and gender become commodities whether explicitly, for sale in tourism (in sex tourism, mostly, but not exclusively, females; and in 'nature' – sun, sea, sand and forest), or as the means of reducing production costs via cheaper labour or by externalising the environmental costs of production.
>
> (Buckingham-Hatfield, 2000: 117)

The 'life-affirming' character of reproductive work is used by Salleh to explain why in Western social theory and history this gendered sphere of activity, and the characteristics and values associated with it, have been downgraded. As she puts it, 'In the Eurocentric tradition, not "giving life" but "risking life" is the event that raises Man above the animal. In reality, reproductive labour is traumatic and highly dangerous . . . birthing . . . is an experience that carves the meaning and value of life into flesh itself' (1997: 39). Hence life-taking and its associated activities and values of violence and war are, within Western culture, seen as an essential aspect of what is distinctively 'human' as opposed to 'nonhuman'. This also partly explains the association of men and manliness (and thus 'humanness') with violence and warfare, something Freud sought to explain, as discussed in Chapter 3. In this way 'human' becomes associated with 'maleness'. At the same time life-giving, birth, reproduction and nurturing are not seen as something that is distinctively 'human', since these biological or 'natural' activities are something we share with the nonhuman world and are consequently regarded as being of lower value.

Materialist eco-feminism is, unlike eco-feminist spiritualism, orientated towards reconfiguring the material basis of human society (covering the formal and informal economy, the nature of work, reproductive relations, and the material exchange between the 'total human economy' and the natural environment). Figure 7.1 outlines the materialist eco-feminist position.

Eco-feminist political economy

The full force of the materialist eco-feminist position demands nothing short of the radical transformation not only of the economy, but also a radically new theory of economics. While Chapter 8 will outline some of the shortcomings of orthodox

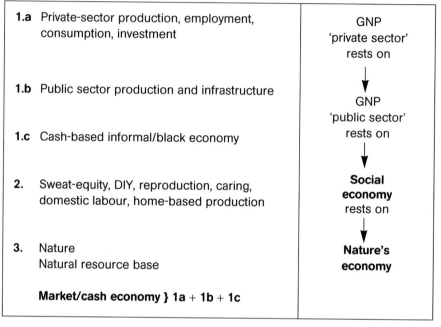

Figure 7.1 The materialist basis of human society
Source: Adapted from Henderson et al. (1986), p.33

theories of economics from a 'green' perspective (which includes some of the insights of materialist eco-feminism), some of the criticisms levelled at these theories can be listed. The changes materialist eco-feminism requires within economic theory include: the creation of meaningful indicators of human 'well-being' rather than abstract measures of 'economic growth'; the reconceptualisation of central economic categories such as 'work' (to include reproductive work); transcending the 'public/private' division by extending the notion of the 'economy' to include the informal, social and domestic economy; and recognising the dependence of the 'total human economy' on the natural environment. The basic eco-feminist political economy position is as follows:

Eco-feminist political economy

1 Sphere of production (industry, formal economy)

rests on

2 Sphere of reproduction (nurturing, informal economy)

rests on

3 Nature's economy (natural resources)

The idea of dependence is central to the eco-feminist materialist position in the critique of orthodox, gender-blind economic theory and practice (and 'malestream'/mainstream political theory and practice). Dependence, vulnerability and the inherent neediness of humans are central ideas and realities of the 'human condition' which have been ignored and/or denied within (male) economic, political and social theory and practice. Because standard or conventional economic theory and practice does not take into account the double dependence of humans (our dependence on each other and on the natural world), the individual and collective vulnerability of human beings is denied.

The result of this is that gender-insensitive and environmentally blind economic theory and practice have resulted in a situation whereby the biological and ecological character of human collective and individual life are simply unacknowledged. As Mellor puts it, 'As a result of women's private and unacknowledged labour we have a *public world constructed on the false promise of an independently functioning individual, with the nurturing, caring and supportive world hidden, unpaid and unacknowledged*' (1992b: 239–40). Thus we have an economy in which childcare considerations are simply not seen as 'appropriate' or central when making economic decisions. These are relegated to the 'private' and supposedly non-economic realm of the home. At the same time the biological and psychological needs of humans are likewise not given the prominence they deserve. An example of this is the argument that the 'twenty-four-hour working day' as an economic ideal is completely out of synch with the biological and psychological needs of human beings, and is not only impossible and undesirable, but is based on a completely false picture of human beings which does not acknowledge their inherent neediness and vulnerability. Finally, orthodox economics has largely ignored the contribution of the natural environment to the human economy and its dependence on that environment.

Eco-feminist materialist political economy stresses the experience and labour associated with *reproduction*, the private, unvalued but fundamental life-sustaining work which women perform. This life-sustaining focus is particularly evident in Salleh's view when she makes clear that 'the embodied materialism of eco-feminism is a "womanist" rather than a feminist politics. It theorises an intuitive historical choice of re/sisters around the world to *put life before freedom.* . . . Eco-feminism is more than an identity politics, it reaches for an earth democracy, across cultures and species' (1997: ix–x; emphasis added). This adoption of a 'womanist' rather than a 'feminist' stance is motivated by a desire to make connections with women in the Southern, developing world, whose concerns, problems and issues are not articulated by the privileged, urban, affluence-based discourses of Northern/Western feminism.

Salleh offers powerful criticisms of Northern, liberal feminism from a materialist eco-feminist standpoint. She criticises Northern, affluent feminist concerns with individual self-realisation, its Eurocentrism and insensitivity to Southern women's concerns, its anti-reproductive bias, and ultimate blindness to its position within global capitalism. Salleh suggests that:

> For too many equality feminists, the link between their own emancipated urban affluence and unequal appropriation of global resources goes unexamined. ... Much of the energy that went into abortion campaigning was clearly a sublimation of this hostility toward the problematic mother. The unreality of mothering experiences to many feminists did not help theorisation. . . . The hope is that feminism's ideological immaturity will be remedied as this generation of career women take up mothering themselves, and draw that learning into feminist thought.
>
> (1997: 104)

She sees Northern liberal/equality feminism as the product of what Marcuse called the 'repressive tolerance' of patriarchal capitalist states, in which feminist issues are 'co-opted' and thus neutralised, and feminist activists become 'femocrats'. Thus materialist eco-feminism is suggested as a maturing or development of feminism both as a form of social theory and a political movement.

One part of this growing eco-feminist political economy which challenges key aspects of the dominant 'malestream' economic paradigm is that of 'subsistence' as developed by Vandana Shiva, Maria Mies and Veronicka Bennholdt-Thomsen. As Bennholdt-Thomsen and Mies put it:

> In the North and, since 1945, increasingly in the rest of the world, everything that is connected with the immediate creation and maintenance of life, and also everything that is not arranged through the production and consumption of commodities has been devalued. This includes activities whose object is self-provisioning, whether in the house, the garden, the workshop, on the land or in the stable. What doesn't cost or doesn't produce money is worthless. . . . How did this alienation between people and their work develop to the point that the most lifeless thing of all, money, is seen as the source of life and our own self-producing subsistence work is seen as the source of death?
>
> (Bennholdt-Thomsen and Mies, 1999: 17)

In opposition to the dominant money-orientated economy and associated modes of economic thinking, they propose the idea of 'subsistence' as a way of expressing an alternative to that economy and way of thinking which encompasses 'freedom, happiness, self-determination within the limits of necessity . . . [and] the historical connection that exists, through colonisation and development, between us in the industrial countries and the countries of the the South. In both cases modern development happened and happens by means of a war against subsistence. . . . The concept of subsistence also expresses continuity with the nature within us and around us' (Bennholdt-Thomsen and Mies, 1999: 19–20). This eco-feminist subsistence perspective also leads to a politics of resistance or 'No substance without resistance' (1999: 213), which we turn to next.

Resistance eco-feminism

While not a 'school' in the sense that essentialist and materialist eco-feminism are, there is a third strand of eco-feminist thought. Overlapping with some of the concerns of the other two, though more on the materialist than essentialist side, it does represent a distinctive 'voice' and perspective within eco-feminism. This final stream of eco-feminism is characterised by its practical political concerns and, while it does have relevance to the developed world, its origins and main focus lie in the 'developing' world.

A key starting point for resistance eco-feminism is the recognition that women are more concerned about the environment than are men, and that women are at the forefront of many environmental struggles. Examples of the latter include the Chipko movement in India, a movement of local women in Uttar Pradesh protesting against commercial logging which was leading to rapid deforestation (Ekins, 1992: 143), and the British Greenham Women's Peace Camp which was a 1980s anti-nuclear movement to remove American nuclear missiles from Britain, and which had strong ecological and feminist aims (Buckingham-Hatfield, 2000). From local community movements against toxic dumping, protests against increased road traffic, resistance to timber-logging, dam-building and other 'mega-developments' in the developing world, women are either in leading positions or make up the bulk of support for these various environmental resistance movements.

For certain environmental struggles, particularly when they have to do with health issues or subsistence livelihoods, it is often the case that women are at the fore-front. In addition, in different environmental struggles or environmental issues, women may have a greater vested interest than men (on the issue of population 'control', for example, or 'toxic shock' syndrome from sanitary products). In other words, there are aspects of environmental protection which are gendered in a way that other areas of environmentalism such as biodiversity loss and global climate change are not. Where environmental problems affect human health such as toxic-dumping, electromagnetic radiation and anti-smog campaigns, women are commonly at the forefront, particularly when it is children's health that is at risk. What is interesting about women's involvement in environmental campaigns is that often it becomes a training ground for more mainstream 'feminist' demands for greater equality, access to employment, wages and general standards

GREEN ECONOMICS

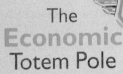

The Economic Totem Pole

What you see is definitely *not* what you get if you look at modern industrial economies in the traditional way. Underlying the visible monetary economy is a whole *non*-monetary area of activity which is both invisible and undervalued. Without the strong shoulders of the bottom characters in this totem pole the market economy at the top would quickly tumble down.

Underground economy

Cash transactions which are hidden to avoid taxes or illegal like drug trafficking, prostitution and pornography.

Mother Nature

The natural resource base is the largest and most basic support for the monetary economy. All economic activity depends on the survival of healthy, natural ecosystems.

Official market economy

All transactions and relationships involve money – including wages, consumption, production, investment and savings.

Government expenditure

Money spent by government on social security, defence, education and infrastructure like roads, bridges, airports, sewers and public transport.

Social economy

All non-market economic activities. Includes subsistence farming, housework, parenting, volunteer labour, home healthcare and DIY. Also includes barter or skill exchanges. In Northern economies the informal economy is estimated to be one-and-a-half times the size of the visible market economy.

Figure 7.2 'The Economic Totem Pole'

Source: Polyp (1996a)

Box 7.1 Summary of materialist eco-feminism

1 Stresses the importance of spheres of production and reproduction, and critical of existing, gender-blind political and economic theories, institutions and structures.

2 Strongly 'feminist' as opposed to 'feminine'.

3 Emphasises the dependence of humans on nature, and also the real work women are socially associated with and which is vital to human society, i.e. reproductive work, caring, nurturing, home-making.

4 The ecological restructuring of the economy requires reconstructing the relationship between the spheres of 'production' (public) and 'reproduction' (private), so that the latter takes precedence.

5 This ecological restructuring requires a new materialist ecofeminist theory of economics, in which central categories of economic thought need to be reconceptualised.

6 Stresses the biological (and psychological) neediness, vulnerability and dependency of humans.

---

of welfare, and political, economic and social respect. As Martin-Brown observes, 'environmentalism has become the "Trojan horse" for the engagement of women in the political process' (Martin-Brown, 1992: 707).

However, as many eco-feminist writers rightly point out, even if women make up the majority of those involved in environmental campaigning on the ground, they are grossly misrepresented in terms of senior positions in environmental organisations. Seager estimated that while in the USA between 60 and 80 per cent of environmental grassroots activity is undertaken by women, there were virtually no women in senior positions in environmental non-governmental organisations (Seager, 1993). It was partly for this reason, as well as the not unjust characterisation of the major US environmental organisations as run by white and middle-class professionals, that the environmental justice movement developed, discussed in the previous chapter, and also that this engendering of the environmental justice movement is not confined to the developed world (Wickramasinghe, 2003) . Thus, although there are obvious connections between feminist interests and those of the environment, including the association of feminism with the peace and anti-nuclear movement as Buckingham-Hatfield points out (2000: 96–97), there is still a long way to go in terms of equality between men and women within the environmental movement. It was only in 1989 that there was a woman head of a major environmental organisation, the Worldwide Fund for Nature (WWF).

Struggles and movements against genetic engineering and biotechnology are another focus for environmental resistance in which women, according to Vandana Shiva have a particular interest. According to her, 'Capital now has to search for new colonies to invade, exploit, and spoil for further accumulation. These new colonies are the interior spaces of the bodies of women, plants and animals. . . . Biotechnology as the handmaiden of capital in the post-industrial era creates the possibility to colonise and control that which is free and self-regenerative' (1992: 13).

It is generally regarded that only by giving women reproductive rights will major environmental and developmental problems be averted in the developing world. Here the traditional feminist demand for women to be given control over their fertility, and environmentalist claims of the positive relationship between population growth and environmental degradation dovetail into one another. On this point the eco/feminist response to the issue of population control completely rejects that proposed by right-wing Malthusians such as Garrett Hardin. It is perhaps on this issue if no other that 'eco-feminism' may be considered as a synthesis of feminism and environmentalism. Studies have demonstrated that there is a causal connection between women's equality in general (and reproductive rights in particular) and population control and environmental protection; equality for women means having control over their own reproduction (Dixon-Mueller, 1993).

Addressing the claims of Malthus, discussed earlier, materialist eco-feminism rejects the idea that population growth is the sole cause of global environmental problems. Many eco-feminists, like Bandarage, see contemporary 'neo-Malthusianism' as the dominant ideological analysis and approach to the global ecological crisis. She suggests that, 'Like Malthus, contemporary Malthusian analysts who work within the population control paradigm advocate population stabilization as a substitute for social justice and political-economic transformation' (1997: 6). However, as she puts it, 'growing global economic inequality, not population growth, is the main issue of our time' (1997: 12).

Bandarage also uncovers the violence underpinning population control rhetoric and practice, noting how 'military metaphors that "declare war", "target" and "attack" "over population" with an "arsenal" of new drugs have become the standard language of global population control' (1997: 65). This gendered rhetoric and way of thinking about population issues is particularly striking in that its 'masculine', not to say 'macho', character is quite obvious. This violent and warlike mode of thinking and acting leads, according to Bandarage, to the situation where 'Aggression and conquest rather than compassion and care drive the population control establishment and the larger model of technological-capitalist development that it represents' (1997: 103).

In many so-called 'Third World' countries, it is women who have to provide, tend and prepare food. With modernisation, imported technologies, increasing immersion into the global market and so on, it is women who lose out more, as they are displaced from subsistence labouring by mechanisation and pesticides. Women in Third World economies are increasingly forced to produce food on marginalised land as previously farm land is intensively cultivated for the production of cash crops for export. This not only increases the burden on them but degrades the surrounding land. Women as the primary 'land managers' and workers in agriculture in the South mean that any attempt to implement sustainable policies needs to take this fact into account. That is, policies aimed at **sustainable development** or environmental protection must be formulated with women and women's needs in mind. According to Martin-Brown, 'Traditionally, that responsibility [of ecosystem management] has fallen to women. Throughout time and around the world, the traditional role of woman has been to *manage* prescribed resources. . . . The complementary historical role of men has been to *enlarge* the available resource base' (1992: 707). From this she takes the view that sustainable development will depend on how women can be empowered to manage ecosystems.

At the same time, in the developing world there is a connection between poverty and environmental degradation. As Buckingham points out, 'The link between women and the environment was consolidated, internationally, at the 1995 4th UN Conference on Women in Beijing. The resulting Platform for Action identified "women and environment" as one of the critical areas of concern . . . this concern [was subdivided] into education, health, marginalized groups, planning, housing and transport, Local Agenda 21, and consumption and waste' (Buckingham, 2004: 148). Yet while this is a generally recognised relationship (Goldblatt, 1996; Doyle and McEachern, 1998: 77), there is also a gender aspect to this relationship which needs equally to be recognised. The fact is that women (and children) are the 'poorest of the poor' in the developing world, and thus suffer more than anyone else from the effects of poverty, poor environments and environmental degradation. Thus within the context of sustainable development in the South, joining development with environmental protection, one needs to be aware of the gendered distribution of environmental and economic burdens/costs, which means that it is women as a group who suffer the most and therefore whose needs are greatest. As Martin-Brown notes, 'Women and the environment are the "shadow subsidies" which support all societies' (1992: 717).

According to Vandana Shiva (1988), women are in the vanguard against Western forms of modernisation and its damaging ecological and socio-economic effects in the South. For her, women in the developing world offer resistance to what she calls the 'colonisation' of the developing world, by the imposition of a

Western and male view of progress, economic vision, institutions and modes of thinking and acting about the natural environment. As Doyle and McEachern note in discussing Shiva's argument, for her, women in 'subjugated cultures have been direct activists in opposing modernisation in parts of the third world' (1998: 51). At a more conceptual level, Shiva advances an eco-feminist critique of the Enlightenment (the original source as it were for modernisation theory and practice). For her:

> Throughout the world, a new questioning is growing, rooted in the experience of those for whom the spread of what was called 'Enlightenment' has been the spread of darkness, of the extinction of life and life-enhancing processes. A new awareness is growing that is questioning the sanctity of science and development and revealing that these are not universal categories of progress, but the special projects of modern western patriarchy. . . . *The violence to nature, which seems intrinsic to the dominant development model, is also associated with violence to women* who depend on nature for drawing sustenance for themselves, their families, their societies. This violence against nature and women is built into the very mode of perceiving both, and forms the basis of the current development paradigm.
>
> (Shiva, 1988: xiv; emphasis added)

Thus this 'resistance' eco-feminism shares some of the concerns of materialist eco-feminism, but its critique of ecological degradation and alternatives to the status quo are premised on practical experience of ecological struggles. In its developing world forms resistance eco-feminism constitutes a rejection of Western forms and models of modernisation.

For Shiva, then, resistance to large-scale projects of modernisation, such as the Narmada Dam in India (Ekins, 1992: 114, Roy, 1999), or protests against the patenting of seeds and genetic information by Western biotechnology indus-tries (Purdue, 1995), may also be seen as part of a wider and deeper resistance to a particular Enlightenment or 'modern' (i.e. Western) way of thinking, valuing and acting. She criticises biotechnology on the grounds that:

> Biotechnology, as the hand-maiden of capital in the post-industrial era, makes it possible to colonise and control that which is autonomous, free and self-regenerative. Through reductionist science, capital goes where it had never been before. The fragmentation of reductionism opens up areas for exploitation and invasion. . . . It is in this sense that the seed and women's bodies as sites of regenerative power are, in the eyes of capitalist patriarchy, among the last colonies.
>
> (Shiva, 1991: 128–43)

Other eco-feminists have also highlighted the gendered dimensions of Enlightenment/Western thinking in relation to the scientific understanding of the environment. Buckingham-Hatfield (2000) suggests that the dominant scientific paradigm is not immune from social, ideological, class or gender influences and that science is not a 'value-free' form of knowledge. For her, 'gender relations (as well as class and race relations) are likely to have a bearing on [science]' (2000: 21), and points out the argument by feminist critics of 'malestream' science (such as Harding, 1990) that if women's concerns were at the heart of the scientific project, rather than male definined scientific issues, then perhaps 'ecological research may well be preferred over military, or research with a social benefit preferred over that which serves solely industrial interests' (Buckingham-Hatfield, 2000: 21–2). Another illustration of the sexist thinking within male-stream science are those studies of the metaphors and descriptions used in scientific analyses which betray a clear gender dimension, such as the common portrayal of the male sperm as 'active' while the female egg is presented as 'passive' (Martin, 1991). Other examples include research into the different results from studies in primate behaviour with male researchers focusing on male primates and coming to the conclusion of the importance of aggression and hierarchy to explain primate social order, while female reseachers studied female primates and demonstrated the centrality of co-operation and sociability (Buckingham-Hatfield, 2000: 22).

Some of the issues she raises here, notably in relation to the threat to 'autonomy' and 'freedom', are remarkably close to the misgivings outlined by Habermas in Chapter 4 on the ethical dangers of biotechnology and genetic engineering. Now while Shiva is raising deeply normative issues, she is also concerned with the negative practical, mainly economic and, in terms of human health and well-being, consequences of biotechnology. Thus her characterisation of the threat of biotechnology in terms of its 'colonising' the bodies of women and sub-sistence agricultural practices means that a political struggle to 'decolonise' and protect them from the predatory motivations of biotechnology corporations is required.

From a Northern/Western perspective there are affinities with resistance eco-feminism and some of the strategies outlined by Val Plumwood, who stresses the need to cultivate what she calls 'counter hegemonic virtues' of sensitivity to others (human and nonhuman), emotionality/emotional intelligence, embodiment and corporeality and attentiveness to our natures as animals, and to cultivate modes of thinking and acting which resist the dominant ideological mechanistic and reductionist worldviews perpetuated by capitalism, patriarchy and technocentrism (Griffin, 2000: 288–9). Equally the strategies and political struggles required to implement the types of structural changes envisaged by eco-feminists such as Mellor and Salleh would be similar and indeed linked to the struggles of women in the developing world.

Other examples along these lines include strategies to reclaim 'citizenship' for eco-feminist purposes. Macgregor (2006) articulates a political project for a *feminist ecological citizenship* informed by feminist critiques of citizenship, and points to and draws on those eco-feminists who have regarded citizenship in theory and practice as an important site for political struggle (Plumwood, 1993; Gaard, 1998; Sandilands, 1999). This project has much in common with both materialist and resistance forms of eco-feminism, and indeed could offer a way of bringing a sense of 'unity in diversity' for different schools of eco-feminist theory. Equally, Quinby's (1990) suggestion that eco-feminism should not unduly concern itself with some finished coherent theory to guide practice, but rather focus on 'decentered political struggle', since such 'grand theorizing' may 'limit political creativity' and replicate 'the totalizing impulses of masculinist politics', not only echoes key aspects of postmodernist thinking discussed in Chapter 9, but is also more in keeping with the eco/feminist spirit of encouraging not discouraging 'unity in diversity'.

Conclusion

Feminism has made the relationship between human society and the natural environment central to its concerns, more so than any other twentieth-century social theory. Eco-feminism, as a sub-branch of feminism concerned with ecological issues, highlights the role of gender in social–environmental relations. Different strands of eco-feminism, such as essentialist/spiritual, materialist and resistance, each offer their own critical analysis of the connection between the oppression of women and the degradation of the natural environment, and their own (sometimes competing) alternatives to what they regard as contemporary anti-women and anti-environmental political, social and economic arrangements.

Summary points

- Exploring the relationship between gender and the environment has been the central contribution of feminist social theory to the study of social environmental issues. Social theorising about the environment is not a 'gender-free' zone.
- Western culture is based on a set of gendered dualisms such that (1) certain values, principles, characteristics and activities are either 'male' or 'female', and (2) those that are 'male' are regarded as both separate and superior to those associated with 'female'.
- Essentialist or spiritual eco-feminism is based on the claim that women are 'naturally' closer to nature than are men, and that if the cause of ecological

problems is men and male culture, then the solution is the creation of a women-centred society.

- Essentialist eco-feminism has been criticised for confusing biological sex (which is 'given') with gender (which is socially constructed), and in being more 'feminine' than 'feminist'.
- Materialist eco-feminism begins from the observation that what connects women and nature is that both are exploited within patriarchal or male-dominated society.
- It stresses the biological embodiedness and ecological embeddedness of human beings, and draws attention to the vulnerability, neediness and dependence of humans.
- It highlights the role of women's unpaid and undervalued reproductive labour in meeting human needs and the ways in which 'malestream' social, political and especially economic theory and practice ignore both this vulnerability and women's work.
- Eco-feminist political economy calls for the radical reconceptualisation of 'malestream' or orthodox economic theory, especially central economic terms, while in practice calling for the radical restructuring of the economy.
- Resistance eco-feminism links the feminist and ecological movement in terms of common political aims, such as defending women's reproductive rights and protecting women from poverty and degraded environments.
- Resistance eco-feminism also builds on the observation that certain environmental issues (population, health-related and children-related, and the link between poverty and environmental degradation) seem to be more gendered in terms of actual social support by more women than men, than others.

Further reading

For a general overview of eco-feminism, see Susan Buckingham-Hatfield, *Gender and Environment*, London: Routledge, 2000; Val Plumwood's *Feminism and the Mastery of Nature*, London: Routledge, 1993, and her more recent book *Environmental Culture: The Ecological Crisis of Reason*, London: Routledge, 2002; Karen Warren (ed.), *Eco-feminsm*, London: Routledge, 1995, and A. Collard, *Rape of the Wild: Man's Violence against Animals and the Earth*, Indianapolis: Indiana University Press, 1988. For shorter overviews see ch. 2 of Timothy Doyle and Doug McEachern's *Environment and Politics*, London: Routledge, 1998; ch. 5 of Andy Dobson's *Green Political Thought*, London: Routledge (2nd edn), 1995; Judy Evans, 'Ecofeminism and the Politics of the Gendered Self', in A. Dobson and P. Lucardie (eds), *The Politics of Nature*, London: Routledge, 1993; ch. 4 of Vernon Pratt, Jane Howarth and Emily Brady's *Environment and Philosophy*, London: Routledge, 2000; Mary Mellor, 'Eco-feminism', in J. Barry and

E.G. Frankland (eds), *International Encyclopedia of Environmental Politics*, London: Routledge, 2001; M.J. Breton, *Women Pioneers for the Environment*, Boston, MA: Northeastern University Press; Susan Buckingham (2004), 'Ecofeminism in the Twenty-first Century', *The Geographical Journal*, 170:2.

On 'essentialist' eco-feminism, see the collection of writings edited by Judith Plant, *Healing the Wounds: The Promise of Eco-Feminism*, Philadelphia, PA: New Society Publishers, 1989; Susan Griffin, *Women and Nature: The Roaring Inside Her*, San Francisco, CA: Sierra Club Books (revised edn), 2000.

For a brief overview of materialist eco-feminism, see my 'The Emergence of Eco-feminist Political Economy', *Environmental Politics*, 7:4 (1998). Fuller elaborations of materialist eco-feminism may be found in Mary Mellor, *Breaking the Boundaries: Towards a Feminist, Green Socialism*, London: Virago, 1992, and her latest book, *Feminism and Ecology*, Cambridge: Polity Press, 1997; Ariel Salleh, *Ecofeminism as Politics: Nature, Marx and the Postmodern*, London: Zed Books, 1997, 'Nature, Woman, Labor, Capital: Living the Deepest Contradiction', in M. O'Connor (ed.), *Is Capitalism Sustainable?: Political Economy and the Politics of Ecology*, New York and London: Guildford Press, 1995, and Salleh (2005), 'Moving to an Embodied Materialism', *Capitalism, Nature, Socialism*, 16:2; Asoka Bandarage, *Women, Population and Global Crisis: A Political-economic Analysis*, London: Zed Books, 1997; V. Bennholdt-Thompsen and M. Mies (1999), *The Subsistence Perspective: Beyond the Globalised Economy*, London: Zed Books.

On 'resistance' eco-feminism, see Vandana Shiva, *Staying Alive: Women, Ecology and Development*, London: Zed Books, 1998, *Monocultures of the Mind*, London and Penang: Zed Books and Third World Network, 1993; and Joan Martin-Brown (1992), 'Women in the Ecological Mainstream', *World Development*, 47: 4.

8 The environment and economic thought

Key issues

- **The environment and the 'economic problem'.**
- **Economics as social theory.**
- **The environment and the history of economic thought: classical liberal political economy.**
- **Land, labour and the enclosure movement.**
- **Material progress, poverty and economics.**
- **Economic theory, science and environmental policy-making.**
- **Economising the environment: the rise of environmental economics.**
- **Ecological economics.**
- **Green political economy.**

Introduction

> Anyone who believes exponential growth can go on forever in a finite world is either a madman or an economist.
>
> (Boulding, 1966)

This chapter takes up some historical themes discussed in Chapters 2 and 3 as the starting point for an examination of the relationship between economic thought and the environment. The reason for focusing on economic thought is defended on the grounds that, first, economic thought *is* a form of social theory. Different forms of economic thought are premised on particular analyses of society and views on alternative social arrangements. Second, different forms of economic theory are based on particular moral principles, including views and conceptions of human nature and the value of the nonhuman world. Third, and this will form the bulk of this chapter, of all forms of social theorising, economics has had the most influence on both how the environment and social–environmental interaction has been conceptualised. Fourth, and following on from the last, economic thought has had (and continues to have) a dominant position within prevailing political and economic institutions which mediate and shape the actual, material relationship between society and environment.

The main aims of this chapter are, first, to trace the changing character and role of nature in economic thought; second, to distinguish political economy from the modern discipline of 'economics' and to argue that this has important implications for the economic analysis of environmental issues; and third, to look at how 'value' in general and in relation to 'nature' in particular is and has been understood within economic thought.

The environment and the 'economic problem'

Economics, both as a discipline as well as a particular form of human practice, may be understood with reference to what has been called the 'economic problem'. The economic problem refers to the fact that human wants are infinite in comparison to finite or scarce resources (or means) to meet those wants (or ends). Hence economics may be defined as how the economic problem is resolved. At this initial stage it is significant to note that there are two possible ways of resolving the 'economic problem': either we expand resources to meet more and more wants (increase the supply), or we can limit wants in relation to (fixed or limited) resources (decrease the demand). While over-simplifying the issue greatly, it is on this basic distinction that the difference between orthodox economic theory and ecological or green critique of economic theory and practice is based. Thus a basic definition of the concerns of economics may be reduced to two: 'the efficient allocation of available resources and the problem of reconciling available resources with a virtually infinite desire for goods and services' (*Hutchinson Dictionary of Ideas*, 1994: 162).

Central to economics then is the notion of scarcity, the simple fact that resources (natural and human-made) are scarce in relation to infinite wants. Basically, humans can potentially have more wants than can ever be met by the resources available to meet them. Now while resources may include such things as 'money', 'labour' and 'capital', it is the case that natural resources (i.e. the useful things humans can get from nature) form the greatest part of the 'economic problem'. In other words, it is the natural environment around us which provides the primary resources upon which the human economy is based. At the same time, the second problem that economics deals with, as indicated above, has to do with the fact that different resources have competing uses. For example, a forest may be logged for the timber it contains, it may be preserved as a 'national park', or it may be developed as a housing estate or an airport. Providing methods to enable us to make such choices is central to economic theory and practice.

Given this stress on scarcity, resource limits and so on, it would seem that economics is close to green or ecological concerns about limits, yet this is not the case. It is also important to note that the etymology of 'ecology' may be defined as the 'economy of nature', that is, 'the study of Earth's household', while economics derives from the Greek term for 'management of the household'. Yet, throughout the modern history of economics, there has been little real sense of the intrinsic connection between the two, and the fact that economics is embedded in ecology has been largely ignored.

Economics as social theory

> It has long appeared to many people that economics is the most successful of the social sciences. It has assumed that people are motivated by money and by the possibility of making a profit, and this has allowed it to construct formal, and often predictive, models of human behaviour. This apparent success has led many other social scientists to cast envious eyes in its direction. They have thought that if they could only follow the methods of economics they could achieve similar successes in their own studies.
>
> (Scott, 2000)

While it may be seen as odd to include a chapter on economic theory in a book about the environment and social theory, I contend that unless one appreciates it as a form of social theory and practice, one cannot fully understand the relationship between the environment and social theory. The main reason for this is that in the modern period, economic forms of conceptualising, valuing and understanding the place of the environment within human society have had the greatest practical as well as theoretical effects on social–environmental relations.

Alongside scientific and technological innovation (with which it is closely associated), economic thought and practice (whether it be the 'positive' economics of this century, or the political economy of the last three) has largely created the modern social world, shaped its view of the natural world and focused attention on 'worldly' affairs. Economists, according to Heilbroner, may be seen as 'worldly philosophers', since 'they sought to embrace in a scheme of philosophy the most worldly of all of man's activities – his drive for wealth. It is not, perhaps, the most elegant kind of philosophy, but there is no more intriguing or important one' (1967: 14). Its importance as a dominant and dominating form of social theory and practice cannot be underestimated. It is also worth noting how radical economics and the economic worldview was in that the view of society which it elaborated, and the particular mechanisms it suggested for holding it together, went against most of the previous history of human societies and received social theory.

One of the central achievements of the 'economic' view of society and social affairs, originating in the seventeenth and eighteenth centuries, was to 'disembed' the economic activities of individuals and groups (production, consumption and distribution) from 'social' or 'political' regulation. That is, modern economic activity and reflection on it was impossible and unnecessary up until the economy emerged as a distinct entity and set of relationships which was separate from religious, political and customary rules, laws, rights and prohibitions. Thus the rise of economic thought must be seen as part of the complex historical processes of change which took place within European societies in the period leading up to, including and continuing beyond the Enlightenment. While not exactly correct, one could say that the birth of economics, the emergence of the 'market economy' governed by prices, voluntary exchange, and not by custom or political-religious rules, may be said to herald the beginning of the modern era:

> The market pattern . . . being related to a peculiar motive of its own, the motive of truck or barter, is capable of creating a specific institution, namely, the market. Ultimately, that is why the control of the economic system by the market is of overwhelming consequence to the whole organisation of society: it means no less than the running of society as an adjunct of the market. Instead of economic relations being embedded in social relations, social relations are embedded in the market.
>
> (Polanyi, 1947: 57)

Thus, the roots of modern economics, as approaches to the age-old 'economic problem' which faces all societies, cannot be separated from the rise of the market society (and later industrial-capitalist society) and its associated theories of **political economy** (including those political economies which were critical of industrial capitalism). As Polanyi puts it, 'Market economy implies a self-regulating system of markets; in slightly more technical terms, it is an economy directed by market prices and nothing but market prices' (1947: 43). And while in the modern world the market and 'market society' would be commonplace notions and practices, in the context of the emergence of political economy in the seventeenth and eighteenth centuries, such ideas were revolutionary and resulted in the transformation of Western societies.

More recently, the 'post-autistic economics movement' has sought to highlight the ideological and dangerous fixations of the dominant 'neo-classical' economics paradigm (Fullbrook, 2003, 2004, 2005) . The post-autistic economics movement is but the latest of a long and distinguished line of dissident or 'heterodox' economic thinking which also includes green/ecological, feminist and socialist economics, elements of which have been discussed in previous chapters. According to Fullbrook:

> For half a century neoclassical economics has hidden its ideology behind the notion that it calls *positive economics*. This is the idea that it contains no value judgements because it mentions none. Of course such a notion belongs to an intellectually more naïve age than today, but it nonetheless persists as an effective tool of indoctrination of undergraduates. The fact that neoclassical economics requires a highly restricted focus in order to maintain its atomist and determinist metaphysics compels it to make many extreme judgements about what is and is not economically important. . . . But one key example is its notion of 'economic man' – an acutely ideological term, as it emphasises some roles and relationships and excludes others; by allowing only decisions based on utility maximisation, it excludes other forms of ethics. As an economic agent, each individual acts in many roles, not just market ones, and is guided by his or her 'ideological orientation'. That orientation may be founded on utilitarianism or not. It may, for example, be based on social and environmental ethics . . . the neoclassical insistence upon the utilitarian ideology legitimises a kind of 'market ideology' and 'consumerism' that increasingly appears dangerous to society, and sidelines the debate about sustainable development.
>
> (Fullbrook, 2005: 98–9)

Although woefully (and without being conspiratorial about it, deliberately) under-represented in the standard courses taught about economics at universities and schools, and within the mainstream economics literature, this heterodox tradition does offer a coherent critique and, more importantly, alternatives (plural emphasised), to the singular and dominant hegemony of the conventional model. Part of the ideological autism of conventional economics, including its simultaneous eradication of value judgements (in its pyrrhic and disingenuous quest to be an 'objective science') yet at the same time being deeply normative and supportive of the status quo, is related to its disciplinary amnesia. That is to say, increasingly the history of economics is no longer seen as a core subject matter, reduced at best to 'soft' optional courses. This not only means students of economics 'forget' that their discipline arose out of a much older disciplinary tradition of natural philosophy, theology and 'political economy', but that such founding fathers of modern economics as Adam Smith would find modern economics incomprehensible. Modern students of economics would perhaps know Smith as an early proponent of the 'free market' economy, as gleaned from selective readings of his *Wealth of Nations*, but such an appreciation of Smith is distorted and one-sided, since a knowledge of his *Theory of Moral Sentiments* is needed to properly contextualise his deeply moral, cultural and politicised mode of economic thinking. But such scholarly standards have long ago deserted neo-classical economics.

An ideological reading of modern economic thinking shows it to construct a fantasy world of monetary valuation and human productive activity which has only tenuous links to the real source of all economic activity – namely the ecological dynamics of nature and the biological and social needs of human beings. The global and globalising economy operates in a fantasy world of '24/7' work and production patterns, and quarterly returns which fly in the face of the ecological time of the seasons and regenerative capacities of natural resources as well as the biological and psychological time human beings need to not only function efficiently but live well and prosper, as eco-feminist and eco-Marxist political economy points out (Benton, 1993; Mellor, 1997). The obsession with speed, rapidity and mobility within the contemporary global economy may be seen in the marketing of 'energy-boosting' drinks and snacks, advertisements promoting remedies for illness which focus not on the illness to the individual, but on the productive days or crucial deals lost as a result of illness. Such a view of orthodox economics clearly places it as a form of social theory in the service of a right-wing political project, elements of which were outlined in Chapter 5. Others, such as the Canadian political theorist John McMurtry, would go further and claim that the orthodox economics taught in universities and used to structure and inform state and business decisions in investment, production and employment and so on is an economics in the service of, in the titles of one of his books, the *Cancer Stage of Capitalism* (McMurtry, 1999). The argument here, and one which echoes elements of the *Limits to Growth* analysis, is that growth for growth's sake is the logic of the cancer cell, and that is precisely what we have today in orthodox economics promoting growth which does not add to quality of life or long-term sustainable development (Barry, 2006a, 2006b).

If one reads the standard literature of neo-classical economics, from undergraduate textbooks to the major economic journals, one would come away with a sense of a discipline and body of human knowledge which simply describes the world as it is, dealing in facts and 'truths'. Economics is often described as the 'queen of the social sciences' (see the quote from Scott above) since it commonly projects itself as the one social science discipline which is held to come close to the standards of enquiry and investigation of the natural sciences. However, not only is conventional economics not an objective, neutral and value-free 'science' but it is deeply value laden and as normative as other forms of knowledge about the human social world.

To test whether economics is a science would simply require us to ask a number of economists to predict, using whatever advanced and computer-aided econometric models of their choice, the rate of inflation or unemployment rate for this time next year and ask them to bet their house on it. Not only would we get different answers but there are serious doubts as to whether any of them would risk their house, thus indicating a less than confident belief in the predictive powers of their models. Economics, even when it surrounds and adorns itself with dense mathematical formulae and ever more complex econometric regression models, cannot hide its unscientific character in not being able to predict the phenomena that are the objects of its enquiry.

In short, in addressing the status of economics as a science, critics argue that we have to say loudly and often, 'the emperor has no clothes' (Barry, 2006c). That is, the dominant economic paradigm is *not* a science but has been falsely trading on its reputation as a 'hard social science', in contrast to the 'soft' social sciences of sociology, cultural studies and politics, and the non-social scientific disciplines of the humanities such as philosophy. This lack of predictive capacity (which I am taking as constitutive for a body of knowledge to count as a 'science' in the manner of the natural sciences) does not mean that I view the knowledge produced by conventional economics as useless. Conventional economics has produced knowledge and findings which have been useful. The point here is that economics cannot persist in the pretence that it is a value-free and predictive social science and that consequently its usefulness and rationale needs to be grounded upon other principles.

That economics is not value free, objective or neutral is a standard argument levelled against orthodox or neo-classical economics. The normative character of conventional neo-classical economics may be gleaned from any undergraduate textbook on the subject. These textbooks talk of a world of 'rational consumers' and 'utility-maximising individuals' with determinate 'consumer preferences' in a world of 'perfect competition' and so on. The '*homo economicus*' that forms the bedrock of much neo-classical economic thinking (particularly micro-economics) is not only a fiction (a fantasy to return to one of the features of autism discussed above) but a deeply normative and not to say ideological fiction (Dryzek, 1996). The human subject of economics, that is, is not described or simply reflected in economics, but conventional economics actively *creates and prescribes* this human subject as an ideal to be attained. In other words, what economics is doing is not describing how the world and human beings are (facts) but mandating or prescribing how the world and human beings ought to be (values or ideology). Now while all forms of human knowledge from physics to philosophy make generalising and simplifying assumptions, within neo-classical economics there is a clear sense that the *homo economicus* and her associated human economic experience discussed here is simply how human beings are and how the economic world is. That is, these models are not simplifying assumptions so much as capturing the essential character, motivation and *modus operandi* of human beings when they enter economic relationships. Human beings as revealed by neo-classical economics qua 'homo economics' or 'economic man' (the gendered connotations of this are noteworthy and of course criticised by eco/feminists as discussed in Chapter 7) are essentially selfish, individualistic, hedonistic and possessed of desires that explain their behaviour, and these desires can never be satisfied.

A key way in which to undermine or demonstrate the 'unreality' of the rationalist model of the self is to highlight the lack of dependence and interdependence within such models. The rational egoist of orthodox, neo-classical economics is not only rational, self-interested and utility-maximising, but independent from social, biological and ecological needs and relationships. As eco-feminists have argued, *homo economicus* is invulnerable, free from biological (socially mediated through gender relations) concerns directly since these are 'taken care' of within the non-economic, non-political domestic or private sphere. To paraphrase Benton (1993), this 'fiction' (which is essential to the dominant economic model) denies that *homo economicus* is 'biologically embodied'. It also denies 'ecological embeddedness' in that there is no sense of the dependence of the human economy upon the wider 'economy of nature'. In short, *homo economicus* exists neither in relationships of vulnerability or responsibility with fellow humans/citizens nor in a relationship of dependence upon the natural world. *Homo economicus* is, to use Kantian terms, a 'noumenal' (abstract) rather than a 'phenomenonal' (real) being. And it is no coincidence that this neo-classical economic, 'unreal' model of the human self is also found in many liberal political theories, particularly in neo-liberal political economic goals.

Ultimately we need to recover *homo sapien* in such a way as to integrate it with *homo economicus* (there is no suggestion that we simply abandon this fiction or banish the economists, as Plato wished to banish the poets in his ideal political society) to create what Dryzek has called *homo ecologicus* (Dryzek, 1996). There is a place for *homo economicus* and economic rationality, but it needs to be placed within political-normative bounds so as to prevent its tendency to 'crowd out' or devalue other forms of behaviour, valuation and interaction. However, we need also to note that *homo economicus* is not just a morally suspect model of human beings, but the behaviour of *homo economicus* (most noticeably in the form of individual material consumption as the 'good life', that is, *homo economicus* qua consumer) is also a constitutive aspect of collective forms of identity (particularly in the 'overdeveloped' world) and a valued and desired activity. Here the 'good life' becomes the 'goods life' (Doran, 2006), and the critique of the simplisitic and one-sided view of the good life for humans has long been a central feature of critiques of the ideological and conservative bias of orthodox economic theory and practice (Barry, 2006c). However, there are signs that those within the mainstream economics discipline have recognised that there are serious problems and some ackowledgement of the discipline's autistism. According to one leading UK-based professor of economics:

Modern economics is sick. Economics has increasingly become an intellectual game played for its own sake and not for its practical consequences for understanding the economic world. Economists have converted the subject into a sort of social mathematics in which analytical rigour is everything and practical relevance is nothing. To pick up a copy of *The American Economic Review* or *The Economic Journal* these days is to wonder whether one has landed on a strange planet in which tedium is the deliberate objective of professional publication. Economics was once condemned as 'the dismal science' but the dismal science of yesterday was a lot less dismal than the soporific scholasticism of today.

(Blaug, 1997: 3)

It is perhaps more than a coincidence that Marx's quip about Malthus that he went in search of human nature and found the English shopkeeper may, with some modification, be said to be as true of today's conventional economists, who also go in search of human beings and come back with *homo economicus*, a dangerous fiction and a one-sided distortion of humanity. Or rather, kept to its proper subject area, say, decision-making with limited income about the choice of this or that piece of house furniture, or making choices about average financial returns from different investment or employment decisions, *homo economicus* is harmless enough and indeed may be said to offer interesting analytical explanations of human behaviour. However, when taken outside of its specific remit, the distortions of *homo economicus* are all too apparent (to everyone it seems except economists). *Homo economicus* is a view of humanity distorted to fit pre-decided abstract categories and modes of analysis rather than being based on analyses of how actual human beings are and actual economic transactions take place. It takes humanity as it ought to be rather than how it is in order to fit the theory. Conventional economics does not take into account psychology, sociology, anthropological studies or politics and is particularly resistant to interdisciplinary approaches. In this sense, the dominant economic paradigm does not reflect or study human economic behaviour, but rather in describing that behaviour seeks to bring it into being. In this manner, the decisive power of the dominant neo-classical economic paradigm and its neo-liberal, free market policy prescriptions lies in its role as a political and ideological project that endeavours to create a social reality which it suggests *already* exists, when in fact that reality does not, in fact, already exist, but is being created (and resisted) all the time, as in the global justice and other anti-capitalist movements discussed in Chapter 6.

The environment and the history of economic thought: classical liberal political economy

In the history of economic thought (that is, reflections on the 'economic problem' as described above, which has a longer tradition than modern economics), one thing stands out about the role of the natural environment: namely, its reduction to a set of resources to be exploited for human economic ends. From an economic point of view, the natural environment has instrumental value; that is, it is useful insofar as it can be exploited in fulfilling human wants. What is problematic about this view is *not* the instrumental valuation of the natural world that economic thought expresses. Rather the problem is the dominance of this view which 'crowds out' alternative non-economic forms of valuation and ways of relating to and thinking about the natural environment. A related problem is the misperception within modern economics that there are no 'natural limits to economic growth'. Finally, there is an equally unjustified assumption at the heart of modern economic thought to the effect that natural resources, the contribution nature makes to the human economy, is a 'free gift'.

One of the first to outline the modern view of the economic problem was Locke. As discussed in Chapter 2, Locke was one of the first social theorists to offer a theory of value in which human labour was central, and also gave the classical liberal defence of private property of land and environment, which was one of the main tenets of the emerging market society. What marks Locke as a political economist is that his economic views were not motivated by or based on any idea that what he was offering was an 'objective' or 'value-free' view of the human economy in relation to the natural environment. His defence of private property in the external natural world was in part motivated by a view in which such exclusive, individual ownership would offer some protection of the individual from arbitrary interference by political authority. He also acknowledged that privately owned land was more productive than unowned land left in a 'natural state'. That is, the unowned, and thus 'untransformed' environment, was 'valueless', since all value derives from human labour. Thus alongside the 'economic' argument that owned land is more productive than unowned land, we need also to see that for Locke (and for the classical liberal tradition as a whole), the defence of private property was also premised on a defence of individual liberty. Thus for Locke, private ownership was not only more efficient, in economic terms able to produce more goods and services, but also a central part in the creation of a sphere of private, individual freedom from political authority. Here, the natural environment (understood principally as ownable 'land') and property relations over it, are defended on economic and political grounds. Particular ownership relations over the external environment in Locke's social thought become central in outlining the classical liberal view of society, namely the importance of private property for social order and material prosperity.

Second, an important aspect of this classical liberal position may be found in the logic of Locke's view of unowned land as 'unproductive' and his explicit defence of material progress as central to the development and advancement of human society. This 'progressive' element in Locke's liberalism (which had Christian aspects) is simply another example of the historical 'spirit of the age' which puts Locke in the company of other writers and thinkers such as Francis Bacon, René Descartes and Isaac Newton. This 'spirit of the age' was the belief in material progress premised on the more efficient exploitation of the natural environment, as a result of the application of science and technology.

Third, given the Christian context within which Locke was writing, it is important to see how the appropriation and transformation of the natural environment could be justified within this Christian and often Puritan context. Within the dominant Christian culture of Europe of the time, it was accepted that the world (including humans) was 'God's Creation'. It is important to realise the theoretical and practical implications of this on regulating how the natural environment was viewed, valued and used. The first thing to notice is that since God (not humans) made the environment, they have no right to destroy it. As Locke himself put it, 'Nothing was made by God for man to spoil or destroy' (*Second Treatise on Government*, IV: 31). So while the natural environment was not the property of humans in the sense that God and not humans had made it, Locke, as we saw above, did develop a justification for humans to claim parts of the natural environment ('land') as private property.

Without going into the precise detail and logic of Locke's justification of private property in the natural environment, what is important to note is that he used a particular reading of Christian virtue in order to ensure that this idea of human private property in land did not offend against the dominant and politically central idea of the natural environment being part of 'God's Creation'. Passmore (1980) suggests that the way Locke (and others after him) could square the circle of radically transforming nature (on the basis of establishing private property in it) was by justifying human transformation of (and property in) the natural environment on the grounds that humans were simply 'perfecting' or 'improving' nature. Human alteration of the world could be justified on the grounds of constituting an 'improvement' of creation for the glory of God (Passmore, 1980: 28–32). In this way, the early economic view of human relations to the natural world (which was still heavily circumscribed by Christian ideals and tradition), as a potentially productive 'resource' which requires land to be transformed into private property which can then be 'improved' (i.e. exploited and used) by humans, was tied up with the whole (Christian and particularly Protestant) notion of 'progress'. And in Locke we see that one of the consequences of 'progress' and social development is an alteration in how the natural environment is viewed and used: that is, the necessity of the external environment becoming a commodity and an economic resource, which like any other commodity could be given a money price and bought and sold on the open market.

In this way we can observe how the 'disembedding' of the 'economy' from other social spheres and regulations (religious, cultural and political), and the emergence of a distinctly 'economic' space and motive, took place. This 'commodification' of the natural environment, which was a central aspect of classical political economy (from Locke to the eighteenth century), was thus a necessary prerequisite for the rise of modern industrial capitalism, as well as being one of the key structural features of that socio-economic order up until the present day.

Land, labour and the enclosure movement

It is as land that the natural environment was equated in the history of economic thought, up until the first stirrings of environmental problems in the 1960s. As it was with land, its transformation into a 'commodity' like any other to be bought, sold and exchanged, that characterises much of the pre-Enlightenment period's concerns. Much of the character of pre-Enlightenment feudal Europe was based on its rural, land-based socio-economic relations (peasant–lord relations, the guild and apprenticeship system), institutions (the monarchy and aristocracy) and customary rules (such as the peasants' right of access and use, not ownership, of common land). All of this was to change both in theory with the emergence of political economic thought, and in practice with the emergence of the self-regulating market or commercial society of the eighteenth century. But it is also important to remember the explicit link between economic theory and actual economic practice of this time. In other words, the pre-eminence of economic thought in influencing political decisions of the day, its role as the legitimating ideology or rationalisation for the tremendous and revolutionary changes in European societies, cannot be overemphasised. Just as a combination of custom, tradition, Christianity and monarchical political requirements furnished the pre-Enlightenment, feudal social order with its legitimating self-understanding, so economic thought functioned in a similar way to justify, defend and legitimate the radically different social and economic order of industrial capitalism. And unlike the later attempt to turn economic thought into an objective 'science' (the idea of 'positive economics'), the early history of economic thought is made up of competing schools of 'political economy', all of which share the same object of study: the political and economic basis of industrial capitalism. Political economy was *the* social theory of this historically unprecedented form of society. It reduced the seeming chaos of this emerging, bustling industrial-capitalist system, the trading, factory system, international trade and exchange, the growth of unknown levels of material wealth alongside great poverty, the urbanisation of society and the decline of the rural character of society, to its barest essentials. And in the works of classical political economy from Locke to John Stuart Mill we may find the clearest expressions of social theorising about this unique, endlessly changing and dynamic form of society.

One of the key ways in which economics captures the character of early industrial capitalism, as well as providing an economic reason for creating that character, is in its attitude to the land. For the market system to work, land, labour and capital had to be 'freed' or 'disembedded' from non-economic restrictions, customs and rules. In short, the industrial-capitalist system required that land, labour and capital be 'free' to move where they were economically required, and where the market dictated they should go. Another way of putting this is that these 'resources', in order to be resources in the sense required by the new economic system, had to be seen as commodities, things that could be bought, sold and exchanged. From the early 'economic' point of view, the natural world was simply a set of resources, and the most important of these resources was land. However, the problem was that land in the pre-Enlightenment period was not viewed solely as an 'economic resource'. Rather what we now call 'real-estate' land as sellable and exchangeable property for money was something alien to feudal, pre-industrial-capitalist society, though of course this did not mean that it was not owned (by an aristocratic landed class) or productively used, as may be seen in the discussion of the landowning classes' reaction to the industrial revolution in Chapter 5. As Polanyi puts it, 'What we call land is an element of nature inextricably interwoven with man's institutions. To isolate it and form a market out of it was perhaps the weirdest of all the undertakings of our ancestors' (Polanyi, 1947: 178). He goes on to point out that traditionally land and labour form part of the same whole; one is inextricably bound to the other:

> The economic function is but one of many vital functions of land. It invests man's life with stability; *it is the site of his habitation; it is a condition of his physical safety; it is the landscape and the seasons.* We might as well imagine his being born without hands and feet as carrying on life without land. And yet to separate land from man and organize society in such a way as to satisfy the requirement of a real-estate market was a vital part of the utopian concept of a market economy.
>
> (Polanyi, 1947: 178; emphasis added)

Thus, according to Polanyi, in the feudal world, 'land' was not viewed the way it is understood in the modern age. Land was 'home', 'place' (as opposed to just physical space or resource). In short, 'land' was the milieu of everyday life in feudal thought and experience. How did this new idea of land as an economic resource to be bought and sold on the market arise? While an incredibly complex issue, one of the key ways in which land was made into an economic resource was by dissolving the cultural and social context within which it was embedded, a major part of which was outlined in the previous section on Locke and early classical liberal political economy. Within the pre-Enlightenment social world, 'land' was not an economic resource in the required sense. This is not to say that it was not used in ways that we would say are recognisably 'economic'. Rather, and this expresses the character of this pre-modern, feudal social order, the economic functions and uses of the land were enmeshed in a whole series of cultural, social, political and religious rules and customs. That is, the idea of land as a distinctly 'economic' resource was unintelligible in the pre-capitalist, feudal era.

In England, for example, while peasants did not own the land, as commoners they had customary rights of access to use common land. What happens in England is that the emergence of 'land' (a particular understanding of the natural environment) in its modern economic sense of sellable, private 'real estate' emerges after a long and bitterly resisted process of 'enclosures' which, over a period of two hundred years or so (from the seventeenth to the mid-nineteenth centuries), transforms the commons from a part of the cultural fabric of rural life into private property and the exclusive economic resource of the landowner. To use an inexact modern expression, the enclosure of the commons was one of the first acts of *privatisation* of previously 'publicly' shared (if not commonly owned) property. Of course, a modern-day example of this is the 'free market environmentalism' discussed in Chapter 5 which promotes 'enclosure' and privatisation as the default and preferred solution to environmental and resource conservation problems.

The enclosures were viewed as a necessary step to take in order for social development, progress and civilisation to proceed. The removal of people from the land, enabling landowners to 'develop' it and thus secure a greater return on their investment, was not only part and parcel of the birth pains of removing the 'land' from tradition and custom, but also in the process created landless peasants who would form the urban working class in the rapidly urbanising areas. At the same time, the 'guild system' and craft-based production were eventually eroded and replaced by the modern factory system, centralised and hierarchical forms of production in which the 'worker' had little direct say in what was made or how it was produced.

Yet, as the history of the enclosures in Britain demonstrates, this radical change of a whole way of life, its set of social, cultural and political institutions, the transformation of land and labour into exchangeable, marketable resources or commodities, did not occur either overnight or peacefully. While such changes to how the natural environment was viewed and used, the creation of new 'economic' modes of acting and behaving towards both land and labour were seen as necessary for social progress, for the majority of the people at the time, these and other changes were 'costs', and they were the main 'losers' from this unprecedented erosion and destruction of a settled, familiar culture and way of life. And faced with such changes, the period leading up to and including the industrial revolution was marked by popular struggles, uprisings and resistance to the creation and maintenance of the emerging industrial-capitalist social system, such as movements featuring the Diggers, the Levellers and the Luddites.

The important point here is to note how different ideas of the natural environment and land played a central part in the transition from the feudal to the modern industrial-capitalist socio-economic order. The main struggles concerned competing ideas of the land as 'private property', a commodity to be bought and sold, versus the claims of those for whom the land was not just a public resource, but also 'home' and an essential part of the social and cultural fabric and a constitutive element of collective identity.

Material progress, poverty and economics

What marks the political economy of the time as a form of social theory is the explicit recognition that the changes within European societies, like all forms of societal change, would produce 'winners' and 'losers'. Much of the political upheaval associated with the emergence of the industrial-capitalist system may be seen as struggles between winners and losers. Landless, homeless peasants having been removed from the land, and forced into the cities with their labour-saving technologies, fought long and hard against the bright new world of early industrial capitalism. These losers smashed machinery (the Luddites discussed in Chapter 2), organised themselves against the emerging modern, urban economy, and, at least initially, sought to return to the security and relative prosperity of their previous rural and feudal way of life. As discussed in Chapter 5, this negative reaction to the modern industrial world was also shared with conservative-minded landed classes. To the economic 'spirit of the age' such disturbances, and the poverty, insecurity and degradation which occasioned them, were a necessary 'price' to be paid for the new capitalist social order. As Heilbroner explicitly notes, 'The market system with its essential components of land, labor, and capital was born in agony – an agony that began in the thirteenth century and did not run its course until well into the nineteenth' (1967: 30).

Some early social theorists of industrial capitalism (e.g. Jeremy Bentham) were explicit. In his book *Principles of Civil Code* he wrote, 'In the highest stages of social prosperity the great mass of the citizens will most probably possess few other resources than their daily labor, and consequently will always be near to indigence' (quoted in Polanyi, 1947: 117). This is a theme also found in Malthus who argued that the 'poor' (the new urban working class) ought to be kept at subsistence wages, since higher wages will only encourage greater increases in population of the 'lower orders' and lead to a catastrophic imbalance between human population levels and agricultural food production, as well as threatening the established social order. Hence the 'naturalism' so typical of other ideological defenders of the *laissez-faire* capitalist system, such as Herbert Spencer and the Social Darwinists (discussed in Chapter 3), had a history in other 'liberal' strands of social thought of the late eighteenth and early twentieth centuries. By naturalism was meant that the market system was not only the *natural* outcome of human behaviour, but also conveyed the claim that this system *naturally* required poverty as part of its own unalterable '*natural* order'.

The link between progress and poverty was for many in the eighteenth and nineteenth centuries regarded as perfectly 'natural', and indeed as Polanyi puts it, 'Poverty was nature living in society' (1947: 84). Thus was the glaring paradox between the unprecedented productive potential of industrial capitalism in creating the most materially affluent societies the world has ever seen alongside the persistence (and indeed deepening) of poverty within such materially affluent societies. For those who either experienced this poverty and its consequences (primarily the urban working class), or who were horrified at the social and environmental harms that industrial capitalism had caused (such as the Romantics discussed in Chapter 2), a return to the material security, and relative environmental harmony of the rural, feudal social order, was an attractive proposition. For many critics of the new capitalist social order, it was evident that it could not eradicate poverty, since the economic system required the poverty of the mass of its population in order for them to engage in wage labour, which for the majority of the newly formed working class was an alien concept and one which they resisted. As Polanyi (1947) explains, although as the industrial revolution and the development of capitalism continued more people were (on some measurements, such as health and not simply financially) better off than before, there were serious costs and socio-economic changes which made people worse off:

> In spite of exploitation, he might have been financially better off than before. *But a principle quite unfavorable to the individual and general happiness was wreaking havoc with his social environment, his neighborhood, his standing in the community, his craft; in a word, with those relationships to nature and man in which his economic existence was formerly embedded.* The Industrial Revolution was causing a social dislocation of stupendous proportions, and the problem of poverty was merely the economic aspect of this event.
>
> (Polanyi, 1947: 129; emphasis added)

In this way, the struggles against the industrial-capitalist order, which were often linked directly or indirectly with the 'defence of nature' or which advocated a 'back to the land' alternative to the urban, industrial-capitalist society, are important features of the conflict between different understandings and aspirations advanced by different classes, ideological positions, writers and activists concerning the relationship between 'environment', social progress and social order. At such times of crises, uncertainty and conflict over social development and change, appeals for a return to a more secure, rural and traditional past were common.

It is interesting to note that such 'back to the land' alternatives, which are an important forerunner of some contemporary 'green' alternatives to 'late' capitalism, continued right through the nineteenth century in utopian, socialist and anarchist political thinking (Gould, 1988), and may also be observed in the mid-war period throughout Europe (Bramwell, 1989). *The important point to note is how in political struggles between defenders of industrial capitalism and those critical of it and proposing an alternative, particular construction of 'environment' and 'nature' as well as justifications for particular sorts of relations to the natural environment are central parts of these political (and cultural) struggles.* The historical argument here thus complements the discussion outlined in Chapter 3 dealing with the different conceptions and ideological uses of 'nature', 'natural' and 'environment' between classical liberalism (such as Social Darwinism) and critiques of liberal capitalism (such as communist or left-wing anarchism).

The following section will discuss other ways in which conceptions of 'nature' and the 'environment' were (and are) used in political economic thought to justify or establish particular arguments, positions or practices.

Economic theory, science and environmental policy-making

In keeping with the general dominance of orthodox economic thought over public policy-making in liberal democracies, it comes as no surprise to see its centrality in environmental policy-making. Of all forms of human thought, it is economics which almost since its birth as 'political economy' and its later transformation into 'positive economics' at the end of the nineteenth century that has had the most lasting effect and hold upon political decision-making. From early forms of political economy, such as English mercantilism and French physiocracy, there has been a close, if not symbiotic, connection between 'orthodox' economic thought (i.e. broadly in support of the industrial-capitalist economy and its political requirements), and the nation-state, its institutions and decision-making processes. Unlike 'speculative philosophy', tradition, custom or religion, which had previously been the main sources of knowledge used by political authorities, economics was always regarded as a 'practical science' perfectly suited to the *realpolitik* of statecraft and political decision-making.

Given its explicit recognition of the disparity between human wants and limited means to fulfil them, economics has always presented itself, and has been perceived as uniquely equipped to deal with 'tough choices', to inform decision-making on mutually exclusive outcomes. And as it moved from political economy to economics, economic thought has sought to refine its scientific character and present itself as the objective, dispassionate study of the 'economic problem' within the context of modern, complex, capitalist societies. Even anti-capitalist political economies, such as Marxism, shared the character of being 'scientific', as discussed in Chapter 3.

Now, with regard to the 'spirit of the modern age' which has been a constant point of reference so far, to be 'scientific' was (and largely still is) to be considered progressive, as well as trustworthy, rigorous, objective and dispassionate. And it is in this privilege accorded to science and the 'scientific method' that we can find roots of the dominance of economic forms of reasoning and thinking within contemporary capitalist liberal democratic nation-states. Its predominance as the central form of knowledge (along with natural science) used by state actors, bureaucracies and leaders to make decisions, implement policies and propose reforms, while of course not excluding other forms of knowledge and political judgement, has had a profound effect on the environment and social–environmental relations. As Francis Bacon, one of the founders of the modern scientific method and worldview, noted, 'Knowledge is power', and this is particularly true in respect of the natural sciences and economic science. At the same time, economic forms of thinking do not simply express themselves within state policy-making, but can also seep into 'ordinary' or 'common-sense' modes of thinking. While the powerful effect of economic reasoning on modern perceptions of the environment and its official (and unofficial) influence on state decision-making which affects the environment cannot be overestimated, I will limit my discussion to a few salient points.

The first and most obvious point is that economics as a self-styled 'science' modelled itself on the physical sciences, particularly physics and mathematics in the late nineteenth century. While it was a 'social science', its rigorous scientific methodology made it a 'hard' rather than a 'soft' form of theoretical enquiry. What is meant by 'hard' and 'soft' is that economics claimed, like the physical sciences, to be able to explain, predict and measure its subject matter, whereas the 'soft' forms of social enquiry, philosophy, sociology, politics and history could at best 'interpret' and give meaning to rather than describe and predict causal relationships with the accuracy of economic methodology. Thus economics became (and even today is often known as) 'the queen of the social sciences'. While it borrowed the scientific method of inquiry, it also absorbed the instrumental view which science had of the natural environment since Bacon and Descartes. As we noted in discussing Habermas in a previous chapter, for him, what marks the scientific method is that it requires an instrumental attitude towards the natural environment in order to produce knowledge which we can use to better understand, explain and exploit it.

A second and equally important issue is how political debate over environmental issues within public policy is heavily influenced by economic forms of reasoning and argumentation. Precisely because of the dominance of economic considerations in public policy-making, environmental issues are often translated into 'economic' problems and courses of action pursued on the basis of the economic costs or benefits of the environmental issue in question. For example, in the classic case of environmental protection versus development, it is very often the case that environmental campaigners have to couch their case in economic terms and language as well as have an economic reason for environmental preservation. From campaigns to save the rainforests on the basis of the unknown medicinal substances or genetic knowledge that may be lost, to anti-roads protesters arguing their case on the basis of the drop in tourism or decline in town shopping, it seems that public policy-making requires participants in the policy-making process to adopt economic forms of reasoning and justification. To base one's case for environmental protection on the intrinsic value of the environmental space or landscape in question (as opposed to some economic-instrumental value it may possess) is to adopt a strategy that would be difficult to persuade or influence the environmental policy-making. In other words, there is a lot of strategic advantage in using economic forms of argumentation in advancing the case for environmental protection, since one is speaking a language which politicians and policy-makers understand. As David Pearce has noted:

> politicians and their advisors are engaged in the activity of trading off environment against economic activity. . . . *Defending the environment means presenting the arguments in terms of units that politicians understand . . .* adducing evidence that the environment does matter in economic terms is important, especially as the record of decision-making in the absence of such valuations is hardly encouraging for the environment.
>
> (1992: 8; emphasis added)

This is not to deny the importance of economic considerations, but simply to note how an economic approach to and understanding of social-environmental problems can (and does) 'crowd out' non-economic forms of environmental valuation and argumentation. The privileged position occupied by economics in environmental policy-making has the effect of drowning out other 'voices', other forms of reasoning, valuing and thinking about the environment.

Slightly over-exaggerating (but then one could say that exaggeration is when the truth loses its temper!), economic reasoning, methodology and forms of valuing the natural environment may be regarded as not simply the *language of power* in policy-making, but the *grammar of power*. What is meant by this is that economic theory functions as the dominant way in which environmental policy-making is debated, thought about and ultimately decided. From the post-structuralist perspective of the influential French thinker Michel Foucault, neo-classical economics becomes a knowledge/power discourse (Foucault, 1980) which shapes the way we think, act and decide policy. In Foucault's thinking neo-classical economics becomes a 'truth regime' and constitutes the very 'rules of the game' in the same way as grammar is the rules for the correct use of language. Thus those who either do not know or refuse to accept this particular grammar (such as non-economic arguments for environmental preservation or those economic perspectives critical of the neo-classical framework) are at a severe disadvantage in trying to influence environmental policy-making within the current institutional and power/knowledge framework. A good illustration of this latter point is the success and rise in prominence of 'environmental economics' over the past decade or so.

Economising the environment: the rise of environmental economics

Environmental economics holds that the environmental problems facing society may be solved by a suitably regulated market and using the tools and reasoning of neo-classical economics. The most prominent exponent of this view is David Pearce and his colleagues, who outlined the environmental economics approach in their widely read book, published in 1989, *Blueprint for a Green Economy* (which was a report commissioned by the then Conservative Secretary of State for the Environment, Chris Patten). This document and subsequent writings from this 'green' form of economics is characterised by its aim to translate environmental problems into economic problems. A central argument of this environmental economics position is that environmental problems arise due to 'market externalities', a market imperfection which means that environmental costs, such as pollution, are outside of the market mechanism. Pollution thus constitutes an 'externality' because it does not have a price (or cost), and as there is no market in pollution, a price is unlikely to emerge. By viewing pollution as an economic problem and using economic techniques to calculate the costs it creates in terms of ill-health leading to fewer working hours, or greater costs to the health services, its negative effects on house prices and so on, environmental economics seeks to find the 'economic price' of pollution. Once this is arrived at, it may be used in environmental policy-making, such as imposing a tax on pollution which reflects its cost. In this way, the 'externality' (a cost which is borne by society as a whole or a local community) through the imposition of a tax can be 'internalised' by the polluting industry or firm. This technique is also extended to calculating the economic benefits of environmental goods and services, viewed as 'natural capital' (Pearce *et al.*, 1989).

Now while clearly not without its merits (and it cannot be over-emphasised how much of an advance environmental economics is over orthodox approaches to environmental problems), it does have many problems. One of these is that it depends on seeing environmental problems as economic ones, and also translating all values into economic costs and benefits. *It tries to deal with social-environmental problems by reducing them to economic ones: it economises the environment rather than ecologising economics.* The environment is viewed as an economic resource, providing economically important environmental goods and services, just as early economic thought commodified the land. As a basis for environmental protection therefore, one has to ask whether this approach provides convincing reasons for the preservation of economically marginal environmental goods, species, ecosystems, and so on. At the same time, there are many who question the whole idea of viewing the environment as 'natural capital' on the grounds that preserving natural capital is *not* the same as preserving the environment. As Holland has pointed out, 'insofar as there is a distinctively *environmental* crisis, it lies in the fact that the natural *world* is disappearing, not in the fact that natural *capital* . . . is disappearing' (1997: 127). Thus the preservation of natural capital will not necessarily lead to the preservation of particular parts of nature, ecosystems, species and landscapes. Hence if one wishes to preserve the natural environment, then arguing for it in terms of 'natural capital' may not be the best way to do this.

A third issue is that the dominance of this economistic form of environmental valuation and reasoning encourages a reductionist and atomistic approach, which is particularly unsuited to the holistic and integrated approach that is needed to deal with most social-environmental problems. Here in terms of environmental policy-making, it is not just the economistic forms of looking at the issue that is problematic, but also the hierarchical, segregated bureaucratic structure of the state institutions which make and implement environmental policy-making. As Dryzek (1987) notes, modern, centralised state institutions and the econo-mistic forms of formulating environmental policy they use are inappropriate to deal with environmental problems which require holistic and integrated solutions. According to Dryzek, what tends to happen in centralised forms of environmental policy-making is a marked tendency towards 'problem displacement' and not 'problem solution'. Take pollution, for example. Solid pollution (domestic and industrial sewerage or garbage) can be transformed into water pollution (by dumping it into rivers or the sea) or air pollution (by incinerating it). But the point is that this simply 'displaces' the pollution, transforming it from one media to another, without actually solving the pollution problem.

Fourth, economic views of environmental issues encourage a short-term perspective on environmental issues (based on the idea of 'discounting' future economic benefits or costs, i.e. a pound today is worth more than a pound next year or in ten years' time). This short-termism within economic forms of looking at environmental problems is often associated with and backed up by a belief that any future environmental problems that may be caused by current decisions will be solved by technological developments. This 'techno-fix' view, as O'Riordan (1981) terms it, means that we do not need to worry about long-term environmental effects (not just because, as J.M. Keynes, one of the greatest economists of the twentieth century, said, 'in the long run we're all dead'), but rather because the future will take care of itself by virtue of the continuing progress in science and technology in identifying, preventing and coping with environmental problems. While there have been examples of successful technological solutions to environmental problems, this does not offer a strong position to suggest that past successes will be repeated in the future to deal with environmental problems of which we are only dimly aware. At the same time, past experience of technological solutions to environmental problems has demonstrated that they often cause other or worse environmental problems ('displacing' rather than 'solving' the problem in Dryzek's terms above). For example, building higher smoke-stacks on fossil fuel-based power-stations on the east coast of Britain may disperse the pollution higher into the atmosphere, but it also has the effect of causing **acid rain** in Scandinavian countries resulting in ecological damage to lakes, wildlife, soils and forests (Elliot, 1997: 25). There is thus a strong argument to suggest that this 'technologically optimistic' underpinning of economic thought, while it cannot and should not be rejected completely, should, at the same time, not be viewed and used as a panacea for all environmental problems. Rather, at times, it has the character of a 'belief' or 'hope' rather than a firm or self-evident proposal, and one that, though it may sometimes be a necessary part of an overall solution, is very rarely a sufficient condition for arriving at one.

Finally, and perhaps most importantly, there is an in-built presupposition for continuous economic growth within most forms of economic theory which goes against long-term environmental sustainability. In this assumption, this idea of 'progress' – as increases in the production and consumption of goods and services – economics is simply articulating the more general and widespread idea of progress and the 'good life' which can be traced back to the Enlightenment as we saw in Chapters 1 and 2. In this way, economic thought may be regarded as the purest and clearest expression of the modern 'spirit of the age': the belief in a particular linear view of progress. The problem with this view of progress is that it has been associated with increasing levels of environmental degradation and the proliferation of environmental problems from local ones such as soil erosion to global ones such as climate change, not to mention the human cost in terms of increased stress, breakdown in traditional family life and cultures.

There is also another reason for the in-built bias towards growth within orthodox economics. This has to do with the social theory of orthodox economics which favours and requires socio-economic inequality. On the basis that it offers 'neutral' or objective analyses of the economy, neo-classical economics simply says that it provides the models and predications of how the economy will or should work; the distribution of income, wealth and so on are not its concern. This leads to a bias towards growth. As Mulberg explains, 'growth becomes a vital issue because of the lack of an adequate (or indeed any) distributional theory within mainstream economics. . . . In practice economic growth has acted to deflect questions of redistribution' (1995: 147; Barry, 1999a: ch. 7).

All this seems to add substance to Milton's (1996) view of the innate conservatism of economics. According to her, 'While sociology and political science (and for similar reasons, cultural anthropology) are inherently subversive, economics, at least in its neoclassical form, cannot be' (1996: 72). However, there has been a recent attempt to make economics more 'subversive', namely ecological economics, to which we turn next.

Ecological economics

Building on, though largely arising out of a critique of environmental economics, has emerged ecological economics. As Juan Martinez-Alier, one of the leading ecological economists, describes it, 'The inability of orthodox economics to cope with green issues has given rise to ecological economics, which is the study of the compatibility between the human economy and ecosystems over the long term' (1995: 22). A short definition of ecological economics is given by Common, who defines it as 'An economics that takes what we know about our biophysical circumstances, and about human psychology seriously – which standard neoclassical economics, including sub-disciplines of environmental and resource economics, does not do' (Common, 1996: 7). A longer definition and context for ecological economics is given on the European Society for Ecological Economics website:

Since the 1970s, researchers from various economic, social and natural science domains have sought to formulate new approaches to questions of economic development in response to environmental challenges, increasingly framed as the problems of sustainable development. This new perspective has become known, since the creation of the International Society for Ecological Economics (ISEE) in 1987, under the name Ecological Economics. Ecological Economics does not constitute a new single unified theory for or of sustainable development. The emergence of this field of activity signals, rather, the need for economic, social and natural science analyses to be brought together in new perspectives, responding to the concerns expressed worldwide for ecological, social, economic and political dimensions of sustainability. It represents a new practice of economics responding to a specific problem domain which may legitimately be addressed in a variety of ways. Ecological Economics thus envisages the use of analytical tools and concepts coming from many different disciplines and fields of experience. Among these, the results and techniques of neo-classical economics can be appropriate if their conditions of applicability and limits are made clear and they are placed in a wider framework of interpretation. At the same time ecological economics insists that economic science needs to open out to the insights and analytical techniques that may be offered from other fields such as the life sciences, the humanities and technology assessment.

The social dimension in Ecological Economics

Proponents of ecological economics in the initial years have, sometimes, tended to neglect the socio-cultural and political dimensions of economic development and change, while focussing on the biophysical analyses of phenomena. The starting point of the ESEE is recognition that economic activities are embedded in and dependent upon the ecosphere. It is necessary, however, to move beyond the simple recognition of biophysical limits to economic growth, in order to explore how, in what ways, and to what degrees the socioeconomic objectives traditionally associated with growth can be reconciled with concerns for environmental quality and preoccupations with social justice and variety of cultural forms.

(European Society for Ecological Economics homepage)

In many respects ecological economics represents a return to the tradition of *political economy*, which orthodox economics has long since abandoned in its aim of being 'objective', 'value neutral' and 'scientific'. That is, ecological economics does not regard the solution of the 'economic problem' to be a purely 'economic' matter (Costanza, 1989). Rather it recognises that the economy is not only a subsystem within a wider ecological system, but also operates within and is influenced by a wider political and cultural context. This systems methodological approach to the study of the human economy is something which

Figure 8.1 'Economic Growth'

Source: Polyp (1996b)

distinguishes ecological economics from neo-classical economics. The economy *ought not* to be disembedded from politics and culture or its ecosystemic and natural bases. Equally, ecological economics not only integrates social and natural science (and in this way 're-embeds' the economy within ecology) but also recognises and acknowledges the normative and ethical as well as political character of the economy and economic decision-making.

Ecological economics seeks to integrate nature's economy (ecosystems) and the human economy, which requires that the latter be seen as dependent upon and a subset within the former. As this is the main focus of ecological economics, it differs from mainstream economics in that it seeks to base its theories and models on the insights of natural science (particularly ecology and thermodynamic theory from physics) as well as having roots in economic science. In its focus on the wider social context of the economy it is an institutional form of economics (Jacobs, 1994), which together with its natural science basis makes it an explicitly multidisciplinary approach to the 'economic problem' introduced at the start of this chapter. As van den Bergh puts it, 'ecological economics should not only search for common elements, theories and approaches in the sciences of economics and ecology. It should try and take a broad view encapsulating economic, social, ethical, historical, institutional, biological and physical elements' (1996: 35–6). Thus ecological economics requires that we develop a new set of terms, models and conceptual tools in order to accurately theorise the interaction beween ecology and economy, such as ideas and the development of measureable indicators of 'ecosystem health' and indices of sustainable economic welfare (Daly and Cobb, 1989). It expands one of the aims of orthodox economics, namely managing the human economy to include managing the natural environment and the human economy together. Of central concern to ecological economics are the issues of the scale of the economy in relation to its ecological basis; the distribution of economic wealth and the need to distinguish quantitive measures of 'economic growth' from qualitative indicators of 'development' and quality of life, issues about which mainstream economics has little to say. As Daly and Cobb suggest, 'Environmental degradation must be shown to result from *the scale of the economy in general*, rather than only from allocative mistakes that can be corrected while throughput continues to grow exponentially' (1990: 368; emphasis added). For example, one of the most powerful ideas of ecological economics concerns the 'mismeasure' of economic welfare at the heart of neo-classical economics.

Ecological economics is an extremely new and challenging development within economic theory, and while it has succeeded in becoming a subdiscipline within economics it is still a minor school of thought. As yet it does not have the same hold on environmental decision- and policy-making as do orthodox economics or environmental economics. It does, however, represent an attempt to bring together the study of the human economy (economics) and nature's economy (ecology) within one multidisciplinary body of knowledge. In evolutionary terms, while ecological economics may still be a 'marginal' perspective in economic thinking, it has inspired and reinvigorated radical political economy alternatives to the dominant model, the most prominent of which is 'green political economy' which we turn to next.

Green political economy

> First, the evidence is conclusive that, beyond a certain point of affluence, the achievement of ever higher levels of material well-being and of income does not lead to increased happiness. The consumer society is chasing a rainbow . . . growth, money and what they can buy us is necessary but very far from sufficient for well-being. 'Having it all' is impossible; the attempt can turn . . . into addiction and compulsiveness.
>
> (Christie and Nash, 1998: 4)

Another closely related development within alternatives to the orthodox neo-classical economic model which overlaps substantially with ecological economics is green political economy which may be distinguished from ecological economics in being much more explicitly political and prescriptive in its analysis and suggested alternatives to the dominant model. In particular, green political economy explictly proposes the achievement of 'quality of life' or 'well-being' as both the basis for its critique of the dominant model and the main objective of an economic system aimed at sustainability rather than 'growthmania' (Daly, 1973). Green political economy is linked to other criticial perspectives on economics, such as 'post-autistic economics' (Fullbrook, 2003, 2004, 2005), eco-feminist political economy (discussed in Chapter 7), eco-socialism (discussed in Chapter 6) and institional economics (Jacobs, 1994), and can be traced to the sentiments of E.F. Schumacher's *Small is Beautiful* economic outlook, that is, in the words of the subtitle of his classic text, *Economics as if People Mattered* (Schumacher, 1973) and other radical thinkers of the late 1960s and early 1970s such as Ivan Illich (1973, 1978, 1981) and James Robertson (1976, 1983).

With ecological economics green political economy argues that the key issue for decision-making is how to achieve sustainable development in terms of determining the sustainable 'scale' of the economy relative to the ecosystem. What is proposed in this shift of economic thinking is a focus on the issue of making *optimality* rather than *maximisation* the main objective of economic decision-making (Barry, 1999a: ch. 6). This is closely associated with eco-feminist ideas of subsistance and related notions of sufficiency and satiety, rather than the imperative for more. In other words, green political economy challenges the basic ideological/value assumption of the neo-classical perspective which proposes a utilitarian imperative that 'more is better'; that is, more production, consumption of commodities and services and so on is the aim of economic life and constitutes the (neo-classical) 'good life'.

Another important feature of green political economy is the idea of the central place of the 'localisation' and self-reliance of the economy in the transition to a more sustainable economy. The slogan 'act local, think global' has long been a rallying slogan for the international green movement, and the policy and practical implications of this principle have been developed within green political economy in terms of its being a guiding principle of a sustainable economy. If the economy's ability to expand is limited by the extent of the market, in ecological terms the smaller and more local the market the less likely it is that the economy will expand beyond its ecological parameters. Lessening dependence upon the whole world as one's 'ecological hinterland' implies a much closer link between economic activity and the ecological conditions which facilitate that activity. This is not an argument for complete self-sufficiency or 'autarky' – that is, a 'closed economy' with no trade whatsoever. The localisation of the economy would lessen the 'ecological footprint' of the economy and thus make it more sustainable. In addition, the localisation of the economy has the political and cultural advantage of making the dependence of the economy on ecological goods and services more visible, since the economy is embedded in local ecosystems rather than using the resources of distant ecosystems.

The 'ecological footprinting' idea is a key analytical tool of green political economy, and has come to be accepted by states and NGOs as an appropriate tool to analyse global sustainability. Ecological footprint analysis approximates the amount of ecologically productive land and sea area it takes to sustain a population, manufacture a product, a particular lifestyle or undertake certain acitivities, by calculating the amount of energy, food, water, building materials and other consumables needed to sustain that population, lifestyle or activity. It is a simple but powerful way to determine relative consumption patterns and use, and is commonly expressed in terms of the number of Earths it would take to support every human living exactly the way one does. Thus, for example, *The Living Planet Report 2004* (WWF, 2005) calculated that humanity is now consuming over 20 per cent more natural resources than the Earth can produce, which is of course unsustainable. More worrying is the fact that humanity's ecological footprint is unequally distributed with the 'minority world' in the North (Europe and North America) having bigger 'footprints' than the 'majority world' of the South. For example, the WWF report calculated that the global ecological footprint was 13.5 billion global hectares – 2.2 global hectares per person – and that the per person ecological footprint of the United States was 9.5 global hectares and that of Britain 5.4. What this means is that the amount of materials, resources and energy needed to sustain the lifestyle of the average Briton requires approximately *two and a half planets*, while that of the average American requires *over four planets*. But of course we do not have four planets! Thus the lifestyles and economies of both Britain and the United States are unsustainable; that is, it is biophysically impossible for the whole of the world's population to live with their current levels of resource use. A corollary of the ecological footprint idea is the concept of 'ecological debt'; that is, the debt those over-consuming nations owe to the rest of the world by consuming more than their fair share of the available finite biophysical resources of the planet. Echoing the debate about 'overpopulation' in Chapter 5, the ecological debt idea turns the overpopulation debate on its head by suggesting that the real issue is the over-use of resources by the minority world and not necessarily the 'overpopulation' within the majority world. Echoing the global justice movement's arguments, the idea of ecological debt means that rather than the Southern world being dependent upon the Northern world, the developed world owes the developing world, since it is using up an unfair and unjust share of the finite resources of the planet. The implications of the ecological debt the minority world owes to the majority world are that what is owed (in terms of reparations and compensation and distribution of wealth and resources) is owed as a matter of *justice* not charity (Dobson, 2003: ch. 3). According to Martinez-Alier (2001), ecological debt denotes an unjust and exploitative exchange between the powerful minority world and the Southern world. As he puts it:

> By ecologically unequal exchange we mean, then, the fact of exporting products from poor regions and countries, at prices which do not take into account the local externalities caused by these exports or the exhaustion of natural resources, in exchange for goods and services from richer regions. The concept focuses on the poverty and the lack of political power of the exporting region.
>
> (Martinez-Alier, 2001: 214)

There are a number of important implications of the ecological footprinting and ecological debt ideas. One is the (obvious) point that what is needed is 'one-planet' thinking; that is, explictly recognising the finite resources of the planet which have to be shared and used equally by the human population as a whole (and the non-human world, one may add) and not unjustly consumed by the most powerful sections of the global population. This has been one of the key principles of green political economy, one shared with the global justice movement, and means that the minority world needs to decrease its ecological footprint, repay the ecological debt owed to the majority world, and for the majority world to increase its ecological footprint through development policies.

Another is the connection between global justice, sustainability and the local-isation of the economy. The upshot of the ecological footprint analysis and associated concept of ecological debt is for a shift towards a localisation of the economy as much as possible – but without ruling out trade and exchange. This self-limiting character of local market economies harks back to an earlier tradition of political economy associated with the ancient Greeks. This refers to the distinction Aristotle made between *chrematistics* and *oikonomia* within political economy. *Chrematistics* is defined as that branch of the political economy relating to the manipulation of property, wealth and currency so as to maximise short-term returns. *Chremastistics*, in short, mistakes a means to an end, and according to Aristotle it is characteristic of this form of acquisition that 'there is no limit to the end it seeks; and the end it seeks is wealth of the sort we have mentioned [i.e. wealth in the form of currency] and the mere acquisition of money' (Aristotle, 1948: 1257b). *Chrematistics* is the money/cash economy which has come to displace the long-term management of the economy in the interests of all and has become separated form the ecological realities. *Oikonomia*, by contrast, is, according to Aristotle, a more limited form of acquisition. Its central concern is the 'management of the household' geared towards long-term maintenance of the welfare of all household members. The limited nature of this form of acquisition is given by Aristotle thus: 'the amount of household property which suffices for a good life is not unlimited' (1948: 1236b). It is clear that what sustainable development requires from a green political economy perspec-tive is the integration of the 'management of the household' with the 'economy of the household'; that is, integrating economy and ecology, which is a central aim of ecological economics. This distinction represents the separation and 'disembedding' of the economy and the 'economic motive' (economic rationality) from social relations (Polanyi, 1957: 54) and the natural world upon which the human economy depends (O'Neill, 1993: 169). The result of this disembedding, according to Hutchinson *et al.* (2002), has been the 'cancerous growth' of the money economy, which is divorced from ecological reality and which now threatens the survival of the human species through undermining the global ecosystem's capacity to sustain human life (Hutchinson *et al.*, 2002: 226–9).

According to Lee (1989), the development of the money economy was central to the modern capitalist economy, and laid the basis for the separation of the human economy from nature's economy. The separation of the economy from its ecological context also meant the increasing separation of the economy from wider non-economic considerations. In ecological terms, John Locke's argument in defence of money (and the inequality that a money economy requires and justifies) permitted the accumulation process that is at the heart of the capitalist market system. Until the creation of money and its widespread acceptance, wealth and accumulation were limited by natural constraints (the limits of a person's stomach or the length of time natural products would last without spoiling). With the widespread use of money as a non-putrefying store of wealth, limits to accumulation could be overcome. As Gorz puts it, 'once you begin to measure wealth *in cash*, enough doesn't exist. Whatever the sum, it could always be larger' (1989: 112; emphasis in original). With money not just as a medium of exchange but now a store of value, what Lee calls the 'organic' basis of human wealth was overcome. With the invention of money, 'Accumulation of this non-putrefying object on the part of the individual can now be limitless and go on for ever, the accumulation process having being emancipated from the workings of Nature' (Lee, 1989: 164). It is money which is at the root of the separation of the formal/cash economy from both the non-money economy and nature, the separation of economic from ecological rationality, and leads to a blindness with regard to the ecological origins of and contributions to human wealth and the non-monetary activity which contributes to and constitutes human well-being and quality of life.

A key feature of green political economy is a non-ecological critique of economic growth which goes beyond the 'Limits to Growth' thesis that exponential economic growth is *impossible*, to one which argues that economic growth is *undesirable*. The basis of this critique of economic growth is twofold, the first related to the way orthodox economic growth depends upon and reproduces socio-economic inequalities, while the second relates to the argument that beyond a threshold, additional economic growth does not add to and may take away from life satisfaction and quality of life.

In general terms, rather than investing energy, time and resources into simply ameliorating the worst effects of inequality, green political economy seek to address the root causes of it. However, given the centrality of the critique of conventional economic growth, green political economy rejects the dominant view that only by redistributing the fruits of a growing capitalist, competitive economy can equality, social inclusion and environmental improvement be achieved, via the 'trickle-down' effect of progressive taxation and social welfare transfers. Rather, the green path to tackling inequality is premised on redistribution (of existing social wealth) without the commitment to unsustainable and undifferentiated economic growth, alongside a radical shift from money and commodity-based measurements of welfare to a focus on well-being, quality of life and free time. In directly linking the achievement of a sustainable society to social justice, greens recognise that redistribution from the better-off to the most vulnerable members of society is absolutely essential, both practically and morally. The basic green argument with regard to lessening socio-economic inequalities is that linking this social goal to redistributing economic growth merely entrenches inequality rather than lessening it. The reason for this is that in societies such as the welfare states of the West, state legitimacy is, in part, dependent upon a commitment to lessen inequalities via redistributive measures. However, co-existing with this is the standard defence of economic inequality which claims that it is necessary for creating the conditions for economic growth. In other words, an unequal distribution of the benefits of socially produced wealth is a necessary condition of a growing, successful economy (less is said about the unequal distribution of the social, economic and environmental costs and risks of economic growth). Wealth and income inequalities are argued to be economic incentives that are absolutely essential for encouraging employment of the best individuals, which contributes to overall economic productivity and growth. The basic argument is that while some gain more than others, everybody does gain in the end. Thus the green critique of economic growth may also be regarded as an argument against the social inequalities that are a structural component of contemporary social and economic policies. The increase in inequality in the United Kingdom and the United States over the past twenty years may be interpreted as evidence of this relationship between economic growth and socio-economic inequality – as the economy has grown (as measured by conventional GDP/GNP accounting), so has inequality. The US is frequently cited as one of the most unequal societies in the world. While having unequal slices of a growing cake means that in absolute terms everyone is gaining – or is 'better-off' using the conventional (and misleading) measure of disposal income – the reality is that the existence of relative inequality detrimentally affects people's quality of life. For example, even though economic growth may mean the car-less can buy more things, and also benefits car-owners by enabling them to buy new and more expensive cars, the reality of persistent relative inequality between the two groups highlights the problem. As Levett points out: 'The more the well-off drive, the more amenities move to sites with plentiful parking, and the fewer remain accessible without a car' (Levett, 2001: 30). Improving the quality of life of individuals and communities requires shifting attention away from income and benefit measurements alone (the fruits of economic growth) towards the non-income (and non-employment) components of quality of life and well-being. As Levett again succinctly puts it: 'The key is to target well-being directly, and stop treating economic growth as a proxy for it' (Levett, 2001: 31). Thus, from a green point of view government social (and public policy) should be aimed at improving the quality of life of individuals, families and communities, and this requires a shift from a focus on ensuring economic growth, competitiveness and employ-ment. The green argument for a 'steady-state economy', or an economy in which maximising output, profits or paid employment would not be the dominant imperative, may thus be seen as a strong egalitarian argument for decreasing social and economic inequality. With an economy not geared towards maximising production, income and formally paid employment, the justification for an unequal distribution of socially produced wealth cannot be that it is required for procuring greater wealth production. In short, with the shift to a less growth-orientated society, the normative basis for social co-operation needs to be renegotiated, as does social policy. The implications of this are dramatic for social policy, given that one of the central justifications for social policy is the lessening of socio-economic inequalities via the redistribution of income, goods and services generated from a growing economy (Huby, 1998: 15). The green argument is that if one wishes to reduce inequalities, abandoning the exclusive focus on a growing economy as the way to do this may be a more realistic way of achieving it, since an inegalitarian distribution of social wealth is less morally and politically justifiable within the context of a non-growing economy. However, such talk of a principled rejection or downplaying of the traditional commitment to economic growth would obviously lead to strong resistance from both labour and capital interests, since it spells nothing less than the radical transformation of industrial society and the welfare state system which has developed alongside and with it (Barry and Doherty, 2001: 603).

Equally, greens are committed to a more egalitarian society on the grounds that this is a more effective way to increase general well-being or quality of life. As Boardman *et al.* note, reducing inequalities:

> not only benefits the poorest in society, but society as a whole, for there is growing evidence that narrowing income inequality in a society adds to the overall social quality of life. The benefits for the poorest households go substantially beyond the direct effects of extra income. There is strong evidence that in the developed world, it is not the richest countries which have the best health but the most egalitarian.
>
> (Boardman *et al.*, 1999: 11)

One interpretation of the green economic critique of orthodox growth and the commitment to establishing a more egalitarian distribution of income and wealth is that economic security rather than economic affluence is important for a more equal social order within modern democracies and is a better way to increase overall social well-being, as suggested above. The green view is that it is the distribution of wealth within society, not the absolute level of wealth, which is important in a democratic political system.

A final and related characteristic of green political economy is the Marxist notion of the need to connect analysis with 'strategies of resistance' (Barry, 2003). That is, political economy is about analysing the current social reality *and* challenging the political status quo; it is not an 'academic' exercise but is connected to political struggle, analysis, agency and strategy (Wall, 2006). For some, such as Martinez-Alier (2001), this means that green political economy should mobilise local populations, classes and groups around unequal and unjust distributional conflicts by highlighting the ecological debt owed by the North to the South, for example, or the ecological debt owed by the rich to the poor within a country. For him, 'To exaggerate slightly, the focus should not be on "environmental conflict resolution" but rather (within Gandhian limits) on conflict exacerbation in order to advance towards an ecological economy' (2003: 257); that is, what we need is to focus on mobilising struggles around the reality of 'ecological and social injustice' and the recognition that history shows that the achievement of justice (ecological or otherwise) will not be achieved without a struggle. In the words of Frederick Douglas, freed slave and anti-slavery campaigner, writing in 1857, 'Those who profess to favour freedom and yet depreciate agitation, are men who want crops without plowing up the ground. They want rain without thunder and lightning. . . . Power concedes nothing without a demand. It never has and never will' (in Barry, 2005: 21). In many respects green political economy echoes the position of the Victorian social critic John Ruskin who noted that 'The real science of political economy, which has yet to be distinguished from the bastard science, as medicine from witchcraft, and astronomy from astrology, is that which teaches nations to desire and labour for the things that lead to life: and which teaches them to scorn and destroy the things that lead to destruction' (in Boyle, 2002: 13). This stress on valuing that which is life-affirming is something that green political economy shares with and owes an intellectual debt to eco-feminist political economy which also stresses the ways in which neo-classical unsustainable economics value and prioritise 'lifeless' measures and indicators of the economy – such as money and capital – or in practice serves to elevate 'life-destroying' human productive activity such as arms manufacturing over life-sustaining reproductive work by women in the domestic sphere. Unlike teachers, nurses or voluntary organisations, when was the last time you saw an army general collecting for a new tank outside a shopping mall?

Conclusion

Of all the social sciences, economics, from its origins in political economy, has perhaps had the most effect on how the natural environment has been viewed, valued and treated in Western societies. As a form of social theory, it has had widespread and far-reaching consequences for how the relationship between society (and the economy in particular) and the natural environment has been thought about and analysed. Historically, political economy and its associated economic worldview was used to legitimate the various changes that constituted the transition from feudalism to an industrial-capitalist market society. In the modern world, the heir of political economy, orthodox 'economics' or positive 'economic science', has a powerful hold on public policy-making, and environmental policy-making in particular.

At the same time the 'economic' view of the natural environment is one which has commonsensical appeal. Most people would go along with the economic view of the natural environment, i.e. that it has only instrumental value to humans and its instrumental value is of an economic form. This economic value of the natural environment is in terms of its functions as a 'resource' or 'input' to the human economy. Economics is thus a major form of social theory and practice which upholds an anthropocentric or human-centred view of nature and human relations to it. Its importance lies in that its area of study (i.e. the human economy) is the one part of human society which has a direct and material interaction with and relation to the natural environment. The economy is where nature and society meet and is increasingly the ideological terrain upon which critics of the contemporary economic system are waging their struggle for a post-capitalist sustainable economic system.

Summary points

- Economics is a form of social theory, indeed perhaps one of the most powerful and important in terms of how the natural environment is viewed, valued and treated.
- One of the central achievements of the emergence of an 'economic' view of society in the seventeenth and eighteenth centuries was the 'disembedding' of the economy from the wider social, political, cultural and religious context.
- The emergence of liberal political economy led to the enclosure movement, the commodification of the land and a view of the natural environment as only of instrumental value to humans.
- There was a historical link between resistance against and criticism of the 'economic worldview', the rise of industrial capitalism, the nation-state system in Europe, and particular traditional views of 'nature' and 'land' and rural ways of life in relation to the land as 'place'.
- There is a strong link between orthodox economic thought and policy-making in industrial societies.
- Environmental economics, in which environmental goods and services, as well as environmental risks and bads, are given a monetary value, is the first systematic attempt to introduce the environmental dimension within mainstream economics.

- Environmental economics may be criticised for economising the environment rather than ecologising economics.
- Ecological economics is a recent development within economy theory which attempts to integrate economy and ecology; that is, it is explicitly based on the dependence of the human economy upon nature's economy.
- Ecological economics is a mutlidisciplinary approach to ecological–economic interaction, which not only integrates natural science into its analysis, but in its attention to the political, cultural and social context of the economy may be seen to represent a return to the older tradition of 'political economy'.
- Green political economy builds on ecological economics accepting most of its insights but is more explicit in its political aims of arguing for an alternative mode of economic organisation and stressing the connections between analyses and political struggle, or theory and practice.
- Green political economy challenges the dominant neo-classical economic view of the 'good life' as one based on ever-growing levels of material consumption achieved through increases in economic growth. It proposes that 'quality of life' and 'well-being' be the aim of sustainable economic policy.

Further reading

For some general reading on the relationship between economics and the natural environment, see Donald Worster's *Nature's Economy: A History of Ecological Ideas*, Cambridge: Cambridge University Press, 1994; P. Mirowski (ed.), *Natural Images in Economic Thought: Markets Read in Tooth and Claw*, Cambridge: Cambridge University Press, 1994; and Jon Mulberg, *Social Limits to Economic Theory*, London: Routledge, 1995.

On the historical relationship between political economy and the natural environment, see A. Clayre (ed.), *Nature and Industrialization*, Oxford: Oxford University Press, 1997.

On environmental economics, see David Pearce *et al.*, *Blueprint for a Green Economy*, London: Earthscan, 1989; and the rest of the 'Blueprint' series published by Earthscan, particularly D. Pearce *et al.*, *Blueprint 3: Measuring Sustainable Development*, London: Earthscan, 1993. For critical assessments of environmental economics, see Michael Jacobs, 'The Limits of Neoclassicism: Towards an Institutional Environmental Economics', in M. Redclift and T. Benton (eds), *Social Theory and the Global Environment*, London: Routledge, 1994; and John Barry, 'Green Political Economy', ch. 6 of *Rethinking Green Politics*, London: Sage, 1999.

On ecological economics, see Herman Daly and John Cobb, *For The Common Good*, Boston, MA: Beacon Press, 1989; Herman Daly, *Ecological Economics and the Ecology of Economics: Essays in Criticism*, Cheltenham: Edward Elgar, 2000; Robert Constanza (ed.), *Ecological Economics: The Science and Management of Sustainability*, New York: Columbia University Press, 1991; Juan Martinez-Alier, *Ecological Economics*, Oxford: Blackwell, 1987; Nicholas Georgescu-Roegen, *The Entropy Law and the Economic Process*, Cambridge, MA: Harvard University Press, 1971; Peter Söderbaum, *Ecological Economics. A Political Economics Approach to Environment and Development*, London: Earthscan, 2000; S. Dovers, D. Stern and M. Young (eds) (2003), *New Dimensions in Ecological Economics: Integrated Approaches to People and Nature*, Cheltenham: Edward Elgar.

On post-autistic economics, see E. Fullbrook (ed.) (2003), *The Crisis in Economics*, London: Routledge; Fullbrook (ed.) (2004), *A Guide To What's Wrong With Economics*, London: Anthem Press; Fullbrook (2005), 'Post-autistic Economics', *Soundings: A Journal of Politics and Culture*, spring 2005.

On green political economy, see John Barry and Brian Doherty (2001), 'The Greens and Social Policy: Movements, Politics and Practice?', *Social Policy and Administration*, 35:5; John Barry (2003), 'Holding Tender Views in Tough Ways: Political Economy and Strategies of Resistance in Green Politics', *British Journal of Politics and International Relations*, 5:4; Barry (2006a), 'Towards a Concrete Utopian Model of Green Political Economy: From Economic Growth and Ecological Modernisation to Economic Security', *Post-Austistic Economics Review*, 38; Barry (2006b), 'Towards a Model of Green Political Economy', *International Journal of Green Economics*, 1:3; M. Kennet and V. Heinemann (2006), 'Green Economics: Setting the Scene', *International Journal of Green Economics*, 1:1/2; D. Wall (2006), 'Green Economics: An Introduction and Research Agenda', *International Journal of Green Economics*, 1:1/2; Jon Mulberg (1995), *Social Limits to Economic Theory*, London: Routledge; Richard Douthwaite, *The Growth Illusion*, Dublin: Lilliput Press, 1999; Douthwaite, *Short Circuit: Strengthening Local Economies for Security in an Unstable World*, Dublin: Lilliput Press, 1996.

9 Risk, environment and postmodernism

Key issues

- Beck's 'risk society' thesis.
- The character of risk.
- The 'precautionary principle'.
- Reflexive modernisation and the redefinition of 'progress'.
- Democracy, democratisation and risk society.
- Postmodernism, social theory and environment.
- Postmodernism, environmentalism and the rejection of modernity.
- Postmodernism and post-industrialism.
- Postmodernism and the social construction of the environment.
- Problems of postmodern environmentalism.

Introduction

Throughout the book so far an important point of reference for analysing the environment and social theory has been the way in which, both historically and conceptually, 'modernity' has affected social theorising about the environment. 'Modernity' may be understood as the (sometimes radical) changes in the organisation and legitimation of 'modern' social, political and economic life and 'modern' ways of thinking and acting, associated with the advent of the 'modern' age in the latter half of the eighteenth century in Europe. The many and complex aspects of these changes in almost all parts of life are central to an adequate understanding of how 'social theory' (itself a product of the 'modern' age), in its different forms, schools and as developed by different social theorists, viewed, valued and thought about the environment.

In this chapter, we discuss two contemporary forms of social theorising about the environment and the environmental crisis, in which 'modernity' and its legacy are central to both. These are the 'risk society' thesis associated with Ulrich Beck, and the relationship between postmodern social theory and the environment. Both may be seen, in sometimes very different ways, to be directly engaging with the problems and potentials of modernity and its legacy.

In the first part of the chapter, Beck's 'risk society' thesis is explored which, in terms of its attitude of modernity, may be broadly viewed as representing an 'immanent critique and reconstruction' of modernity. That is, 'risk society' may be seen as an attempt, in Beck's own terms, at a 'new modernity' (which is the subtitle of his famous book *Risk Society*, published in English in 1992). Thus while Beck does criticise modernity, particularly in terms of the rise in potentially catastrophic environmental risks, he does suggest that modernity has within itself the ability to solve the problems it produces.

The second half of this chapter looks at postmodern social theory and its treatment of the environment and associated phenomenon such as the environmental crisis and environmental politics. Postmodern social theory as its name suggests, unlike Beck's social theory, sees itself as 'beyond modernity'. While no such thing as *a* singular or homogeneous postmodern social theory exists (there are many varieties of postmodern social theory), what the different varieties of postmodernism share is an extremely critical and radical assessment of 'modernity'. In terms of 'modern' forms of knowledge (from natural science to social thought), and modern political, social and cultural institutions (such as the nation-state, liberal democracy, the nuclear family), to 'modern' alternatives to liberal capitalism (particularly Marxist ideology), postmodernism suggests that the various problems associated with modernity (including environmental ones) cannot be solved within modernity, but require a postmodern solution.

Beck's 'risk society' thesis

Stated bluntly, Ulrich Beck's 'risk society' thesis suggests that what we are witnessing in contemporary Western societies is the emergence of a politics concerned with the interpretation and distribution of social and ecological 'bads' rather than 'goods'. The politics which characterised 'industrial society' centred around the production and distribution of 'goods' such as wealth, income and formal employment. As Beck puts it, 'What was at stake in the older industrial conflict of labor against capital were positives: profits, prosperity, consumer goods. In the new ecological conflict, on the other hand, what is at stake are negatives: losses, devastation, threats' (Beck, 1995a: 3). For Beck, 'risk society' refers to the recent transformation of Western societies, with specific focus on environmental issues. It concerns the health, socio-economic, cultural and environmental effects of 'social progress' in general, and scientific and technologically based production in particular.

In industrial society, politics is largely concerned with the distribution of the benefits of society (wealth, income, jobs), hence the dominant political conflict is between 'capital' and 'labour', and political life assumes a 'left–right' character. In risk society, which is located between 'industrial' and 'post-industrial' stages of social advancement, the dominant focus of politics is the distribution of the 'costs' and 'risks' of socio-economic development, dominated by the emergence of unexpected ecological and health hazards.

In this new historical epoch, politics, to use a famous green slogan, is 'beyond left and right'. Beck's ecologically informed analysis of risk and his account of the emergence of a 'post' left–right political landscape is something shared with Giddens, discussed in Chapter 4. Also in common with Giddens (and Habermas), Beck's 'risk society' thesis has both descriptive and prescriptive dimensions. While he is *describing* recent social changes in Western societies, he also *prescribes* ways of dealing with these changes and promotes a particular vision of an 'ecologically rational' or 'ecologically enlightened' society.

There is some empirical evidence to support Beck's thesis that Western societies are (or are becoming) 'risk societies', both in terms of greater awareness and sensitivity of risk, and 'at risk' from their own development paths. In 1995 a MORI poll in the United Kingdom asked the following question: 'Do you think that the kind of world that today's children will inherit will be better or worse than the kind of world that children of your generation inherited, or about the same?' (Jacobs, 1996: 3). The results were:

Better	12%
Worse	60%
About the same	25%
Don't know	4%

What are the reasons for this extremely pessimistic assessment of the future? Why do people think the future will not be better than today? The number and range of possible risks or bads we can think of include: the increase in crime; a decrease in personal safety; rising unemployment levels; declining job security and prospects; the breakdown of families and communities; the risk of contracting diseases/illnesses; the public controversies about the decline in food safety; and rising pollution levels and environmental degradation. Across these and other issues, one can witness the growing risk sensitivity of Western publics due to experience of one 'scare' or 'risk' after another: such as the 'mad cow' disease (BSE) in the United Kingdom, the outbreak of E Coli food poisoning, as well as environmental dangers such as global warming and ozone depletion, and environmental-related illnesses such as the dramatic rise in childhood asthma, linked to the increase in car pollution.

Another form of empirical evidence of the 'riskiness' of modern societies is the rise in the quantity and quality of insurance cover that individuals, groups, firms and corporations have taken out to insure against the various forms of risk outlined above. For example, the risk of global warming, its potentially damaging effects on agriculture, water generation and distribution, and the likely devastating effects of rising sea-levels for low-lying human settlements, may be observed in the quantity of 'global warming'-related insurance being taken out by companies and being held or insisted upon by the insurance industry. Or observe the rise in the private health insurance market in Britain in recent years (though this may be explained in part as a consequence of the decline of the National Health Service, as much as an increase in people's sensitivity to health-related risks).

In Beck's terms, these 'risks', 'bads' or 'dangers' are the side-effects (the costs) of the particular 'development path' or type of modernisation which characterises modern societies. Basically, his point is that the costs of modernisation are beginning to outweigh the benefits. As Beck puts it, '"Risk society" means an epoch in which the dark sides of progress increasingly come to dominate social debate. What no one saw and no one wanted – *self endangerment and the devastation of nature – is becoming a motor force of history*' (1995b: 2; emphasis added).

The character of risk

We can divide risks into a number of categories:

1 *Ecological risks*: global warming, biodiversity loss, ozone depletion, ecosystem destruction.
2 *Health risks*: health risks due to genetically altered foodstuffs, skin cancer, food safety scares ('mad cow' disease), 'avian flu', SARS, pollution-related illness such as asthma, cancer, heart disease.
3 *Economic risks*: unemployment and decline in job security.
4 *Social risks*: decline in personal safety, rise in crime and breakdown of community, rise in divorce and separation.

The character of risks is important in understanding them. First, they are often 'intangible' and remote, either in space (risks and their consequences happen 'somewhere else') or time (the effects of risks will be felt in the future, not now). Second, they are characterised by uncertainty and unpredictability which makes them difficult to insure against. Third, they are often unknown, that is, those who are subject to specific risks are often ignorant of them. Yet there is, according to Beck's thesis, a growing fear from risks in general. Due to both of these issues (uncertainty and knowledge/ignorance of risks), of crucial importance in analysing risks is knowledge and its communication which identify, inform and define the shape and content of risks. Of particular significance for Beck are science and the media, and one of the main aspects of his thesis is the increasing gulf between 'science and technology' and 'society' in general, and the latter's distrust of the former in particular.

Risk, trust and science

According to Beck's 'risk society' analysis, science (and technology) are increasingly seen as the causes of modern environmental (and other) risks rather than the solution. What marks 'risk society' from the previous 'industrial' stage of social development is that whereas in the latter science and technology were seen as positive forces for social progress, a view common to nineteenth-century social theory as discussed in Chapter 2, in 'risk society' this equation of scientific and technological advancement and social progress is broken. Risk society describes a modern sense of fear, distrust and unease about scientific and technological developments. However, for Beck this distrust is not confined to science and technology, but may also be seen in the erosion of 'trust' in dominant social and political institutions, such as industry and government. Here Beck's 'risk society' argument complements other theories which analyse the 'legitimation crisis' of Western societies, such as the early work of Habermas (1975), and other neo-Marxist social theorists such as Claus Offe (1984). Where once people had unquestioning faith and belief in scientific knowledge and technological developments, the experience of the negative side-effects of science and technology has led Western publics to be less trusting of 'the scientific establishment'. The advent of 'risk society' is therefore the advent not just of an 'ecological crisis' but also as Irwin puts it, 'the "environmental crisis" is in essence a social crisis for our institutions and for our own existential beliefs (that is, who we think we are)' (1997: 220).

This development, however, does not mean the simple rejection of science, for as Beck suggests, 'public risk consciousness and risk conflicts will lead to forms of scientization of the protests against science' (Beck, 1992a: 161). What he means by this is that scientific knowledge is no longer a unified body of knowledge but fragmented, and 'risk society' implies the breakup of a homogeneous understanding of 'science' as speaking with one voice. It terms of the association between expert knowledge and power, it is important to stress that the authority and legitimacy accorded to scientific knowledge was in part related to its unified, internally coherent character. Once it begins to fragment and no longer be a unified whole, its ability to command automatic authority and public trust begins to wane.

First, in various environmental controversies which rely on 'scientific evidence' both to 'identify' the problem as well as to suggest solutions, we increasingly find government or industry scientists facing those of environmental **NGOs (non-governmental organisations)**, such as Greenpeace, in presenting scientific evidence to establish or prove that a particular environmental problem exists and to assess its severity. Indeed, there is evidence to suggest that the public distrusts government or business scientists and scientific evidence and has more trust in the scientific arguments of environmental NGOs and other civil society organisations (Bocking, 2004). This is of course directly related to the discussion in Chapter 5 of the 'science wars' and the Lomborg controversy which illustrated the ideological battle between right-wing 'contrarian' rejections of the scientific evidence of environmental problems such as climate change. As Bocking, echoing Beck, asks rhetorically, 'where, in the continuum between hard data and messy questions of responsibility, does science end and politics begin?' (2004: 20).

Second, there are issues on which science itself is divided, such as certain illnesses, and ecological risks such as global warming (though this links with the previous point in that those scientists who dispute that global warming is due to human carbon emissions, who are in a minority, are largely funded by what is known as the 'fossil fuel' lobby, corporations such as car manufacturers, and the oil industry, who clearly have a lot to gain from discrediting global warming). Beck sees this pluralisation or breaking up of a unitary science as something positive. The reason for this is that 'Only when medicine opposes medicine, nuclear physics opposes nuclear physics . . . can the future that is being brewed up in the test-tube become intelligible and evaluable for the outside world' (Beck, 1992a: 234). In other words, having as many scientific voices as possible debating among themselves and with the public (and thereby educating the public) is the best way to identify and possibly cope with or solve the various ecological and health risks that face modern societies.

Developments in science and technology have profoundly affected their relationship with society, and between society and nature according to Beck. On the one hand, scientific and technological innovations have turned society into a laboratory. The proposed development and release of genetically modified organisms into the environment and food chain, or the 'cloning' of organisms, with their unknown effects, may be taken as an example of how, in Beck's terms, we live in an 'experimental' society. Society is being subjected to experiments over which it has no direct control and often unknown to it.

On the other hand, the large-scale, global effects of advanced scientific and technological development has resulted in what McKibben (1989) has called the 'Death of Nature'. What is meant by this is not the death of the natural world as such, but rather the end of nature as an independent entity from human intention and activity. With the advent of human-induced global warming and climate change, as a negative and unintended 'side-effect' of human productive activity, we have what may be called the 'humanisation of nature'. However, unlike the humanisation of nature that Marx advocated, the current process of the human-isation of nature is unintended, undemocratic and not based on the principles he advocated. Indeed, it is truer to say that what we are witnessing is not the human-isation of nature, but the 'capitalisation of nature' (Barry, 1995c, 1998c). Global warming and climate are largely the result of globalised processes of capitalist production, while at the same time it is not the human species as a whole which is to blame, but disproportionately the advanced capitalist nations of the world (the 'minority world'), and large corporate multinationals.

On the other hand, technological developments and innovations such as bio-technology, especially genetic engineering, cloning and the creation of genetically modified organisms, are other examples of the 'experimental society'. For some, these parts of the experimental society are something to be feared. As Levidow and Tait put it, 'In parts of the popular imagination, biotechnology is feared as a violation of nature, which may then go out of control' (1995: 123). While for some people risks are exciting challenges to be overcome, and there are those who positively enjoy taking risks, for most people risks are to be minimised, managed or avoided if at all possible.

The 'precautionary principle'

While Beck does not talk directly of the precautionary principle, it is clearly consistent with the main thrust of his thesis, and constitutes an important aspect of the relationship between social theory and environmental risks. According to Beck, 'insisting on the purity of scientific analysis leads to the pollution and contamination of air, foodstuffs, water, soil, plants, animals and people' (1992a: 62). In other words, in the face of uncertainty and high levels of scientific disagreement, waiting to 'scientifically prove' a particular ecological risk (for example, the link between cancer and nuclear power plants) means the burden of proof is on those who wish to prevent or stop some industrial process, rather than on those who promote it. Meanwhile the negative consequences of the process continues (i.e. people still get cancer).

The precautionary principle, which has come to be a central principle of environmental risk assessment and management in recent years, holds that in the

Figure 9.1 'We're All Gonna Die'
Source: Polyp (1996c)

context of uncertainty it is rational to be prudent and not proceed with a particular action if there is a risk of its resulting in future significant danger or harm. For O'Riordan and Jordan:

> At the core of the precautionary principle is the intuitively simple idea that decision makers should act in advance of scientific certainty to protect the environment (and with it the well-being interests of future generations) from incurring harm. . . . In essence it requires that risk avoidance become an established decision norm where there is reasonable uncertainty regarding possible environmental damage or social deprivation arising out of a proposed course of action.
>
> (1994: 194)

Core aspects of the precautionary principle include a willingness to act in advance of scientific proof and, for example, to deliberately hold back from certain forms of development on precautionary grounds. It also implies a reassessment of the belief that technology can solve all environmental problems, a belief, which as indicated above and in Chapter 2, that was typical of social beliefs and social theory of 'industrial society'. Following on from these two, the precautionary principle seeks to avoid irreversible forms of environmental intervention. That is, the precautionary principle suggests that large-scale developments which cannot be later reversed (such as destroying a particular ecosystem or perhaps damming a river), and which are risky should not proceed, as the costs would outweigh the benefits. There is also a deeply normative dimension to the precautionary principle in that, in seeking to prevent certain forms of 'risky' development, it raises fundamental issues about the direction of social development and the type of society in which we wish to live.

As such, the precautionary principle fits with Beck's view that environmental risks and dilemmas are not simply technical 'problems' to be 'solved' by scientific and/or economic applications. Rather, in 'risk society' the precautionary principle acknowledges the normative character of environmental risks – that is, they are moral questions about right and wrong and not simply about costs and benefits or technical 'problems' and 'solutions'. The normative character of the precautionary principle may be seen if dealing with environmental risks is viewed in the context of avoiding unnecessary harm to future generations and the non-human world. In this way, the precautionary principle suggests a shift from a 'polluter pays' to a 'polluter proves' principle. That is, those proposing large-scale developments with large and unknown or unquantifiable environmental effects, or developments resulting in irreversible environmental degradation (such as dams, road-building schemes, the creation, release and use of genetically modified organisms), should demonstrate that no long-term or irreversible ecological, health or other harms will result. In conclusion, the precautionary principle is in many respects simply common-sense prudence and caution, a virtue all the more pressing in the complex, uncertain world of contemporary social–environmental relations. The spirit of the precautionary principle is nicely expressed by St Thomas Aquinas' wise opinion that 'It is better for a blind horse that it is slow'.

The effects of a widespread application of the precautionary principle would be potentially radical as it would result in a qualitatively different form of development from the standard 'modernisation' model. At the very least it would mean a stricter and more political regulation of economic development, in general, and in particular, forms of modernisation which either rest on innovations and applications of science and technology and/or require the 'use' or consumption of the nonhuman environment. This different form of ecologically informed and sensitive modernisation, consistent with Beck's 'risk society' position, is discussed next.

Reflexive modernisation and the redefinition of 'progress'

For Beck, what may be called 'simple' or 'industrial/economic' modernisation is the form or model of social development associated with the standard 'Western' view of progress and development. This modernisation model is associated with the historical evolution of 'Western' societies over the past two hundred years. Some of the main features of this modernisation model include: the industrialisation of the economy; urbanisation and the emergence of a 'post-agricultural' commercial society; the creation of a nation-state; and above all else the equation of 'progress' with continuous increases in the production and consumption of material goods and services. With the collapse of communism, this 'Western' model has in recent years evolved to include the importance of 'free markets', private property and international free trade. For Jacobs, this 'dominant model' of progress may be understood as meaning that:

> the principal purpose of economic activity is to raise incomes. Income growth makes people better off: it enables them to consume greater quantities of both material and non-material goods, and through taxation enables governments to provide essential public services such as education, health care and social security. Given technological improvements to productivity, annual economic growth not only generates jobs, but is required to sustain them. The motor of growth is free trade: as import duties, foreign exchange controls and other forms of protection are lowered, goods, capital and labour flow to where they are most productive, and more wealth is generated . . . free trade and growth lead to 'modernisation' – the increasing productivity of agriculture, the movement of people into towns and cities and the transformation from traditional to modern cultures.
>
> (1996: 8)

Now, for Beck, the advent of 'risk society' marks the threshold beyond which ecological and other risks outweigh the benefits of further economic growth associated with the model of industrial or simple modernisation. According to Beck, what is now required is a qualitatively new model of modernisation, what he calls 'reflexive modernisation'. The basics of reflexive modernisation are that 'Modernisation *within* the path of industrial society is being replaced by a modernisation of the *principles* of industrial society' (Beck, 1992a: 10). As the term 'reflexive' implies, what Beck (in agreement with Giddens who also focuses on the 'reflexivity' of social institutions) suggests is that modernisation should mean that society as a whole increasingly reflects upon its own development and the institutions which further and/or realise that development. It is important to stress that Beck is *not* arguing from a 'postmodern' perspective, but is arguing instead for a new type of modernisation. This new model of modernisation, reflexive modernisation, may be seen as a form of social learning, an attempt by society, through its political institutions (especially through democratic politics), to deal with and/or cope with the ecological and other risks and dilemmas that have arisen as a result of simple or industrial modernisation.

A central part of reflexive modernisation is the redefinition of 'progress'. According to Beck, '"Progress" can be understood as legitimate social change without democratic political legitimation' (1992a: 214). In this way, 'progress' as orthodox economic growth (in accordance with the 'standard' or 'dominant' model) is giving way, in Beck's analysis, to 'social progress' (1992a: 203). By the latter, Beck means *institutionalised self-criticism* (reflexivity) and increased opportunities for individuals to be conscious of and deliberate upon, and democratically pass judgement on the *principles of modernisation* and not just specific policies associated with it. In other words, by reflexive modernisation, Beck means a redefinition of social progress, a central aspect of which requires the separation of 'techno-economic' progress and 'social' progress. Additionally, and radically, what reflexive modernisation implies is that society democratically make decisions on its development path; that is, democratically 'regulate' social progress. The politics of 'risk society' thus concerns both the *direction* and the *substance* of social progress, and thus of social organisation as a whole.

Beck's reflexive modernisation argument holds that current ecological and other risks will only be resolved if we begin from the question, 'How do we wish to live?' In democratically regulating social progress, what Beck is saying is that we choose to live in a different type of society, one more open to popular, democratic control and accountability. This democratisation process for Beck extends into those areas of social life, such as science and technology, which can dramatically affect people's lives, yet over which they currently have little or no control.

Democracy, democratisation and risk society

Risk society, although it denotes social change in which job insecurity, social unease and ecological hazard figure large, thus also presents opportunities for greater democratic accountability and institutional innovation. Indeed, one can say that Beck's position is that such democratisation, the opening up of previously 'hidden' areas to democratic accountability, is not just desirable, but absolutely necessary to deal with the problems of risk society. Allied with his critique of the various ecological (and other) risks associated with industrial society, Beck also criticises the latter for its limited and limiting form of democracy. According to him, '*Industrial society has produced a "truncated democracy", in which questions of the technological change of society remain beyond the reach of political-parliamentary decision-making.* As things stand one can say "no" to techno-economic progress, but it will not change its course in any way' (Beck, 1992b: 118; emphasis added). In order to rectify this, dealing with ecological risks calls for more not less democracy, more public openness, democratic accountability and popular participation in decision-making in science and technology. The simple question Beck asks is this: 'Why should democracy end at the laboratory door?', especially since the consequences of what is decided in the laboratory are potentially far-reaching. However, the recent establishment of a neo-liberal project of globalisation represents a serious challenge for this democratic imperative. As he notes:

the world market eliminates or supplants political action – that is, the ideology of rule by the world market, the ideology of neoliberalism. It proceeds monocausally and economistically, reducing the multidimensionality of globalization to a single, economic dimension that is itself conceived in a linear fashion. If it mentions at all the other dimensions of globalization – ecology, culture, politics, civil society – it does so only by placing them under the sway of the world-market system.

(Beck, 2000: 9)

In seeking to extend democratic control and regulation over 'techno-economic progress', Beck may be said to be simply asserting the classical defence of democracy; that is, those who are affected by decisions ought to have some say over how those decisions are made. As Levidow puts it, 'Risk assessment serves as an implicit ethics, and even as a potential means to democratize industrial development. . . . Risks lie across the distinction . . . between value and fact and thus across the distinction between ethics and science' (1995: 186).

Examples of this extension of democratic accountability include more 'right to know' information about state and corporate business decisions which affect the environment and people's lives; more opportunities for citizens to influence decisions which affect them; more democratic regulation of 'unaccountable' market decisions of firms; greater popular participation in public inquiries (from proposed new road developments to genetic engineering), and more overall democratic regulation of science and technology (Fischer, 1990).

Another reason for the extension of democratic norms of accountability and openness to science and technology relates to the point made above that according to the 'risk society' thesis, environmental risks are not simply technical matters to be dealt with by 'experts', but are also acknowledged as moral issues which society needs to debate and resolve. Thus, the recent controversy over genetic engineering may be seen as a classic example of what Beck means. Public disquiet and fears over cloning, genetic engineering and the creation of 'new' organisms is in large part motivated by a moral sense of unease with this particular form of human technology. While people are worried about the possible health and safety implications of genetic technology, there is also a deeply moral dimension to the issue, with some people thinking it is not right for humans to 'play God' in creating new forms of life, while others feel that the development of such technology indicates a violation and lack of respect for the intrinsic value of the natural world.

Beck's argument that 'risk society' requires the extension of democracy to deal with and possibly prevent environmental risks from arising in the first place (by the application of the precautionary principle, for example) may be viewed in terms of the two 'imperatives' of modernity discussed in Chapter 2. Recalling the

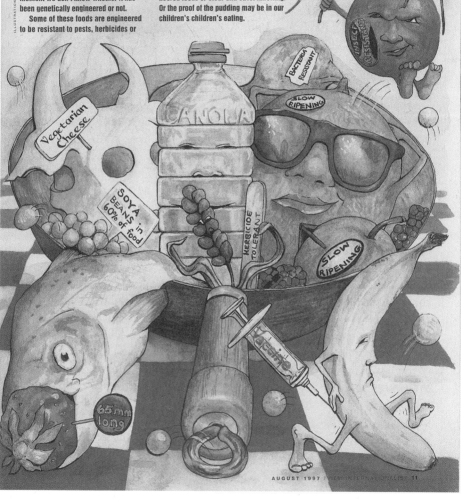

Figure 9.2 'Guess What's Coming to Dinner'

Source: Cakebread (1997)

shorthand description of modernity as being made up of the French and industrial revolutions, where the former represents the democratic imperative of modernity, while the latter represents its quest for material comfort and wealth, one could say that with 'risk society' we come full circle. In many respects Beck may be said to be using the democratic imperative of the French Revolution to control the potentially negative effects of the industrial revolution and its legacy. That is, 'risk society' and its model of 'reflexive modernisation' may be seen as requiring enhanced democratic control of and public accountability for industrial progress, a central part of which leads to the democratic 'redefinition' of what constitutes progress.

Another recent version or movement within social theory, which like Beck's 'risk society' focuses on the relationship between 'modernity' and environment, is postmodernism and the call for a 'modernity without illusions' (Bauman, 1993), which we turn to next.

Postmodernism, social theory and environment

Postmodern social theory (if one can say such a single 'school' may be said to exist) is one of the most recent developments in Western social thought, one of the many characteristics it shares with ecological or green moral and political theory. The aim of this section is to explore the extent to which environmental issues and problems may be said to be 'postmodern' in some sense, and examine the contribution of postmodern social theory to the analysis of environmental issues.

According to Docherty, the term 'postmodern' 'hovers uncertainly . . . between – on the one hand – extremely complex and difficult philosophical senses, and – on the other – an extremely simplistic mediation as a nihilistic, cynical tendency in contemporary culture' (1993: 1). Postmodern social theory has many roots and crosses many disciplines, and its origins may be found in such thinkers as Friedrich Nietzsche, Martin Heidegger and Sigmund Freud. In social theory one can point to the Frankfurt School of critical theory as a key origin, particularly Horkheimer and Adorno's *Dialectic of Enlightenment*, discussed in Chapter 4.

In pointing out the dark side of 'modernity', its various costs, risks and dangers, Adorno and Horkheimer created the theoretical ground from which postmodernism began. Indeed, in the *Dialectic of Enlightenment* one can also see the connection between modernity/modernisation, postmodernity and ecological concerns, as evident in Adorno and Horkheimer's statement that 'The fully enlightened earth radiates disaster triumphant' and 'What men want to learn from nature is how to use it in order to dominate it and other men' (Horkheimer and Adorno, 1973: 3).

Postmodernism is often associated with the abandoning of 'grand narratives' (Lyotard, 1984), such as 'progress', or at least the Western views of industrial and economic progress discussed in Chapters 2 and 8. Other 'grand narratives' which postmodernism rejects or suspects include Marxism and its 'myth' of a worldwide post-capitalist, communist society. Postmodernism also rejects or is sceptical of the grand narratives of science and scientific forms of enquiry which purport to reveal the 'truth' about the human or nonhuman worlds. It also focuses on the self, the individual subject, both the creation of this 'social category' within modernity as well as its problematic status. At the same time some have said that postmodernism is more concerned with the 'body' than with the 'body politic'. The latter is particularly evident in the work of Michel Foucault. Most post-modernists see themselves as against 'totalising' (and therefore potentially oppressive) forms of discourse and power and celebrate diversity, tolerance, difference, fragmentation and self-realisation. Above all else however, post-modern social theory takes a social constructionist approach, seeing 'nature' and the 'environment', for example, as socially constructed categories, brought into being by the operation of discourses and power, something which is also evident in Beck's theory.

Postmodernism, environmentalism and the rejection of modernity

According to Gare, 'It is the idea of environmental crisis as "Enlightenment gone wrong" which has encouraged the view that postmodernism represents a solution to the environmental crisis through a rejection of the modernist project' (1995: 5; Gandy, 1996: 26). Eder, in similar terms, declares that 'Modernity represents a culture which has reduced the material appropriation of nature to the exploitation of nature and thus provoked the ecological crisis' (Eder, 1996: 23). In simple terms, if the ecological crisis is caused by 'modernity' and its legacy, then the solution must be 'post-modernity' something other than modernity as we know it. The ecological crisis as a 'side-effect' of modern societies and the underlying logic and practices of the 'modernist project' thus forms a strong link between postmodern social theory and the environment. As leading green theorist and activist Charlene Spretnak put it, 'Green politics goes beyond not only the anthropocentric assumptions of humanism but also the broader constellation of values that constitute modernity' (1996: 533).

Postmodernism's critique of 'scientific rationality' and the scientific paradigm (part of its wider critique of Enlightenment reason and rationality) is similar to and overlaps with radical green critiques of how science has been central in 'dis-enchanting' nature and transforming it into a collection of resources, whose value is determined instrumentally by human considerations of nature's 'usefulness' as means to human ends. Science strips the natural environment of beauty, power and purpose, and reveals nonhuman beings as 'mere entities' or raw materials. An early exponent of this view of Enlightenment science and technology is the German philosopher Martin Heidegger, who not only prefigured aspects of postmodern theory, but is also seen as a central link in the relationship between postmodernism and environmental concerns (Zimmerman, 1992, 1994). According to Heidegger, modern science and technology offer a misguided and dangerous view of the natural world, and the place and relation of humans in and to it.

At the same time postmodernism, again building upon critical theory, suggests that the domination of nature, which is at the heart of modernity, results in the domination of human beings. Thus the emancipatory aim of the Enlightenment, namely to free humanity from ignorance, fear, squalor and domination, actually creates a new system of ignorance, fear and domination. In this way, postmodernism challenges the idea of modernity being 'progressive' or better, and suggests instead that modernity and modern society are simply 'different' from other possible forms of socio-economic development and organisation.

Thus from a postmodern perspective the Western model of progress and development cannot be said to be 'better' or 'more advanced' than non-Western societies and non-Western models of development. This rejection of the simplistic equation of 'modern' (i.e. Western/Eurocentric) values, principles and socio-economic models with 'progress' is something postmodernism shares with radical green critiques of Western forms of development being imposed upon the non-Western world. While the Enlightenment sought to universalise 'modern' values, principles and practices, in effect what this implied was the 'Westernisation of the world' (Latouche, 1993). This then led to the creation of a hierarchy in which non-Western societies, values and practices were, by definition, 'non-progressive' and 'pre-modern'. In this way, for postmodernism 'modernity' and its allied concepts and practices of 'modernisation', 'progress' and 'development' are and have been used to create a situation in which the 'otherness' and 'difference' of non-Western societies leads them to be seen as unequal and inferior to Western societies and Western values and norms. Postmodernism's view is that 'difference' and 'otherness' (central valued categories within postmodern theory) demand respect and equality and are thus against the creation of a universal set of standards, values and practices which has the effect of homogenising both the human nonhuman worlds and erasing 'difference' and 'otherness'. A more recent instance of this has been the postmodern critique and rejection of globalisation, which from its perspective basically implies the Westernisation of the world and the universalisation of the 'Western' model and standards of 'development' and what constitutes 'progress'.

The ecological sensitivity or potentiality of postmodern social theory is nowhere more evident than in its defence and valuation of otherness and difference. This is because the nonhuman natural environment is the ultimate 'other' for humans, since while our increased scientific knowledge of the natural world means we can learn more about it, ultimately we will never know the natural world fully, since a scientific understanding of the world is only one form of knowing. That is, from a postmodern perspective, the nonhuman world will (or ought) always to remain, at least in essence, strange to us, mysterious and wonderful. We will never be able to possess or grasp it in its entirety. Thus while Donna Haraway, one of the leading postmodern theorists in this area, holds that 'Nature is . . . that which we cannot not desire' (1995: 69), in keeping with postmodern theory, she also notes that a dominating attitude and treatment of the natural world can obscure as much as it reveals. Her view is that 'We must find another relationship to nature besides reification, possession, appropriation, and nostalgia' (1995: 70), one based on a respect for the nonhuman world's very strangeness and alienness. However, basing how we ought to view and treat nature on a recognition of its alienness and otherness is not particular to postmodernism (see O'Neill, 1993).

Postmodernism and post-industrialism

This idea that postmodernism represents a solution to the ecological crisis has expressed itself in many ways, particularly when postmodernism itself is related to 'post-industrialism'. Postmodernist social theory describes the current state of society, not only as 'postmodern' in cultural terms, but, related to that, also as post-industrial in terms of its economic system. Postmodernism in the words of Frederic Jameson (1991) is the 'cultural logic of late capitalism', where 'late capitalism' can be roughly equated with a 'post-industrial' stage of capitalism. The term 'post-industrial' conveys the idea that 'advanced' Western societies are now in a qualitatively different developmental stage than the 'industrial' one, which approximately spans from the beginning of the industrial revolution to the 1980s. The post-industrial hypothesis is that as industrialised societies progress they move away from economies based on large-scale, factory-based, industrial manufacturing towards service-based, hi-tech forms of employment, production, distribution and consumption. One shorthand way of understanding post-industrialism is to say that it implies a shift in industrial society away from making things to providing services. Insofar as postmodernism implies post-industrialism, and insofar as green politics is a post-industrial politics with a vision of a post-industrial society, then to that extent postmodernism and green politics have much in common. The important point here is that 'post-industrialism' is typically argued to require a less exploitative use of the natural environment. A largely service-based economy, with modern electronic communications and information-processing networks and technology, in which information rather than 'raw materials' is the central economic resource, is argued to create a more sustainable economy.

The connection between postmodernism and green politics may also be seen in those analyses which suggest that the non-instrumental appreciation of the natural environment which is central to green politics requires a certain degree of material affluence. In this way 'post-industrialism' is intimately related to 'post-materialism'. According to Ronald Inglehart's famous 'post-materialism' thesis, a profound shift took place in Western societies in the 1960s and 1970s, which helps explain the rise of green politics and a non-instrumental concern with the natural environment. Inglehart (1977), echoing some of the themes within Giddens' thought discussed in Chapter 4, suggested that the experience of peace, rising levels of material affluence, employment, mobility and education in Western societies since the Second World War had led to a profound cultural change in those societies. Basically his thesis was that having secured high levels of material satisfaction and standard of living, many people in Western societies became aware of a lack of 'post-material', qualitative goods and experiences. Chief among these post-material interests was a concern with the preservation of the natural environment which led to a rise in green politics. In this way, then, postmodernism is related to post-materialism and the consequent connection between the latter and ecological politics, interests and movements.

However, the equation of a concern with the environment and environmental politics as 'post-material' is something that can be criticised, particularly when there are many environmental issues and movements which are characterised by 'material' as opposed to post-material interests and concerns. There is a very 'material' basis for environmental concerns and green politics, based on the relationship between environmental degradation, poverty, ill-health and social injustice, aptly expressed as an 'environmentalism of the poor' by Guha and Martinez-Alier (1997).

Postmodernism and the social construction of the environment

A final and arguably more contentious area of postmodern engagement with the environment involves its 'social constructionist' approach to environmental problems (Hannigan, 1995), and its deconstruction of such key terms as 'environment', 'nature' and the 'natural' (Robertson *et al.*, 1996; Andermatt Conley, 1997).

According to Hannigan, 'environmental problems are not very different from other social problems such as child abuse, homelessness, juvenile crime or Aids' (1995: 2), in that what we should focus on is not the 'problems' as such, but how the 'problems' come to be 'constructed' by the interaction of the social actors, the types of knowledge deployed and the power relations between these factors. An extreme version of the social constructionist perspective would be the postmodernist Jean Baudrillard's idea that environmental 'problems' are 'created' by images, the media and means of communicating/constructing them, not that there are objectively existing environmental problems 'out there' in nature for humans. While most social constructionist approaches, such as postmodernism, would not go so far as to deny the reality of environmental problems, the important point they raise is that in analysing environmental issues one must be aware of the different actors, claims, types of knowledge, communication and cultural contexts in which these problems are articulated, contested, presented and re-presented. The key to understanding the attitude of postmodernism to environmental issues is to remember that for it 'knowledge is power'. Hence from a postmodern perspective what is of primary importance is to identify whose or what knowledge is the dominant or most powerful in the social construction of the problem and suggestions for possible solutions.

A theory of 'post-ecologist' politics based on this social constructionist perspective has been advanced by Blühdorn. What may be termed the 'happy nihilism' of the postmodern 'post-ecologist' can be seen in his view that 'The important question is not the reality of ecological damage, degradation, human-induced health problems etc., but rather for what reasons and to what extent such phenomena . . . can be conceptualised as problems and crises' (2000: 10). He criticises existing green politics for wanting to change the world, rather than taking the world and its underlying stuctural dynamics as given and therefore beyond change. That is, greens should simply accept the 'sociological reality' of the modern world ('get with the programme' as it were), and attempt to work with, rather than against, the grain of modern market-based consumer society.

The basic logic of the argument seems to be that:

- The dynamics of development of modern society are going in an anti-ecological direction.
- Ecologically committed thinkers are therefore naïve and unrealistic in thinking that they can alter this.
- Therefore they should abandon 'castle – building' and get 'sociologically real'.
- Seek to accommodate and tailor the 'ecological project' to the 'reality' of an advanced, market-based, high-consumption society.

Thus in the face of continuing and increased ecological destruction, environmental injustice, global economic and environmental inequality, Blühdorn appears to claim that green thinkers should abandon their critical stance, adopt a sociologically grounded and informed viewpoint, explore and discover the dynamics of modern social development, and try to see where they can fit into or attach their ecological aims on to these, so that this social development is 'greened'. It is little surprise then to find that the conclusion he reaches is that the best green politics can hope for, by way of influencing modern society, is through a technocratic, elitist, 'reformist' policy of ecological modernisation (2000: 191–200). That is, to put it bluntly, the best green politics can (and ought to) do is 'green capitalism'.

Depending on one's view, this is either a profoundly conservative (in that it effectively 'naturalises' the current development path of society, so that its underlying principles and dynamics are simply beyond human collective control), and thus could be used to butress right-wing and contrarian arguments as discussed in Chapter 5, or at best indicates an extremely modest reformist agenda (weak ecological modernisation), in which green politics is forced to limit itself to what is currently politically and culturally possible.

The conservative interpretation of Blühdorn's view may be seen (like much of post-structuralism and postmodernism) in his positive endorsement of Hegel's view that 'human rationality is more suitable for interpreting *that which is* than for issuing instructions on how the world *ought to be*' (2000: 5). Blühdorn's postmodern/social constructionist thesis (largely based on Niklas Luhmann's work) in which the environmental crisis is 'entirely internal', and which substitutes human communication about the environment for actual human-induced effects on and vulnerability to changes in the physical environment, at the centre of the environmental debate. From this perspective, the environmental crisis may thus be simply resolved (or as he later puts it 'dissolved') by reminding us that it is all in our heads. Again to put it bluntly, the advice Blühdorn gives is that 'since we cannot change the world, what we ought to do is change ourselves and our perceptions about this unalterable world'. This of course sounds like the logic of 'there is no alternative'. His analysis says nothing about dominant political, cultural and economic forces which underpin the 'ecologically hazardous dynamic' of modern societies, since, as noted above, it 'naturalises' this dynamic as something we cannot control or alter. Since we cannot control or change this dynamic, the best we can do is adapt to it.

According to Blühdorn, green political theory expresses the view that there is 'one nature, one global ecological equilibrium, one human interest of survival, one intrinsic value of nature and one human rationality' (2000: 36). While there are those within the ecological movement and green political theory, such as deep ecology, and those for whom 'wilderness preservation' is a central aim and normative grounding, the majority of green political theorists do not hold this view. This is extremely irritating and academically negligent, in that it reduces the plurality and variety of green political theory to 'eco-fundamentalism'. Green politics, while obviously concerned with human–nature relations, is not reducible to this concern, and is certainly not reducible to a 'fundamentalist' search for one, singular conception of what the 'proper' or 'only' relationship between humans and nature should be. However, it is interesting to note that others have also highlighted the potential 'fundamentalism' of green thinking such as the Jungian psychologist Andrew Samuels, who has noted that:

> Running an analyst's eye over the information and education material put out by organisations like Greenpeace and Friends of the Earth, I am struck by the one-sided censorious portrait of humanity that is presented. . . . The result of too much self-disgust may be the cultivation of a deadening cultural depression that would interfere with the imaginative energy required for environmental action.
>
> (Samuels, 1993: 211)

A major claim which Blühdorn makes is stating that all green political theory is 'fundamentalist' in valorising and privileging some notions of the natural and nature, even if they are concerned with sustainable development or biodiversity. For example, he claims that 'They have found substitutes like sustainability, biodiversity and the like. However, these concepts can hardly be substantiated – and the related practical guidelines justified – without making reference to nature and the natural' (2000: 76). Yet at no point does Blühdorn demonstrate this link, the essential 'eco-fundamentalism' of all green politics, and not just deep, ecology-like ones. Thus a fatal problem with Blühdorn's analysis is his reduction of all green political theory to 'deep, ecological-type' environmentalism. For example, the 'abolition of nature' (taken to mean the abolition of any stable, unitary conceptualisation of nature and human relations to it) is only a threat to those views of green politics for whom this totemic conception of 'nature' is central. It is not a feature of all green theories. For example, pluralist, non-fundamentalist views of nature may be found in eco-socialist/Marxist and eco-feminist views (such as one finds in the works of Ted Benton, Ariel Salleh, Caitriona Sandilands, Mary Mellor, James O'Connor or David Harvey).

By making 'sustainability' a 'code' just like 'nature', Blühdorn voices his central assertion that all green politics is ultimately 'eco-fundamentalist' in character. He claims that 'If we assume that the most fundamental distinction of ecological communication is the distinction *natural–unnatural or sustainable–unsustainable*, all statements which are intended to be a contribution to the ecological discourse must be derived from and reducible to this distinction' (2000: 139). Yet it is difficult to substantiate the claim that *all* green politics is reducible to this master distinction. Sustainable–unsustainable is not the same as natural–unnatural, since, among other things, the former carries with it a whole range of arguments, debates and positions on non-ecological issues from human rights, democracy, equality, gender relations and the organisation of the economy, in a way 'natural–unnatural' resolutely does not.

Blühdorn's 'post-ecologist politics' is a descriptive sociological analysis of ecological politics not an ecologist ideology. Building on Luhmann, 'post-ecologism' is not only 'ecology without identity', it is also 'post-humanist'. Thus, rather than the subject of analysis being the actual, 'real', 'material' degradation of the environment, and the variety of social, political, economic and cultural causes and consequences of this, the subject becomes the disintegration of the ecological issue as a linguistic/communicative problem, as in the common post-modern claim of the arbitrariness of the distinction between 'environmental change' and 'environmental degradation'. That is, when is an environmental problem a problem? Or rather we need to examine who is making this claim, for what reasons and based on what sorts of knowledge claims.

However, as a descriptive sociological analysis rather than a critical account of green ideology, there may be more to Blühdorn's 'post-ecologist politics' in terms of the limits of 'simulative politics' and half-hearted attempts to make 'green consumerism' coupled with encouraging 'recycling' and other forms of 'environmental governmentality' the main ways of realising sustainability (Luke, 1999b). As Blühdorn puts it:

> For late-modern societies, the question is not how (nor on the basis of what kind of frame of mind) the capitalist growth economy can be made socially just and ecologically sustainable. *Instead, the question is; How we can generate the illusion that social justice and ecological integrity are on the agenda while at the same time perpetuate a system and a frame of mind that is inherently socially exclusive and ecologically unsustainable? Put differently, the fundamental challenge for late-modern society is the management of unsustainability and its implications, rather than its removal.* In this situation, late-modern society has developed strategies of simulation that stabilize the normative basis of society and pacify potentials for social conflict that emerge from inherently exclusive and unsustainable practices of growth, accumulation, and consumption.
>
> (Blühdorn, 2002: 22–3; emphasis added)

In a related vein Luke describes the popularity of 'green consumerism' and recycling as the 'domestication' of environmentalism, a way of 'disciplining' the radical potential of the ecological critique, and which locates the major source of responsibility for unsustainability in the wrong places with individual consumers and citizens rather than in industrial, corporate and state-managed modes of technologically organised production (Luke, 1997: 115–36). This 'eco-governmentality', which also includes the new emerging power/knowledges of 'managing unsustainability' through technical and expert means, threaten the emancipatory and radical democratic potential of green/ecological politics, but at the same time are also attractive modes of strategic thinking and acting for the green movement since it offers them a real chance of being taken seriously by state and corporate interests. For Luke, what we are seeing is the realisation of Beck's analysis of 'sub-politics' with a vengence, that is, the 'de-democratisation' of ecological issues away from the public debate about politics to the closed, expert 'sub-political' realm, hidden from democratic oversight (Luke, 2002: 308).

The emergence of 'ecological problems' in the 1970s is for Luke best analysed through Foucault's notion of state policy towards 'population' as a discrete object of analysis. For Focault, modern ecology is a continuation and development of 'one of the great innovations in the techniques of power in the eighteenth century . . . the emergence of "population" as an economic and political problem' (1978: 25). By 'population' is meant the ensemble of relations between 'people' and 'environment' viewed from the perspective of state power, a form of knowledge/power of which Malthus was one of the first to articulate, but which may also be found historically and today in state regulation of population (whether through encouraging/enforcing population control – as in China – or encouraging increases in population – as in some European countries), sexuality, and the productive interaction between economies and the natural environment, as articulated by 'political economists' such as Marx in the nineteenth century or contemporary free market environmentalists. Discourses of 'sustainable development' (and discourses critical of it, such as in the anti-environmental backlash discussed in Chapter 5) are merely modern ecological readings of this Foucauldian power/knowledge discourse of 'population', state, 'political economy' and expertise for Luke. According to Luke:

> Because governmental techniques are always the central focus of political struggle and contestation, then interactions of populations with their natrual surroundings, in highly politicized economies compel regimes to constantly redefine what is within their competence through the modernization process. To survive in the fast capitalist world of the 1990s, it is not enough for territorial states to maintain legal jurisdiction over their allegedly soverign territories. . . . As ecological limits to growth are either discovered or redefined in sustainability discourses, states are forced to make good on an almost impossible obligation: namely, guaranteeing their populations' fecundity and productivity in the total setting of a global political economy by becoming

'environmental protection agencies' . . . green governmentality [means that] economic growth must not be limited, but rather become sustainable.

(Luke, 1999b: 145)

From this Focauldian 'eco-governmentality' perspective, sustainable development is viewed as a state and expert-led exercise in 'greening capitalism' rather than challenging or transforming it, and the expert and knowledge aspects of this 'geo-power' have echoes of the eco-authoritarian Ophuls' suggestion of the need for a 'priesthood of responsible technologists' (1977: 159) to govern in the interest of social survival.

At the same time, building on the critique of 'modernist' science in particular, and theories of knowledge in general, postmodern social theory has questioned the status and meaning of central concepts such as 'nature', the 'natural' and 'environment' as well as 'human' and 'nonhuman'. In many respects the input of postmodern social theory to the analysis of environmental issues highlights the way in which these concepts have multiple meanings and senses. In this way they are always open to alternative meanings and understandings. What 'nature' and 'environment' are and mean varies from culture to culture, group to group and means different things at different historical periods. As Gandy notes:

> The most important lesson to emerge from any serious engagement between postmodernism and environmentalism is that we cannot understand changing relations between society and nature by relying on ahistorical and positivist modes of explanation which refuse to engage with the social and ideological dimensions of environmental discourse . . . our knowledge becomes open to negotiation and interpretation.
>
> (Gandy, 1996: 156)

A good illustration of this is Lyotard's prescient statement that under neoliberal conditions governments and companies have foresaken 'idealist and humanist narratives of legitimation in order to justify the new goal: in the discourse of today's financial backers of research, the only credible goal is power. Scientists, technicians, and instruments are purchased not to find truth, but to augment power' (Lyotard, 1984: 46). This offers another good analytical/interpretive perspective on the 'science wars' discussed in Chapter 5, as well as revealing the political economy of knowledge production, that is, the economic and ideological forces which determine reseach agendas, funding and publication.

The multiple meanings of such terms as 'nature' and the 'environment' are by no means limited to postmodernism; indeed, the contested character of these terms was raised and discussed in Chapter 1. However, postmodernism has arguably gone further in exploring the multiplicity of meanings of concepts such as 'nature', 'environment' and the 'natural' (Cronon, 1995). For example, what does 'nature' mean in an age when nature as an external, autonomous, independent entity is increasingly affected by human activity? What do 'natural' and 'unnatural' mean to different people? Why, in the Western world, is such a premium placed on all things 'natural' (goods, experiences, commodities, foodstuffs) precisely at a time when 'nature' is perceived to be under threat or disappearing? As Cronon puts it:

> What happens to environmental politics, environmental ethics, and environmentalism in general once we acknowledge the deeply troubling truth that we can never know at first hand the world 'out there' – the 'nature' we seek to understand and protect – but instead must always encounter that world through the lens of our own ideas and imaginings? . . . Much of the moral authority that has made environmentalism so compelling as a popular movement flows from its appeal to nature as a stable external source of nonhuman values against which human actions can be judged without much ambiguity. If it now turns out that the nature to which we appeal as the source of our own values has been in fact contaminated or even invented by those values, this would seem to have serious implications for the moral and political authority people ascribe to their own environmental concerns.
>
> (Cronon, 1995: 25–6)

Problems of postmodern environmentalism

Postmodern 'deconstruction' and its view of the 'postmodern' social world as inherently fragmented, chaotic and unstable seems to fit with certain strands in modern scientific enquiry, such as chaos theory, which also see the natural world as unpredictable, chaotic and unstable. However, some schools of postmodern thought which reject the independent 'reality' of the 'real world are deeply problematic from an ecological point of view which is premised on the non-discursive, non-negotiable existence of independent ecological, material limits on human action' (Gandy, 1996: 36).

As Zimmerman in his study of postmodernism and radical ecology notes, 'radical ecologists criticize *some* aspects of modernity, while appropriating and transforming *other* elements of its emancipatory vision' (1994: 4). That is, critical social theory based on ecology and motivated by a concern for environmental destruction need not necessarily require the abandoning of the 'project of modernity'. Like Achilles' lance, modernity (as theorists such as Beck suggest with his theory of 'reflexive modernisation', and also the theories of Habermas and Giddens) is argued to be able to heal the ecological (and other) wounds it inflicts. Thus there is no *necessary* connection between environmental concerns and politics and postmodernism.

The limits of postmodern analysis of environmental issues are no more clear than at the level of global environmental degradation. For some critics, postmodernism is unable to articulate a political economy of global environmental destruction, which leads to an inadequate analysis of power and social agency in a meaningful, useful and political sense. This 'apolitical' or 'anti-political' charge is a common one made against postmodernism. According to Gare, 'Postmodern culture is the culture of a society in which politics has become a farce, where rational critique and protest have become impossible' (1995: 34), and goes on to state that 'Decrying the quest for political power as the problem, they [postmodernists] have handed over responsibility for their fate, and the fate of the world environment, to the economic rationalists, to the new international bourgeoisie and the international market' (1995: 34).

At the same time, focused as postmodernism is on the 'discursive' aspects of environmental politics and debates leaves it unable to deal with the ecological, non-discursive, material dimensions of environmental problems (Gandy, 1996).

Postmodernism offers interesting and sometimes illuminating 'deconstructions' of how terms such as nature, environment, and related concepts such as human, are inherently *relational* in that one cannot speak of 'human' without also speaking of the 'nonhuman' or the 'natural' without its corresponding opposite of 'unnatural'. However, for many, its relativism, ambiguities, lack of a clear conception of political power and agency, its drift into a celebration of 'late capitalism' and its consumer culture, individualistic focus and stress on aesthetics and lifestyle means it is inadequate for the full range of issues around the relationship between human societies and their environments.

Conclusion

This chapter has surveyed two recent social theories which analyse or may be used to analyse the environment and the relationship between environment and society. What Beck's 'risk society' thesis and postmodernism share is that they are both orientated around the Enlightenment or modernity, and both explore how central to modernity and modernisation is the change that takes place in how society views, values and uses the nonhuman environment. However, while 'risk society' and the idea of 'reflexive modernisation' may be broadly seen as within the 'project of modernity', postmodernism, as the name suggests, is more critical of modernity, modern ways of thinking and acting in general, as well as in relation to the environment.

Summary points

- A central point of reference in looking at the environment and social theory is modernity and its legacy.
- Beck's 'risk society' thesis both *describes* the current Western world as one in which people and politics are more concerned with the distribution of 'bads' (such as environmental and other risks) rather than 'goods' (income and jobs), and also *prescribes* a more ecologically enlightened social alternative.
- A central part of Beck's analysis focuses on how risk society demands what he calls a process of 'reflexive modernisation' and the redefinition of what constitutes social 'progress'.
- Central to the latter is the necessity (as well as desirability) of spreading democratic control and accountability to more areas of decision-making which affect people's lives, particularly scientific and technological developments.
- Whereas Beck's 'risk society' thesis argues that modernity and modern society based on its principles can solve its own environmental and other problems, postmodern social theory rejects the idea that modernity can do this, and instead calls for 'postmodern' solutions and approaches.
- Postmodernism is particularly good at indicating the multiple and contested meanings of such social constructions as 'environment', 'nature' and 'natural'.
- While postmodernism highlights the importance of identifying the discourses and powerful actors in environmental issues and disputes, it has an 'apolitical', 'anti-political'or conservative character which is a serious flaw.
- Blühdorn's 'post-ecologist politics' viewed as a postmodern analysis of the 'simulative' politics of environmental rhetoric offers an interesting perspective on environment and social theory, as does Timothy Luke's 'eco-governmentality' critique of the 'normalising' power/knowledge discourse of 'official' environmentalism.

Further reading

On Ulrich Beck's 'risk society' thesis and 'reflexive modernisation' see his *Risk Society: Towards a New Modernity*, London: Sage, 1992; *Ecological Enlightenment: Essays on the Politics of the Risk Society*, Princeton, NJ: Humanities Press International, 1995a; *Ecological Politics in an Age of Risk*, Cambridge: Polity Press, 1995b; 'Risk Society Revisited: Theory, Politics and Research Programmes', in Barbara Adam, Ulrich Beck and Joost van Loon (eds), *The Risk Society and Beyond: Critical Issues for Social Theory*, London, Sage 2000.

An excellent critical analysis of Beck may be found in ch. 5 of David Goldblatt's *Environment and Social Theory*, Cambridge: Polity Press, 1996. Other sources include: Scott Lash, Bron Szersynski and Brian Wynne (eds), *Risk, Environment and Modernity: Towards a New Ecology*, London: Sage, 1996; A. Kerr and S. Cunningham-Burley (2000), 'On Ambivalence and Risk: Reflexive Modernity and the New Genetics', *Sociology*, 34:2; T. Benton, 'Beyond Left and Right? Ecological Politics, Capitalism and Modernity', in M. Jacobs (ed.), *Greening the Millennium? The New Politics of Environment*, Oxford: Blackwell, 1997; Alan Irwin, 'Risk, the Environment and Environmental Knowledges', in M. Redclift and G. Woodgate (eds), *The International Handbook of Environmental Sociology*, Cheltenham: Edward Elgar, 1997; Pat Caplan (ed.), *Risk Revisited*, London: Pluto, 2001; B. Wynne (2002), 'Risk and Environment as Legitimatory Discourses of Technology', *Current Sociology*, 50:3; P. Rutherford, 'Ecological Modernisation and Environmental Risk', in E. Darier (ed.), *Discourses of the Environment*, London: Sage, 1999.

On postmodernism and environmental issues see: Arran Gare, *Postmodernism and the Environmental Crisis*, London: Routledge, 1995; Michael Zimmerman, *Contesting Earth's Future: Radical Ecology and Postmodernity*, Berkeley: University of California Press, 1994; Klaus Eder, *The Social Construction of Nature: A Sociology of Ecological Enlightenment*, London: Sage, 1996; William Cronon (ed.), *Uncommon Ground: Towards Reinventing Nature*, New York: W.W Norton, 1995. The work of Donna Haraway is particularly good as an example of how aspects of postmodern social theory may be used to analyse the environment, and social–environmental relations. See her *Simians, Cyborgs and Women: The Reinvention of Nature*, London: Free Association Books, 1991, and her contribution to the Cronon edited volume above, 'Universal Donors in a Vampire Culture'.

Other postmodern approaches to theorising the environment include: Vera Andermatt Conley, *Ecopolitics: The Environment in Poststructuralist Thought*, London: Routledge, 1997; G. Robertson *et al.* (eds), *FutureNatural: Nature, Science, Culture*, London: Routledge, 1996; and John Hannigan, *Environmental Sociology: A Social Constructionist Perspective*, London: Routledge, 1994; Ingolfur Blühdorn, *Post-ecologist Politics: Social Theory and the Abdication of the Ecologist Paradigm*, London: Routledge, 2000, and his (2002) 'Unsustainability as a Frame of Mind – and How We Disguise It: The Silent Counter-revolution and the Politics of Simulation', *The Trumpter*, 18:1; T. Luke, *Ecocritique: Contesting the Politics of Nature, Economy, and Culture*, Minneapolis: University of Minnesota Press, 1997; Luke, 'Training Eco-managerialists: Academic Environmental Studies as a Power/Knowledge Formation', in F. Fischer and M. Hajer (eds), *Living with Nature: Environmental Discourse as Cultural Politics*, Oxford: Oxford University Press, 1999a; Luke, 'Environmentality as Green Governmentality', in E. Darier (ed.), *Discourses of the Environment*, Oxford: Blackwell, 1999b; Luke, 'The People, Politics, and the Planet: Who Knows, Protects, and Serves Nature Best?', in B. Minteer and B. Pepperman Taylor (eds), *Democracy and the Claims of Nature: Critical Perspectives for a New Century*, Lanham, MD: Rowman & Littlefield, 2002; Luke, 'Environmentality as Green Governmentality', in E. Darier (ed.), *Discourses of the Environment*, London: Sage, 1999.

10 Ecology, biology and social theory

Introduction

Having introduced some of the ways in which the environment has been conceptualised with some central strands of twentieth-century social thought, this chapter continues this exploration, but widens out the issue beyond the 'environment' (both natural and urban), to look at approaches to the study of the natural world and humans in relation to and a part of that world, and how natural sciences (such as biology, ecology and ethology) have influenced social theorising and/or been used themselves as the basis for social theory.

For many social theorists, the social-scientific study of the relationship between society and the natural environment requires that we address the following question: 'How do we open up to investigation the relationships between humans and the rest of nature, without letting in the "Trojan horse" of biological determinism?' (Benton and Redclift, 1994: 4). That is, how are we to study the ecological relationship of the human species to its natural environment (or eco-logical niche) while avoiding thinking about and studying humans as if they were simply another species and thus explain human behaviour using the same biological and ecological models and conceptual analysis used in the case of other species? At the same time it is quite obvious that humans are a particular species, living in particular environments like other species on the planet, and thus it seems odd to reject a biological or ecological approach to the study of human society and behaviour. So why are we worried about 'biological determinism' if humans are a biological species, and the behaviour of other biological species can be explained and theorised on the basis of their biological characteristics within the context of their environment? But are humans the same as other species? Surely, some would say, humans are 'special', 'different' in some important way from other species, which makes the application of biological and ecological models inappropriate to them? This chapter seeks to discuss these questions and examine the various ways recent theorists have sought to integrate biological and ecological insights into the study of human behaviour and relations between human society and the environment.

Sociobiology: genetics, 'human nature' and ecology

While Benton and Redclift (1994) are weary of adopting a 'biological determinist' approach in studying human–environment relations, for some this question is unproblematic. For some a 'biological determinist' approach is absolutely central to understanding not only human relations to our environment, but also to explaining our social relations to fellow members of our species. The school of thought which best expresses this 'biological determinist' view is sociobiology, which as its name suggests is a composite theory of human behaviour in which biological theories are applied to the human social world. As the *Hutchinson Dictionary of Ideas* puts it, sociobiology is the 'study of the biological basis of all social behaviour, including the application of population genetics to the evo-lution of behaviour' (1994: 483). For sociobiology, to deny our biological make-up, our 'selfish genes' in Richard Dawkins' (in)famous term, in our analysis of human society (and especially in our proposals for social change) is a dangerous mistake.

Sociobiology as a form of 'biological determinism', basing its claims on the direct application of genetic and evolutionary theory to humans, has a long pedigree, and has clear antecedents in Malthus's 'theory of population', 'Social Darwinism' and Herbert Spencer's ideas, as discussed in Chapter 3. A flavour of its biological-ecological determinism may be seen in Wilson's claim that 'it is entirely possible for all known components of the mind, including will, to have a neurophysiological basis subject to genetic evolution by natural selection' (1978: 10). Taking human beings and their social and individual behaviour to be just like the behaviour of other animals, sociobiology has sought to explain society, politics, culture, and the relationships between men and women in terms of genetics and evolution by natural selection.

In short, sociobiology advances a particular idea of 'human nature' and uses this to explain human social life. As a form of social theory, sociobiology aims to explain social phenomena by insisting that humans are not just *like* other animals and therefore analogies may be drawn between humans and comparable species, but that we *are* a species of animal, with our own natural characteristics, and to deny that animal nature is mistaken. As Gregory explains, 'sociobiology tells us much about the inherited behavior of animals; and it tells us in a new way that we ourselves are animals' (1978: 293–4). Its view of human nature as a 'fixed' essence and largely immune to the effects of socialisation or 'nurture' shares many of the aspects of the traditional conservative political view in which a similar 'negative' view of human nature is used to criticise political programmes which place great faith in the 'innate goodness' of humans and/or the capacity of altered social practices and institutions to make people less selfish, violent, competitive and so on. In this respect, sociobiology often seems to echo Malthus' in his critique of 'progressive' Enlightenment thinkers of his time. Like Malthus, sociobiologists speak of natural constraints (especially the evolutionary, genetic make-up of humans) which they argue undermines attempts to make the world a better place through changing socialisation patterns or the social environment of individuals. Whereas for many socialists, people can be encouraged to behave in a more co-operative manner by changing their social environment, for socio-biologists what is more important is the evolutionary and genetic aspects of human behaviour (which in part is based on the 'original', prehistoric natural and social environment of the evolving human species).

For Dawkins, another early proponent of sociobiology, human behaviour is explicable in terms of the 'selfish gene'. Dawkins' claim is that human individuals are simply carriers of genetic information, and since the principal aim of genes is to reproduce themselves, all of human behaviour, social rules, taboos, traditions, institutions may be explained by reference to the 'selfish gene'. Biology determines society, and in this way culture is reduced to human nature. One of the interpretations of the modern ecological crisis, from a sociobiological perspective, is that our genetic coding, or instincts evolved in the 'hunter-gatherer' stage of human evolution, are not suited or do not exhibit 'inclusive fitness' in the modern situation. Our advancement, technologically, scientifically, economically and socially, in radically altering our relationship to the natural environment, has 'outstripped' our instinctual ability to cope with these changes (Berry, 1986).

The idea of human nature within sociobiology comes down to the 'instinctual' basis of human action, that these are somehow 'given' and therefore 'beyond change'. The import of this, as Berry has noted, is that 'Humans as humans have a limited range of options open to them. The presence of this limited range means that there is a constancy and predictability about what humans will do in specific situations' (1986: 104). Often sociobiological arguments express themselves not so much in the idiom of 'biological' as 'psychological determinism'. According to Davies (1976), the Russian Revolution (or at least Lenin's part in instigating it) may be explained along the following lines: 'As a very young man, Lenin experienced the death of his father, his brother, and his sister, for reasons attributable to the Tsarist government. Decades later, these experiences became part of the fuel of a revolution' (1976: 104). Such controversial social theorising is part and parcel of sociobiology, controversial in reducing social events, such as revolutions, to matters of individual psychology and personal experience. The point is not that the latter are not important in explaining social action, but rather one cannot *reduce* the latter to the former.

For some sociobiologists, the 'fact' that humans (or at least males) are 'naturally aggressive' and 'competitive' is something that is fixed and 'given' and therefore unchangeable. As these behavioural traits are genetically 'hard-wired', the best society can do is either to minimise their negative impacts or use them to produce/ motivate socially desirable outcomes. Echoing aspects of Freud's work, socio- biology suggests that since war and aggression is 'natural to humans' and instinctual, the best we can do is sublate or channel these potentially harmful impulses into socially beneficial or harmless pursuits. It is from this type of argument that we come to the popular sociobiological explanation for competitive sports in human cultures as evolutionary adaptations to innate human aggression, group loyalty, group and individual competitiveness and so on. Football, on this view, may be explained as a cultural phenomenon in which natural (male) impulses for male bonding, competitive and aggressive territorial behaviour can be exercised and (neutralised) on (but increasingly spills over and off) the pitch.

At the same time, sociobiology can also give an evolutionary-genetic reason for the predominance of males in warfare. While the orthodox sociobiological explanation for this is on the basis of the superior physical strength of (young) men as compared with women – that is, men are 'adapted' for warfare – there is another less 'sexist' interpretation. On this view (partly based on anthropological evidence), the reason why young men have traditionally been the 'warriors', and have monopolised (and still do) warfare and the military institutions of society (that is, the institutions of organised violence and aggression) is that unlike women, young men are 'superfluous' from the genetic and evolutionary perspec- tive. That is, sending a large proportion of young men off to war (with a good chance of their being killed) did not significantly affect the genetic evolution of the species. At the same time, anthropological evidence seems to suggest that in 'hunter-gatherer' societies (in which humans have spent most of their evolution), it was women who provided most of the protein in these societies (from gathering vegetables and rudimentary cultivation) and not the occasional and infrequent contributions of meat from male hunting (Diamond, 1991: 34).

Many have criticised some sociobiological arguments as based on ideological rather than scientific theory. Sociobiology is held to offer 'scientific' legitimacy to right-wing, conservative and 'free market' ideology (Alper, 1978; Berry, 1986: 109). One can see here in the language of the 'selfish gene' the innate competitiveness and aggression of human beings, an echo of the 'invisible hand' idea of the eighteenth century, where the self-interest of each person will, in market transactions, result in socially beneficial results. F.A. Hayek, one of the foremost theorists and defenders of a libertarian, free market perspective, represents a good example of this ideological appropriation. As Dickens (1992) points out, Hayek's evolutionary view of human beings leads him to see the modern capitalist system, the free market and private property as adaptive forms of what he calls 'the extended order', a cultural form in which civilisation, social order, human progress and development have been achieved. Thus Hayek's position is close to the 'Social Darwinian' perspective discussed in Chapter 5. Hayek's 'organic' view of human society, in which society is seen as a 'super-organism' in relation to its environment, places this strand of right-wing thought within strands of conservatism, which also have an 'organic' view of society. This 'naturalistic' view of society is used to combat those (usually from the left of the political spectrum) who argue that conscious collective action, altered social institutions and practices can create a new type of society and human being. For the conservative, such 'social engineering' is both impossible and dangerous, since the essence of human beings, their nature, is fixed and immune from deliberate social changes.

Sociobiological arguments have caused great controversy, particularly when used in arguments about whether 'nature' or 'nurture' determines human behaviour. The political importance of this debate should be obvious. If 'human nature' is fixed genetically and behaviour reduced to genetic causes, then there is little point in trying to create a 'social environment' aimed at encouraging people to be less selfish, competitive and so on. Trying to make people less selfish is seen by those who subscribe to the sociobiological/conservative view as 'unrealistic' and 'utopian', in attempting to 'go against the grain' of human nature.

A particularly contentious example of sociobiological theorising is its use to scientifically 'prove' innate, 'natural' differences between different racial or ethnic groups. Particularly in America there have been attempts to prove that Afro-Americans are genetically less intelligent than whites or Asians. This is the basic argument of a book published in 1994 by Richard Hernstein and Charles Murray, *The Bell Curve*, which was criticised (like much crude, deterministic versions of sociobiology) for being right-wing and reactionary, and for justifying racism and racist attitudes and finally for misusing science in order to do so. It thus stands as another example, along with the Lomborg controversy discussed in Chapter 5, of the role of values, ideology and political power in the generation and dissemination of scientific knowledge, as well as Beck's idea of the 'breakup' of science in that it no longer speaks with one voice.

Sociobiology and the biophilia hypothesis

There have been attempts, notably by E.O. Wilson, to establish a genetic or evolutionary reason/explanation for what he calls 'biophilia' or love of nonhuman life. Now while there are many genuine questions one may ask concerning the evidence for this genetic, hard-wired concern for nonhuman life, whether Wilson is talking about 'biophilia' or mere 'bioconnection', and whether this genetic 'ecological sensitivity' is relevant to modern, urban dwellers (see Baxter, 1999: ch. 3), it still remains the case that biophilia is a sociobiological attempt to explain why we do have a (genetic) concern for the nonhuman world. According to Wilson:

> *Biophilia, if it exists, and I believe it exists, is the innately emotional affiliation of human beings to other living organisms.* Innate means hereditary and hence part of ultimate human nature. Biophilia, like other patterns of complex behavior, is likely to be mediated by rules of prepared and counter-prepared teaming – the tendency to learn or to resist teaming certain responses as opposed to others. From the scant evidence concerning its nature, biophilia is not a single instinct but a complex of learning rules that can be teased apart and analyzed individually. The feelings moulded by the learning rules fall along several emotional spectra: from attraction to aversion, from awe to indifference, from peacefulness to fear-driven anxiety.[1]

In other words, biophilia is a love of, or concern with, the nonhuman world which is rooted in human nature. At root, the genetic explanation for something like biophilia is, as Baxter (1999: ch. 3) explains:

> human groups whose individual members possess something like an epigenetic reverence for nonhuman nature, at least to a degree, may produce better than groups whose members possess a purely instrumental attitude towards it. . . . Evidence also compatible with it may be drawn from the fact that even in our modern artifactual world, where individuals may live their entire existence in

an almost artificial environment, people show a yen for, and exhibit moral attitudes towards, the nonhuman natural world. They make gardens, keep house plants and domestic pets, go for country walks, watch natural-history films and so on.

According to Wilson there are evolutionary grounds for establishing 'kinship' with other nonhuman species:

> Other species are our kin. This perception is literally true in evolutionary time. All higher eukaryotic organisms, from flowering plants to insects and humanity itself, are thought to have descended from a single ancestral population that lived about 1.8 billion years ago. . . . All this distant kinship is stamped by a common genetic code and elementary features of cell structure. Humanity did not soft-land into the teeming biosphere like an alien from another planet. *We arose from other organisms already here, whose great diversity, conducting experiment upon experiment in the production of new life-forms, eventually hit upon the human species.*[2]

This appeal to there being 'something' in human nature (its genetic and evolutionary history), which means that having a relationship to the natural environment is important, is something shared with other 'naturalistic' approaches to social theory discussed below.

For Wilson, biophilia can be proved to have a genetic basis. For example, he claims that fear, and even full-blown phobias, of snakes and spiders (but also other phobic elements of the natural environment such as dogs, enclosed spaces, running water, and heights) are quick to develop with very little negative re-inforcement, while more threatening and risky modern artefacts, such as knives, guns, electric wires and cars, do not usually evoke such a response. Humans find trees that are climbable and have a broad, umbrella-like canopy (such as oaks, poplars) more attractive than trees without these characteristics (such as firs and pines), since the former resemble the ecological niche of early human evolution. Finally, people would rather look at water, green vegetation or flowers than built structures of glass and concrete.

For Wilson (and an ecologically orientated sociobiology), the biophilia hypo-thesis, if substantiated, provides a powerful argument for the conservation of biological diversity, on the anthropocentric (human-centred) grounds that the maintenance of a diverse and richly varied biological range is genetically important for human (psychological, emotional and perhaps spiritual) well-being, as well as being important for providing the ecological resources for society. As Wilson puts it, 'The one process now going on that will take millions of years to correct is the loss of genetic and species diversity by the destruction of natural habitats. This is the folly our descendants are least likely to forgive us' (Wilson, 1984: 121).

In this way the biophilia hypothesis is a potentially important argument in the politics of nature conservation and protection of biodiversity, in linking human emotional and psychological health and well-being to the existence and experience of other nonhuman living creatures. As Milton suggests:

> The biophilia hypothesis provides nature protectionists with a potentially powerful argument. It suggests that nature, and particularly the presence of other living things, is important for our emotional health, that the destruction of nature deprives us of countless opportunities for emotional fulfilment, and that the extinction of other species is, in some ways, the extinction of our own emotional experience.
>
> (Milton, 2002: 61)

The appeal of human nature in social theory

The appeal of using 'human nature' in social theory is understandable. Human nature is universal; that is, it is something shared by all human beings regardless of culture. Thus any assertion or claim based on an appeal to human nature is universalisable to humanity as a whole and not just to one particular society in a particular historical period. Hence the political importance of 'human nature' is that it is at *one and the same time a descriptive and a prescriptive concept*. That is, when one says, for example, that it 'goes against human nature' for women to work outside the home and have careers rather than raise children, one is both describing (explaining) the gendered division of labour between the sexes, but also saying this particular division ought to be. As I am sure most readers will recognise, in popular consciousness and popular argumentation there is often an appeal to 'human nature' either to explain or justify some particular position or to refute competing arguments. Indeed, the appeal to human nature is perhaps the most ubiquitous form of justification and/or explanation within debates about a variety of social phenomena.

As Berry notes, 'the notion of human nature, with its assertion of the givenness of man, which serves to demarcate the boundary between the supposedly reme-diable human world and the supposedly irremediable natural world, is essentially an ideological construct' (1986: 105). The problem with sociobiology as a form of social theory is that it attempts to explain everything about human social life in terms of genetic and evolutionary factors. For Gregory, 'One of the problems sociobiology encounters in seeking genetic determinants of behavior is that it must explain everything or else it explains nothing' (1978: 286). Hence socio-biology's attempt to explain everything from human warfare, the social division of labour, to individual psychological problems in terms of genetics, 'adaptive fitness' and so on. And across a wide variety of social phenomena its explanatory powers have, to say the least, been found wanting (Dickens, 1992; Rose *et al.*, 1984). It is not that sociobiology is necessarily wrong in wishing to break down the barrier between the 'social' and the 'natural', but rather how it simply *reduces* the former to the latter on the basis of a very deterministic account of human nature.

However, to reject the biological or genetic determinism of sociobiology does not mean that we need to deny that human beings are 'natural' or indeed that 'culture' and human cultural practices are somehow not natural. Sociobiology in some senses has to be recognised as representing one of the main attempts to transcend the 'nature/culture' dualism, though misguided in simply reducing culture to nature in general and evolutionary genetics in particular.

Human embodiedness and embeddedness: from a dualistic to a dialectical culture/nature relation

In this section, we look at an alternative view of the relationship between social theory, biology, ecology and the environment. Over the past twenty years or so Ted Benton has sought to integrate biological and ecological ideas into social theory from a broadly left-wing perspective. His starting point is one which recognises the perils of what he calls an over-naturalistic perspective in which 'the concepts of ecology as a biological science apply in an unqualified way to the human species, whilst the philosophical principles underlying ecology are generalizable as a set of norms for human conduct' (1994: 39). He rejects Malthus' population theory (and sociobiology, discussed below) for applying unmodified the concepts of natural science to the study of human society. At the same time, he does not reject the importance of biological and ecological considerations. As he puts it:

> We can, and, I think, we should, continue to view humans as a species of living organism, comparable in many important respects with other social species, as bound together with those other species and their bio-physical conditions of existence in immensely complex webs of interdependence, and as united, also, by a common coevolutionary ancestry. To say this much *is* to be committed to a naturalistic approach, but not necessarily to a reductionist one. It is to be committed to recognizing the *relevance* of evolutionary theory, physiology, genetics and, especially ecology itself, as disciplines whose insights and findings are pertinent to our understanding of ourselves.
>
> (Benton, 1994: 40; emphasis in original)

Thus Benton argues that we should not confuse the relevance of the natural sciences in investigating social phenomena in the human world with these sciences offering a full explanation of those human phenomena. This is the mistaken path taken by sociobiology. Although sharing with nonhuman species similar problems and similar characteristics (e.g. biological needs for food), humans are *not* the same as other species. Our particular natures, needs and modes of flourishing are such that we are different from other species. But being different does not mean that we as a species are somehow radically separate or 'superior' to other species. In this way, Benton, like others in this chapter, seeks to transcend the dichotomy between 'environment' and 'society' as an important step in developing a more ecologically sensitive form of social theory. Yet as many feminists have noted, 'biology is not destiny' and neither is ecology for that matter, and Benton is aware that one must equally recognise that there is both a *difference* and a *connection or commonality* between the social and natural worlds. One way of looking at this is to see that *human beings are a part of, as well as apart from, the environment* (Barry, 1995c).

This latter injunction, to see humans as at one and the same time a part of the natural order (and hence like other species) as well as being apart from it (and hence not being like other species), has also been taken up by others such as Hayward (1995), Dickens (1992) and Brennan (1988). Brennan neatly sums up one of the main reasons for adopting a naturalistic approach: 'in order to discover what sort of human life is valuable we must first consider what kind of a thing a human being is. Although there is, in my view, no complete answer to this question, we can . . . grasp one important aspect of human nature by reflecting on what are essentially ecological considerations' (1988: xii).

An example of this position, and one which illustrates how Benton and others have sought to integrate biological and ecological considerations into social theory (as opposed to reducing social theory to these natural sciences) is the human requirement for food as a condition for survival and flourishing. While, like other species humans require food, in the case of humans it is not food per se that we need, but food collected, prepared and consumed in particular ways. As Benton puts it, 'Proper human feeding-activity is symbolically, culturally mediated' (1993: 50). The importance of this point is its sensitivity to ecological and biological conditions for human need fulfilment. A graphic illustration of this is the title of Lévi-Strauss' famous book *The Raw and the Cooked*, the 'raw' referring to the mode of food consumption within the nonhuman world, and the 'cooked' to the mode of consumption within human societies.

On this view, culture may be regarded as our species-specific mode of expressing our nature, or 'species-being'. As it is continuous with our nature as social beings, human culture does not represent a radical separation from nature, but rather may be viewed as our 'second nature' (Bookchin, 1986); that is, emerging from, but situated within, the natural order. The importance of this has been expressed by Kohák, who notes that 'Were culture a negation of nature, no integration of humans and nature would follow' (1984: 90). Benton seeks to overcome the 'nature/ environment' versus 'human/culture' divide by articulating a naturalistic account of human beings in which human needs, conditions of flourishing and human culture may be understood 'naturalistically'. Benton's naturalistic social theory begins from his observation that 'Humans are necessarily *embodied* and also doubly, ecologically and socially *embedded*, and these aspects of their being are indissolubly bound up with their sense of self and with their capacity for the pursuit of the good for themselves' (1993: 103; emphasis in original). In drawing attention to the 'embodiedness' of human beings, Benton is close to the materialist eco-feminist arguments outlined in Chapter 7, in which human embodiedness is used to highlight and criticise the 'sexist' assumptions of social theories which deny the neediness, vulnerability and dependency of human beings.

Benton criticises what may be called the foundational separation between 'culture' and 'nature'; that is, the basis of social science (as distinguished from natural science) in general, and social theory in particular. The clearest example of this separation which Benton criticises is that between humans (e.g. their unique capacities, needs and forms of sociality) and animals. While not for one moment denying the uniqueness of the human species, Benton does point out that across a broad range, the specific capacities and qualities often assumed to be unique to human beings may also be found in other species. Thus, for example, he states that 'The capacity for and disposition to social co-ordination of activity as such is not a distinctive feature of our species' (1993: 36), since many other social species, such as primates, also exhibit often highly complex forms of social co-ordination. Challenging one of the central claims of contemporary social theory, which not only separates human society from nature but also denotes human capacities and behaviour as 'higher' or superior to that associated with nonhumans, Benton suggests that 'those things which only humans can do are generally to be understood as rooted in the specifically human ways of doing things which other animals also do' (1993: 48). In this way, as pointed out above, like other forms of organic life humans need to eat food, but unlike these other species we have a variety of ways in which food is 'cooked' (culturally mediated) rather than consumed 'raw' (non-culturally mediated).

Another example of the needs and activities of human beings as 'differentiations' rather than unbridgeable 'differences' in comparison with nonhuman species is reproduction and sex. According to Benton, 'continuous sexual receptivity [suggests] an evolutionary significant role for sexual activity in human social life to some extent dissociated from the immediate requirements of reproduction' (1993: 53). Thus unlike other species, in which the female of the species is sexually receptive for limited periods, and in which sexual activity is orientated towards reproduction, humans have a much more complex orientation towards sexual behaviour. The complexity of sexual activity within human society may be taken in part as indicating that sex for humans is not simply driven or motivated by reproductive concerns (an instrumental orientation to sex), but has a deeper evolutionary significance in that it is invested with cultural or social meanings. Sex in human societies is governed by normative and socially authoritative sanctions, rules, taboos, often, as eco-feminists have pointed out, gendered in ways that control or marginalise women. One of the most obvious examples of the 'rule-governed' and socially regulated character of sex within human societies is the existence of an incest taboo in almost all human cultures and the 'private' character of sex within human societies. One can think of many other ways in which sex is *meaningful* for humans in ways which it is not for other species, where it is (by and large) simply the *means* by which reproduction is achieved. However, in this case, one could suggest that the difference between humans and nonhuman animals is a *difference in kind rather than degree*. However, this does not undermine the naturalistic basis of Benton's approach, in which the important point is that sex for the human species is not, from an evolutionary and naturalistic perspective, solely (or perhaps even primarily) explicable in terms of reproduction. On the other hand, the *centrality* of sex (and its various manifestations of sexuality and sexual activity) within human cultures *is* something that is shared with other species where sex (qua reproduction) is a central organising need and activity (even if only infrequently but regularly practised).

Transforming nature: the creation of our ecological niche

From a naturalistic perspective, human culture may be seen as a collective capacity of the human species to adapt to the particular and contingent conditions of their collective existence, including, most importantly, the natural environments with which they interact and upon which they depend. Thus culture is in part the particular mode by which humans adapt to their 'ecological niche', but not simply in the sense that cultures are somehow 'determined' by environments. Rather, in the additional sense that the mode of human adaptation to their 'ecological niche', and the expression of their 'species being', involves the *active transformation of their environment*. As the biologist Lewontin puts it:

> We cannot regard evolution as the 'solution' by species of some predetermined environmental 'problems' because it is the life activities of the species themselves that determine both the problems and the solutions simultaneously.
> . . . Organisms within their individual lifetimes and in the course of their evolution as a species do not *adapt* to environments: they *construct* them. They

are not simply *objects* of the laws of nature, altering themselves to the inevitable, but active *subjects*, transforming nature according to its laws.

(Quoted in Harvey, 1993: 28)

The dialectical character of the relationship and interaction between society and environment may be seen to be consistent with the 'new biology' which unlike earlier biological theory and sociobiology does not take there to be two separate entities to be observed: the 'organism' and the 'environment'. As Lewontin puts it, 'it is impossible to describe an environment without reference to organisms that interact with it and define it. Organism and environment are dialectically related. There is no organism without an environment, but there is no environment without an organism' (1982: 160). The point here is that there is no 'raw nature' to which organisms adapt. Rather the relationship is reciprocal or dialectical. Organisms change and make their 'natural environment', and at the same time the natural environment influences their behaviour and sets limits and opportunities on what organisms can do.

This dialectical relationship between organism and environment helps us to understand why there is no one 'ecological niche' for humans, as can be readily seen in the success of our species' colonisation of even the most inhospitable parts of the Earth's surface. As a species nature did not specialise; our ability to transform the environment means that we are unique in the range of ecological niches in which we can flourish, current ecological problems notwithstanding. That is to say, the 'environment' for humans is not some 'raw' or untransformed one as given by nature, but rather a humanised one. Like other species, humans are not simply faced with the problem of how to 'fit' a particular 'given' environment. The environment is not 'given', fixed and something external to 'society', but nor is it completely 'materially (and socially) constructed', that humans can ignore and/or make for themselves. A dialectical understanding of the relationship between society and environment sees both as interdependent, the environment, in part, transformed (as opposed to 'made' from scratch) by human activity, and the environment in turn providing opportunities and constraints for human activity.

Human flourishing and the natural environment

Another aspect of this naturalistic social theory is Benton's idea that human flourishing has an ecological basis (but not simply in the sense that the natural environment transformed and worked upon by humans provides food, clothing, shelter and so on). At the same time, humans also need to relate to their environment culturally and symbolically as a condition for human flourishing. According to Benton, 'our interaction with and symbolic investment in our external environment are essential to the formation and maintenance of a stable personal identity' (1993: 181). From this perspective, alienation or estrangement from the natural environment can lead to a crisis of personal identity, mental breakdown and psychological distress. This point has also been made by environmental psychologists such as Kidner (1994, 2001), and green thinkers such as Goodin (1992: 37–41) who stress the 'human need' to find meaning by conceiving of themselves as part of a 'larger order of things'. It is also part of Wilson's 'biophilia hypothesis' discussed above. The argument is that one's 'sense of self' is in part tied up with one's 'sense of place'. In this way, human personal identity and psychological well-being have an ecological basis. Thus, for Benton the natural environment becomes an essential part of a naturalistic understanding of human well-being and a condition for human flourishing. As he puts it:

> These features of the physical and social world which enter into and constitute our sense of self are not dispensable features which we may not choose to value or assign significance to. They are features which are basic in the sense that only in virtue of their presence can we hold on to a sense of ourselves as choosers, valuers and assigners of significance at all.
>
> (Benton, 1993: 184)

Hence the contributory factor of alienation from the natural environment in the high proportion of mental (and other forms of) illness and psychological distress in heavily urbanised areas. This is particularly the case in certain forms of land use and urban developments such as high-rise tower blocks and suburban housing estates. The importance of access to the natural environment for human psychological health may also be demonstrated negatively by reference to the individual experience of nature as a central therapeutic stage in the treatment of some forms of psychological distress and illness. It is important to notice here that these same sources of ecologically based psychological harm are particular to humans (though perhaps shared to some degree with apes and chimpanzees). Unlike other species, humans do not simply need an 'ecological niche' or 'habitat', but over and above that humans require a 'sense of place' as a condition for their flourishing. To put it another way, situating humans within a suitable ecological niche (though this niche is not 'given' by the natural environment) is a necessary but not a sufficient condition for their flourishing. Humans may survive within such a natural environment, but without a sense of place they will not flourish. Part of this distinction between 'survival' and 'flourishing' (which of course is not limited to the human species) rests on the ways in which our species relates to the natural environment in cognitive or symbolic-cultural ways. Such cognitive appropriations of nature include the scientific and aesthetic appreciation and experience of the natural world, and other cultural modes of apprehension, valuing and experiencing the natural environment.

At the same time, from the perspective of the late twentieth century, the natural environments faced by human societies cannot be taken to be completely and purely 'natural', i.e. independent of human influence, as the discusion of Giddens in Chapter 4 demonstrates, or as postmodern theorists of 'technonature' argue. In most cases, they have been transformed by past and current human behaviour. There is a real, material basis to the claim that the environment is 'socially constructed'. It is not just that our understanding of the environment is mediated by human social relations and culturally symbolic meanings, but the environment faced by human culture is often partly the 'product' of previous social modification. It is frequently difficult to maintain a strict division between a 'natural' environment and one which is the outcome of human purposive action in conjunction with that natural or given environment. The ecological niche for humans is as much a 'humanised' as a 'natural' one, as Lewontin suggests above. A naturalistic social theory views this transformative activity as central to understanding human nature and human culture. The significance of this cultural dimension is that the human 'ecological niche' is *both* culturally *and* biologically ecologically determined. While like other species we must use the natural environment in order to live and to flourish, we do not react uniformly to it, and we are unique in our species' adaptability to different ecological conditions, and above all in our species' ability to transform the natural environment to our particular needs. As Woodgate and Redclift put it:

> we are uniquely equipped to regulate and refashion the environment in ways that make it more suited to our requirements. *Thus, there is no single way in which we, as human beings, relate to external nature.* Acceptance of the complex, interactive character of social and environmental change, means that simple distinctions between 'social' and 'natural' sometimes become untenable.
>
> (1998: 8; emphasis added)

This echoes Benton's suggestion that:

> *there is no 'natural' mode of human relation to nature.* No original, ecologically 'harmonious' golden age or state of grace from which we have fallen. Humans have no single, instinctually prescribed mode of life, but a range of indefinitely variable 'material cultures'. . . . The forms of human ecology, as culturally mediated relations to physical, chemical and biological conditions, are both limitlessly variable *and* ecologically bounded.
>
> (Benton, 1994: 43; first emphasis added)

There is no single way in which humans relate to the natural environment, as even a cursory examination of the variety (though this variety is not unlimited) of past and present modes of human interaction with the environment will illustrate. Thus there is no 'species-specific' or 'naturally given' manner in which the human species *does and ought* to interact, both materially and symbolically with the natural environment. The importance of this is that, unlike other species, one cannot specify what the 'proper' mode of human interaction with the natural is or ought to be. The mistake of reductionist accounts of human social behaviour such as sociobiology, but going back to Malthus and Spencer, is that the relationship between other species and their environments cannot be used to 'read off' or 'determine' the relationship between human society and its environment.

Population–environment relations in terms of the 'carrying capacity' of the natural environment to support some level of population is different in the human and nonhuman cases. As Benton points out, 'Human inventiveness, with respect to our powers of intentional modification of our environments through normatively ordered social practices, renders quite illegitimate any attempt to read off from a specification of the bio-physical environment what its "carrying capacity" might be for human populations' (1994: 42–3). Benton's reticence about using the carrying-capacity idea may also have something to do with the fact that its origins are ideological rather than scientific. According to Bandarage:

> The concept of 'carrying capacity', for instance, was first put into use by French and British colonial scientists and administrators seeking to estimate the minimum amount of land and labor needed by local people to meet their subsistence needs so that what was deemed in excess of that could be taxed by the colonial state and appropriated for export production.
>
> (1997: 127–8)

However, Benton's point is that in order to find out the 'carrying capacity' of a particular environment for humans one must find out about the social relations, moral codes, cultural values and practices, economic arrangements, agricultural practices, property relations, forms of scientific and technological capacities and so on. Thus while a particular environment may support a particular human population, under changed social, economic, political and cultural conditions, this same environment may support a higher or lower population level. Our 'second nature' (to use Murray Bookchin's (1986) term for human culture/society) modifies or transforms the opportunities and constraints that 'first nature' presents to us.

At the same time, this 'indeterminacy' and variety it gives rise to in the way in which humans use and interact with the environment is in part due to the fact that we not only transform the environment (i.e. materially interact with it). Human relations with the natural world are also, in part, regulated by 'symbolic' interaction with and conceptions of the natural environment. We are unique in having a variety of 'cultural' and 'moral' norms, rules, taboos and so on which influence our material transformation and use of the natural environment.

Human categorisation of the environment: resources, non-resources and proscribed resources

Although an over-simplification, generally speaking human beings and societies divide 'the environment' into three broad categories: 'resources', 'non-resources' and 'proscribed resources'. Other species divide their environment into the first two, basically that which they can eat and/or use in some way, and that which they cannot eat and/or use. Humans, on the other hand, while fulfilling their needs (many of which they share with other species, such as the need for food, safety and reproduction) by using the environment do adopt a discriminating attitude to the natural world on the basis of 'non-instrumental', cultural or symbolic grounds in removing parts of the environment which could be used as 'resources' into the category of 'proscribed resources'. So, for example, we find human cultures in which some animals, which are otherwise perfectly suitable as sources of protein or hides, are not considered as such, as in Hindu cultures where the cow is seen as 'sacred' and thus placed in the 'proscribed resources' category. Other examples include the ubiquity within human cultures of denoting some nonhuman species as 'vermin', a cultural category which only makes sense against a background of there being 'non-vermin' species and 'non-vermin' modes of treatment. According to Diamond, 'the notion of vermin makes sense against the background of the idea of animals in general as not mere things. Certain groups of animals are then signalled out as *not* to be treated fully as the rest are, where the idea might be that the rest are to be hunted only fairly and not meanly poisoned' (1978: 476; emphasis in original). The important point is to recognise that the human species is set apart from other species by the ubiquity of this category. That is, in all human cultures we find the category of 'proscribed resources'. Figure 10.1 gives a brief overview of this simple model outlining the different categories which are important in the metabolism between species and the environment.

The usefulness of this model may be seen to outline the main issue at stake in current debates about biotechnology and genetic engineering. In terms of the tripartite division of the natural world, the technological advances represented by biotechnology may be seen to raise problems and opportunities which cut across all three categories. For those in the biotechnology industry one could say that what they are doing is moving genetic information which was previously in the category of 'non-resource' (since we did not have the technological capacity to use genetic information) into the category of 'resource'. At the same time, many of those who raise moral or prudential reasons against genetic engineering may be seen either to wish genetic information to be placed in the 'proscribed resources' category (that is, we have the technological potential to use this information as a resource, but choose not to) or advocate the abandoning or scaling

	Rats	Humans
Resources	Most organic matter, including waste and inorganic nest-building materials	Organic and inorganic material and energy
Non-resources	Materials not biologically-suited (e.g. inorganic matter)	Not 'given', i.e. dependent upon human biology (depends on technological factors)
Proscribed resources	Not applicable	Cultural/moral variations (e.g. Hindus and cows, vegetarians, vegans). Genetic information as standing between 'resource' and 'proscribed resource'

Figure 10.1 Species metabolism with the environment
Source: Author

down of biotechnology and genetic engineering, thus maintaining genetic information as a 'non-resource'. In this way, this discriminating attitude which humans adopt towards their environment (an environment that is often transformed to suit their own particular purposes) illustrates the importance of the symbolic, cultural or moral dimension to the social–environmental metabolism.

The simple tripartite model in reference to the inclusion of a cultural or moral aspect to social–environmental relations also illustrates another weakness of sociobiological arguments which depend on 'hard-wired' or an 'instinctual' basis for human behaviour. According to Benton:

> Lacking what Mary Midgley has called 'closed instincts', humans depend on their capacity to identify and meet their full range of needs upon the conceptual resources and normative rules which constitute their local culture. . . . Specifically culture, identity, self-realization and aesthetic needs interact with and complement organic needs for food and shelter in ways which figure less, if at all, in the ecological requirements of other species.
>
> (Benton, 1994: 42)

Unlike other species, human beings do not have a predetermined set of needs that must be fulfilled for them to 'flourish'. While they have in common with other animals and biological entities 'basic' needs to meet (food, reproduction, shelter, security, sociality), there is no universal list of other needs which humans have to meet in order to flourish. First, humans as a species are different from other species in having particular needs and conditions for flourishing. Second, some of these needs are culture-specific and cannot be generalised to all human beings.

However, this stress on the uniqueness of the human species from other species should not be read as simply reasserting the divide between 'society' and 'culture'. We share many characteristics and needs with other animals. Thus, while we share with other social animals a need for sociality, we are different from other species in the *forms* our sociality takes, not that we are the only species in which sociality exists.

In conclusion then, a naturalistic approach to social theory, as proposed by Benton and others, may be simply stated as indicating, first, human society and social practices by noting the commonality between the human and the natural world, and the untenability of a rigid, dualistic separation between 'society' and 'nature'. Second, a naturalistic perspective highlighting the centrality and constancy of the following in human life: birth, sex, death and collective subsistence, features of the 'human condition' which we share with other species, but which we fulfil and invest with normative/cultural significance in our own particular ways, ways which are not shared with other species.

In terms of the 'big issues' in human life, those aspects which have been constant features of human life around which humans have created moral codes, laws, institutions, taboos and customs, those of birth, sex, death and collective subsistence may be said to constitute a 'core' range of needs, conditions of flourishing which are central to understanding any recognisably human society. These core needs are part of life, any life whether human or nonhuman, but what marks us out is the variety of meanings we attach, the ways we make these 'natural events' or 'needs' meaningful, by marking them in ways not shared with other species. This does not deny that other species can 'grieve' for dead fellow members in ways that are intelligible to us, but even those social species which do grieve do not have anything remotely comparable to the variety of often elaborate and culturally specific ways in which humans express their grief. These cultural and morally important ways in which death, sex, birth and collective subsistence are dealt with, expressed, and the needs associated with them fulfilled, are in fact what defines the 'human' and 'human society' as both a part of as well as apart from the nonhuman world.

Conclusion

In many respects the right and left versions of how ecological and biological insights can explain (and justify) social relations is at one level another rehearsing of the old 'nature/nurture' debate, with left-wing social theories insisting on the primacy of 'nurture' and right-wing theories stressing 'nature'. The importance of these distinctions cannot be over-emphasised, since they have far-reaching consequences for the prospects for social change. If one takes the view that 'nature' (whether 'selfish genes' or the hunter-gatherer 'nature' of human beings) explains human behaviour, then the scope for changing behaviour on the basis of social change is limited. For example, if there is a genetic basis for particular types of behaviour (sexual orientation, violence, altruism and so on) then simply altering social rules or institutions will not change the genetic causes of behaviour though it might constrain it to some extent. This is because if something is 'natural', to do with 'human nature', there is very little that can be done to alter it. This type of argument is common in everyday conversation where often people put down particular types of behaviour (usually, though not always, 'bad' behaviour) to 'human nature'. 'It's only human nature isn't it?' is an all too common refrain in our daily discourse, and its significance is that it serves to both explain and close any discussion, since if it's 'human nature' then there is nothing that can be done to alter it, it just *is*. This type of argument is very common in conservative and right-wing thought, which usually starts from a negative view of human nature to defend social institutions as a way of 'managing' or perhaps lessening, but not changing or eradicating, the fundamentals of human nature. However, this deterministic view is usually modified to permit some personal agency and responsibility for individual action and behaviour. Hence, a typical conservative view of crime would see it as down to the individual and having little to do with the individual's social conditions, or how they were brought up.

Against this view we have left-wing perspectives which stress social conditions, and the individual's 'social environment' including how he or she was 'socialised' or 'nurtured' as an explanation of behaviour such as crime. On this view, creating a less unequal society, one in which socio-economic opportunities were more evenly distributed, would greatly lessen criminal behaviour (though not fully eradicate it). It is from this type of thinking that British Prime Minister Tony Blair formulated his famous statement about 'Being tough on crime, and tough on the causes of crime'. Starting from a less 'negative' view of human nature and human beings, the standard left-wing position is that individuals are basically 'good' and want to be 'good' but that social and especially economic conditions prevent this from being realised, and individuals are led, encouraged or otherwise motivated to do 'bad' things. So the solution from a left-wing view is to create the social and economic conditions (or 'social environment'), by altering the present institutions of social order, to remove the socio-economic causes of criminal behaviour. Thus we come to another meaning of 'environment' which is important in social theory, environment as the 'socialising' conditions within which individuals are 'nurtured', develop their personalities, views and values. Important here is the early socialising and experience of young children, the centrality of which has been stressed by various branches of psychological theory and practice for at least a century. Part of the reason for this is that unlike most other animals, human infants are vulnerable and dependent on adults for a much longer period. This relatively long period of dependence in which the human infant is dependent on others to feed, cloth, protect it, is also the period within which the human child learns to walk, speak, think, develop the capacity for abstract thought and its sense of self. According to Dickens (1992), there is an important analogy between the 'nurturing' relationships that exist between the human infant and other humans (especially parents), and a particular 'nurturing' view of the relationship between humans and nature. For Dickens the human mind has evolved in such a way that we are 'predisposed' to thinking about the natural environment as something upon which we depend. However, as I point out in Chapter 11, the environment is also something which depends on us and can be put at risk from our actions and inactions.

Summary points

- The relationship between knowledge of the natural environment (natural sciences) and the human social world has a long and complex history.
- Sociobiological theory is an attempt to explain human social behaviour, practices and institutions based on genetics, evolutionary theory and biology. It works with a particular notion of 'human nature' in which selfishness, competitiveness and individualism are 'given' (i.e. unalterable) features of humanity.
- For critics, sociobiology is an ideology rather than a science, given its close current and historical connection with right-wing and conservative political positions. This is particularly so with sociobiological attempts to establish the 'superiority' of some races over others.
- E.O. Wilson's 'biophilia' hypothesis claims that there is a genetic or evolutionary basis for human concern for the nonhuman world, which provides a sociobiological reason for conserving biological diversity.
- The concept of 'human nature' is central to social theory, and is a common reference point in arguments about how society is and ought to be. That there is no one accepted account of it illustrates its 'political' or 'ideological' character, such that one can often indicate a social theory's position on the political spectrum by its view of human nature.

- 'Progressive' social theories usually have a positive account of human nature and stress the importance of 'nurture' or the 'social environment' in explaining behaviour, while 'conservative' social arguments are usually based on a 'negative' view of an unchangeable human nature.
- A less deterministic and reductionist way of integrating biology, ecology and social theory is one which recognises that humans are both part of and apart from nature, and that humans are biologically embodied and ecologically embedded.
- We should not confuse the relevance of the natural sciences in investigating social phenomena in the human world with these sciences offering a full explanation. We cannot reduce social behaviour to the explanations of natural science.

Further reading

A gentle (and short!) introduction to some themes in evolutionary theory is Stephen Jay Gould's *Adam's Navel*, Harmondsworth: Penguin, 1995. An excellent overview of different accounts of 'human nature' and how the latter is used in different political arguments is Christopher Berry's *Human Nature*, London: Macmillan, 1986.

Accounts of sociobiology include: Richard Dawkins' *The Selfish Gene* (2nd edn), Oxford: Oxford University Press, 1989; M. Gregory *et al.* (eds), *Sociobiology and Human Nature*, San Francisco, CA: Jossey-Bass, 1978; E.O. Wilson, *Sociobiology: The New Synthesis*, Cambridge: Belknap Press, 1980. Matt Ridley's *The Red Queen: Sex and the Evolution of Human Nature*, Harmondsworth: Penguin, 1995, and his *Nature via Nurture: Genes, Experience, and What Makes Us Human* London: HarperCollins, 2003; and Jared Diamond's *The Rise and Fall of the Third Chimpanzee*, London: Vintage, 1991, present a less reactionary account of sociobiological theory.

On the biophilia hypothesis, see E.O. Wilson, *Biophilia*, Cambridge, MA: Harvard University Press, 1984; Stephen Kellert and E.O. Wilson (eds), *The Biophilia Hypothesis*, Washington, DC: Island Press/Shearwater, 1993, and E.O. Wilson, *In Search of Nature*, London: Allen Lane, The Penguin Press, 1997. See also Kay Milton's *Loving Nature: Towards an Ecology of Emotion*, London: Routledge, 2002, and David Kidner's *Nature and Psyche: Radical Environmentalism and the Politics of Subjectivity*, Albany, NY: State University of New York Press, 2001.

On naturalistic social theory, see Ted Benton, *Natural Relations: Ecology, Animals and Justice*, London: Verso, 1993; Peter Dickens, *Society and Nature: Towards a Green Social Theory*, Hemel Hempstead: Harvester Wheatsheaf, 1992, and his *Reconstructing Nature: Alienation, Emancipation and the Division of Labour*, London: Routledge, 1996; Mary Midgley, *Animals and Why They Matter*, Harmondsworth: Penguin, 1983, and her *Beast and Man: The Roots of Human Nature* (revised edn), London: Routledge, 1995; and John Barry, *Rethinking Green Politics: Nature, Virtue and Progress*, London: Sage, 1999.

Notes

1 URL: matu1.math.auckland.ac.nz/~king/Preprints/book/diversit/restor/bph1.htm# anchor89361 (emphasis added).

2 URL: matu1.math.auckland.ac.nz/~king/Preprints/book/diversit/restor/bph1.htm# anchor89361 (emphasis added).

11 Greening social theory

Introduction

This chapter has a twofold purpose. The first is to try and bring together some of the main themes and issues of the book. The second is to discuss reasons for the necessity and desirability for the 'greening' of social theory. The latter is developed through a brief discussion of the emergence and main principles of green social theory, which is then used to suggest ways in which social theory can or ought to be 'greened'. Reference will be made to green social, political and moral theory, which should be understood as a particular approach to the greening of social theory. That is, green social theory is only one particular way social theory can be greened; others are always possible (indeed desirable).

The call for putting the natural environment and social–environmental relations on the agenda of social and political theory is one that is due in large part to the theoretical and practical impact of green politics and philosophy over the past thirty years or so. However, as suggested in Chapters 2 and 3, we can also find some origins and antecedents of these 'green' claims which pre-date the rise of the modern green movement and its political, scientific and moral claims.

Origins of green theory

Often a 'green' or 'environmental' perspective is caricatured as something that is 'counter-cultural', hippy, and out of touch with the 'realities' of modern, late twentieth-century life (see Figure 11.1). In modern Britain 'green' is most vividly associated with groups such as 'New Age Travellers' and the animal rights and anti-roads protest movement, in North America with the 1960s 'counter-culture' and 'hippies'. However, the focus here is not public perceptions of 'greenies' or a sociological analysis of green groups and movements, but rather origins and principles of green social, moral and political theory.

For reasons of space, some origins of green social theory are listed below:

1   the 'romantic' and negative reaction to the industrial revolution;
2   the positive reaction to the French (democratic) Revolution;
3   a negative reaction to 'colonialism' and 'imperialism' in the nineteenth and twentieth centuries;
4   the emergence of the science of ecology;
5   growing public perception of an 'ecological crisis' in the 1960s, claims of 'limits to growth' in the 1970s, and the emergence of 'global environmental problems' in the 1980s and 1990s;
6   transcending the politics of 'industrialism' (organised on a left–right continuum) by a politics of 'post-industrialism' (beyond left and right);
7   increasing awareness of and moral sensitivity to our relations with the non-human world (from the promotion of 'animal rights' to ideas that the Earth is 'sacred');
8   the integration of progressive social, political and economic policies with the politics of transition to a sustainable society, principally the universal promotion of human rights, socio-economic equality, democratisation of the state and the economy.

Of particular importance is the central concern of green theory and practice to overcome both the separation of 'human' from 'nature' and also the mis-perception of humans as above or 'superior' to nature. Green social theory may be seen as an attempt to bring humanity and the study of human society 'back down to Earth'. The science of ecology played an important part in arguing that humans as a species of animal (that is, we are not just *like* animals, we *are* animals) are ecologically embedded in nature, and exist in a web-like relation to

Figure 11.1 Caricature of greens

Source: Fardell (1992)

other species, rather than being at the top of some 'great chain of being'. It is crucial to note the significance of green social theory having a strong basis in the natural sciences (mainly ecology, evolutionary and environmental psychology, the biological life sciences and thermodynamics), because, as will be suggested below, this gives us a strong indication of what the 'greening' of social theory may involve.

A second and related point is that green social theory, in transcending the culture/ nature split, begins its analysis based on a view of humans as a species of natural being, which like other species has its particular species-specific characteristics, needs and modes of flourishing. Central to green social theory, unlike other forms of social theory, is a stress on the 'embodiedness' of humans.

A third issue which green social theory raises is the ways in which social– environmental relations are not only important in human society, but also *constitutive* of human society. What is meant by this is that one cannot offer a theory of society without making social–environmental interaction, and the natural contexts and dimensions of human society a central aspect of one's theory. In its attention to the naturalistic bases of human society, the green perspective is 'materialistic' in a much more fundamental way than materialism within Marxist theory. Unlike the latter, green social theory concerns itself with the external and internal natural conditions of human individual and social life, whereas the 'material base' for Marx is economic not natural. At the same time, this material- ist reading of green social theory questions the 'post-materialist' character often ascribed to green politics and issues, as given by Inglehart's popular explanation for green politics as 'post-materialist' (Inglehart, 1977) and mainly a middle-class phenonmeon. This middle-class characterisation of green politics is one that Marxists have drawn attention to and used to demonstrate the 'anti-working-class' interests of green theory, as discussed in Chapter 6. However, both this Marxist critique and Inglehart's thesis fail to explain the 'environmentalism of the poor' (Martinez-Alier, 2001), the class, ethnic and race dynamics of the environmental justice movement (Schlosberg, 1999), or 'resistance eco-feminism', as discussed in Chapter 7. The Eurocentric perspective of Inglehart's analysis is of course limiting, as is the empirically weak connection he makes between wealth/income levels and post-materialist values (Cudworth, 2003: 71) and his limiting of 'environmental concern' to aesthetic/amenity rather than material or productive interests which people have with their environments.

A fourth issue to note about green social theory is its moral claim about our relationship to the natural environment. What makes green moral theory distinc- tive is that it wishes to extend the 'moral community' beyond the species barrier to include our interaction with the nonhuman world as morally significant. In part, this moral concern with the nonhuman world is what gives green politics its self-professed character as 'beyond left and right'. As Lester Brown puts it, 'Both capitalists and socialists believe that humans should dominate nature. They perceive nature as a resource base to be exploited for the welfare and comfort of humans' (1989: 136).

Some of the basic principles of green social theory are listed in Box 11.1.

Box 11.1 Some basic principles of green social theory

Overcoming the separation between 'society' and 'environment' (which includes extending environment to include the human, built environment).

Appreciation of the *biological embodiedness* and *ecological embeddedness* of human beings and human society.

Views humans as a species of natural being, with particular species-specific needs and characteristics.

Accepts both internal and external natural limits, those relating to the particular needs and vulnerable and dependent character of 'human nature', and external, ecological scarcity in terms of finite natural resources and fixed limits of the environment to absorb human-produced wastes.

As a critical mode of social theory, green social theory criticises not just 'economic growth' but the dominant industrial model of 'development', 'modernisation' and progress.

Claims that how we treat the environment is a moral issue, and not just a 'technical' or 'economic' one. This ranges from claims that the nonhuman world has **intrinsic value**, to the idea of animal rights.

Prescriptive aspects: restructuring social, economic and political institutions to produce a more ecologically sustainable world.

'Act local, think global': ecological interconnectedness and interdependence which transcends national boundaries.

Futurity: time-frame of green social theory is expanded to include concern for future generations.

Scientific: based on ecological science (but also other natural sciences such as biology and physics).

Green social theory: from 'development' to 'sustainable development' and beyond

As explained in Chapter 1, modern social theory as a body of knowledge begins as the systematic study of modern or industrial society, explaining its emergence from a pre-industrial stage, and analysing its internal dynamics and processes. In this way social theory is intimately connected with the theory and practice of the 'development' or 'modernisation' of modern societies. That is, modern social theory takes the processes of development as its object of study and aims to provide a critical analysis of *what* development consists of, *how* it occurs, *who* or what are the main features or actors of development, and *where* and *when* it occurs or has occurred. As such, the increasing concern with 'sustainable development' within political and economic theory and practice (particularly since the 1992 Rio 'Earth Summit' conference organised by the United Nations Conference on Environment and Development) presents social theory (and the social and natural sciences more generally) with an opportunity (some might say obligation) to expand its parameters to include key aspects of this 'sustainable development' agenda (which is based upon, but not co-extensive with, the 'green' political agenda).

The essence of sustainable development is that it integrates a concern for the environment and environmental protection with obligations to current and future human generations. In terms of its most famous definition, contained in the Brundtland Report *Our Common Future*:

> Sustainable development is development that meets the needs of the present without compromising the ability of future generations to meet their own needs. It contains within it two key concepts: – the concept of 'needs', in particular the essential needs of the world's poor, to which overriding priority should be given; and the idea of limitations imposed by the state of technology and social organisation in the environment's ability to meet present and future needs.
>
> (WCED, 1987: 43)

Sustainable development is thus development that is ecologically sustainable; that is, development that is consistent with external, natural ecological constraints and limits. Another way of looking at it has been advanced by Jacobs (1996):

> The concept of 'sustainability' is at root a simple one. It rests on the acknowledgement, long familiar in economic life, that maintaining income over time requires that the capital stock is not run down. The natural environment performs the function of capital stock for the human economy, providing essential resources and services. Economic activity is presently running down this stock. While in the short term this can generate economic wealth, in the

longer term (like selling off the family silver) it reduces the capacity of the environment to provide these resources and services at all. *Sustainability is thus the goal of 'living within our environmental means'. Put another way, it implies that we should not pass the costs of present activities on to future generations.*

(1996: 17; emphasis added)

The discourse (or rather discourses) of sustainability and sustainable development acknowledge:

1 human dependence upon the natural environment, i.e. that the human economy is a subset of ecological systems;
2 the existence of external natural limits on human economic activity;
3 the detrimental effect of certain industrial activities on local and global environments;
4 the fragility of local and global environments to human collective action;
5 that one cannot talk about 'development' without also linking it to the environmental preconditions for development;
6 following on from 4, development decisions now may have environmental (and thus development) consequences for future generations and those living in other parts of the world.

In this way, green social theory, together with the emerging centrality of the theoretical and practical dimensions of sustainable development, are suggestive of why and how social theory can be 'greened'.

Towards the greening of social theory

Green social and political theory, expressed in part through the discourses and practices of 'sustainable development', presents at least five issues for social theory.

The first is the one at the level of the knowledge approach to the study of society. Green social theory suggests that not only should any social theory have 'social–environment' interaction as a central object of study, but that the greening of social theory requires the adoption of explicitly multidisciplinary, inter-disciplinary or transdisciplinary approaches and methodologies. This is because social–environmental relations and the natural contexts of human social life are so complex and involve factual as well as normative issues, which no one discipline can hope to monopolise.

The second issue concerns the temporal frame of social theory: green social theory, as expressed in its central concern with ecological sustainability and sustainable development, suggests the integration of a concern for the future

Figure 11.2 'Under New Management'

and future generations. Greening social theory requires that the future be included as an explicit, rather than implicit dimension of social theory.

The third issue has to do with the brute fact that ecological problems do not respect national or cultural boundaries. Pollution problems such as global warming and climate change are transnational and global in scope. Thus social theory must also be transnational and global in scope and approach. It can no longer (as if it ever could) be solely concerned with a particular society independent of other countries and international processes.

The fourth issue, perhaps most contentious of all, suggests that the greening of social theory requires that social theory can no longer remain within the species boundary; that is, being solely or primarily concerned with human social relations and phenomena.

The fifth issue, equally contentious, turns on the connection between theory and practice; that is, the obligation for social theory to move beyond theoretical reflection, abstraction and 'discourse' and to seek to apply that knowledge in the 'real world' to improve both human society and the nonhuman world. Does social theory in general, but particularly green social theory, have an obligation and responsibility to 'contribute to creating a fairer and more sustainable world'? (Buckingham-Hatfield, 2000: 123).

Greening social theory: beyond 'environment' versus 'society'

The greening of social theory involves the necessary and desirable bridging of the gap between 'society' and 'nature' and also between the 'social' and 'natural' sciences. For Hayward:

> *The most distinctive green idea is that of natural relations.* These are of numerous kinds: there are natural relations of biological kinship between humans, on which familial and social relations are supervenient; between humans who are not kin, too, relations are naturally mediated, for instance in the sense that reproductive and productive activities occur in a natural medium; such activities normally involve modifying the natural environment in some way, and all humans, individually and collectively, have relations to their environment.
>
> (1996: 80; emphasis added)

This suggests that greening social theory requires a naturalistic perspective which has two main components. First, a naturalistic social theory recognises the natural environmental contexts, preconditions, opportunities and constraints on human activity. Second, a naturalistic social theory recognises the centrality of internal human nature, of seeing humans as natural beings with particular modes of flourishing, like other natural beings.

For Ted Benton, discussed in Chapter 10, a recognition of the problems with the separation of 'society' and 'environment' and the other related distinctions between 'mind' and 'body', 'human' and 'nonhuman' (as discussed by eco-feminists in Chapter 7) requires the integration of the biologically based life sciences and the social sciences. For him, 'The task for any proposed realignment of the human social sciences with the life sciences can now be seen as providing conceptual room for organic, bodily, and environmental aspects and dimensions of human social life to be assigned their proper place' (1991: 25). This integration, as he is quick to point out, *does not* mean the reduction of one to the other. At the same time, as materialist eco-feminists point out, there needs to be a recognition of the biological reality and needs of human beings, a central part of which would oblige the reorientation of social theory towards issues around reproduction and not just production (which is the narrow focus of orthodox neo-classical economic theory as discussed in Chapter 8, and most of the history of pre-Enlightenment and Enlightenment political and social theory: Chapters 2 to 4).

The integration of biological and ecological insights into social theory would produce a form of social theory which began from acknowledging human biological embodiedness and ecological embeddedness. The greening of social theory focuses on an explicit recognition of the human body and its organic needs, and fully acknowledges human limits, dependency and neediness (in part, along the lines suggested by materialist eco-feminism in Chapter 7, and non-dualistic views of the relationship between culture and nature discussed in Chapter 10). Such a social theory would see the bodily vulnerability of humans as an essential context within which to assess human mental, conceptual and non-bodily achievements and processes. Rather than seeing humans as essentially abstract, disembodied centres of reason, thinking and acting, the greening of social theory requires re-embodying the human self as a biological entity, with non-voluntary impulses, needs, instincts and feelings.

In terms of ecological embeddedness, the greening of social theory involves the acceptance of ecological limits and parameters to collective human activity. As Lee puts it, given the ecological facts of the world and our dependent relationship with it, 'Any adequate social/moral theory must therefore address itself to these characteristics [of the world] and the character of the exchange [between humans and nature]. If it does not, whatever solution it has to offer is of no relevance or significance to our preoccupations and problems' (1989: 9).

A good example of the latter (from within sociology) is Catton and Dunlap's (1980) call for the development of a 'post-exuberant sociology' (Box 11.2). What they suggest is that the dominant paradigm within sociology (though the argument could be extended to other disciplines within the social sciences and humanities) is a 'human exceptionalist' one based on the dominant 'Western worldview' which they claim is out of keeping with the ecological reality, context and limits of human societies.

Ecology: connecting the natural and social sciences?

Ecology or the 'ecological paradigm' has, for some, since its emergence in the 1960s, promised a 'unified science of nature and society'. Ecology and its aim of a 'naturalistic' account of the human condition, while originating as an empirical natural science dealing with the relationship between species and their environments, has also become a form of social and moral theory. According to Vincent,

---

Box 11.2 Catton and Dunlap's call for a 'post-exuberant' sociology

Assumptions of the dominant 'human exceptionalist' paradigm

'1. About the *nature of human beings* – people are fundamentally different from all other creatures on earth, over which they have dominion.

2. About the *nature of social causation* – people can determine their own destinies, can choose their goals and learn whatever is necessary to achieve them.

3. About the *context of human society* – the world is vast and provides unlimited opportunities for humans.

4. About the *constraints on human society* – the history of human society is one of progress, there is a solution to every problem, and progress need never cease.'

(Catton and Dunlap, 1980: 34)

Assumptions of the 'new ecological paradigm'

'1. Humans have exceptional characteristics but remain one among many in an interdependent global ecosystem.

2. Human affairs are influenced not only by social and cultural factors but also by the complex interactions in the web of nature, so that human actions can have many unintended consequences.

3. Humans are dependent on a finite biophysical environment which sets physical and biological limits to human affairs.

4. Although the inventiveness of humans and the power derived therefrom may seem for a while to extend carrying capacity limits, ecological laws cannot be repealed.'

(Catton and Dunlap, 1980: 34)

'ecology has had moral and religious import for humanity' (1992: 213), noting however that 'there is a tangled and uneasy relationship between those who perceive ecology as an established science and those who mesh the scientific findings with strong doses of normative theory' (1992: 209).

However, the science of ecology (i.e. dealing with facts and the way the natural world is) has tended to go hand in hand with normative claims (i.e. dealing with values and how the world should be, and how we ought to treat and use nature), and has found it difficult to maintain a strict and lasting separation between 'facts' and 'values'. Eroding this strict distinction has placed ecology in a unique position as a 'science', as a form of knowledge which seems to bridge the natural and social sciences. This may be seen in how 'environmental studies' or 'environmental management' programmes in universities necessarily have to straddle the social and natural sciences, reflecting the interdisciplinary character of the forms of knowledge appropriate to articulating social–environmental relations. This of course goes against Habermas' ideas discussed in Chapter 4, in which only natural science and an instrumental/technical relationship to the environment was deemed to be the most 'fruitful' approach to take.

However, if we examine the writings of classic ecological writers such as Eugene Odum, one of the giants of the field in recent decades, we find an almost 'natural' extension of scientific ecological research leading to explictly political, ethical and economic areas and proposals. For example, writing in 1971, Odum claims that 'Sociologists, anthropologists, geographers and animal ecologists first developed an interest in the ecological approach to the study of human society. Now, as we have seen, nearly all the disciplines and professions in both the sciences and humanities are eager to find a common meeting ground in the area of human ecology' (Odum, 1971: 510). He went on to conclude that the ecological management of the environment required planning, policies for optimum population levels, internalisation of ecological costs of production, and 'Development of a "spaceship economy" in which the emphasis is placed on the quality of the capital stock and human resources rather than on rates of production and consumption as such, and a shift from quantity to quality' (Odum, 1971: 516). These proposals, and the others he suggested, could be easily inserted as part of the programme for government of one of the world's myriad Green Parties; or the basic principles of ecological economics or indeed of green political economy – both discussed in Chapter 8 – yet it is important to remember that Odum's suggestions were coming from ecological science not political theory or economics. Thus Odum can rightly be seen as a pioneer of interdisciplinarity as a necessary feature of green social theorising, thus making the study of ecology itself a form of social theorising.

One way in which this conjoining together of knowledge bases and disciplines has been expressed historically in Marx's and contemporary eco/Marxists' attempts to construct a 'unified science of humanity and nature'.

From a unified science of humanity and nature to interdisciplinarity

Greening social theory seems to take us in the direction of challenging strict disciplinary boundaries, not just within the social sciences, but also between the social and the natural sciences. According to Dickens:

> I would argue that a new paradigm is now being forced by environmental issues and related matters such as human health, animal welfare and the application of new reproductive technologies to the human species. *Such a paradigm rejects the distinctions between, for example, the life sciences, the physical sciences and the social sciences.* It is nevertheless a combination of these apparently alternative ways of viewing the social and natural worlds, within a coherent epistemological framework.
>
> (1992: 2; emphasis added)

This idea of a unified and integrated science of humanity and nature is something that may be found in the thought of Karl Marx. For him, 'The idea of *one* basis for life and another for *science* is from the very outset a lie. . . . Natural science will in time subsume the science of man just as the science of man will subsume natural science: there will be *one* science' (Marx, 1975: 355).

While the greening of social theory may not (and perhaps should not) move in the direction of the emergence of a unified, homogeneous and singular 'science', it is clear that overcoming the culture/human versus nature/environment dichotomy also takes us in the direction of weakening and loosening the boundaries between different forms of human knowledge. As Hayward (1995) notes, given that the economy is the most important and visible aspect of human society which interacts with the natural environment, it is clear that the greening of social theory will necessarily have to have an economic dimension. As he points out, 'If a unified theory of economics and ecology is to be possible, it will neither hypostatize (*sic*) an opposition between economy and ecology nor posit a straightforward identity of the two' (1995: 116). Rather, in the spirit of returning to the older tradition of political economy, the greening of social theory will, in part, be a theory of political ecology within which is located the study of political economy, along the lines suggested by the new 'interdisciplines' of ecological economics and green political economy (discussed in Chapter 8).

In many respects the emergence of interdisciplinary reseach on environmental issues, of which the greening of social theory is but one expression – involving collaboration between social science disciplines as well as between the social and natural sciences – offers a positive reply to C.P. Snow's lament of the widening gulf between the 'Two Cultures' of the natural sciences on the one hand and the humanities and social sciences on the other (Snow, 1959). Snow's basic point was that the lack of integration between the different branches of knowledge was not just regrettable but positively harmful in terms of closing off possible forms of knowledge needed to deal with the problems which humanity faced. This movement towards reintegrating academic and disciplinary specialisms, or at the very least opening up lines of communication between them, is the hallmark of recent interdisciplinary research and teaching in the environmental area and elsewhere (Brockman, 1995).

One of the obvious characteristics of research into the relationship between human societies and the environment is that there is no one 'master' discipline which can encompass the multi-faceted, complex and complicated dimensions of that relationship – though, as suggested in Chapter 8, neo-classical economics often assumes this priviledged position, particularly in terms of presenting itself as the most appropriate form of knowledge for decision and policy-making. Barnett *et al.* (2003) have usefully summarised some of the main elements of interdisciplinarity and interdisciplinary research:

- a problem focus, whether the problems be applied, theoretical or methdological;
- but also wariness of the dangers of capture by singular or partial policy objectives;
- a critical capacity, including recognition of normative elements of theory and practice;
- openness to other disciplines, theory, method and arenas of inquiry;
- a 'systems' orientation, in terms of appreciating the whole rather than only selected parts (and encompassing both quantitative and qualitative constructions of systems);
- a close appreciation of multiple and dynamic spatial and temporal scales, including a capacity to account for historical determinants of modern situations;
- appreciation of the personal/group qualities required to undertake inter-disciplinary work, and the balance of risks and rewards in disciplinary boundary transgression.

(Barnett *et al.*, 2003: 70)

To this list we could add the need for interdisciplinary undergraduate and post-graduate programmes – especially postgraduate research training (more common in North American and Australian universities than in the UK and Europe); team-based research (more common in the natural than social sciences); solutions-focused research; and the development of 'collaborative research' between 'experts' and the 'public' as in calls for 'civic science' (Lee, 1993) and more participative forms of scientific research, particularly in relation to public decision-making processes (Fischer, 1990, 2000). Good examples of interdisciplinary research may be found in the area of climate change where both the scientific dimensions of the issue (climate modelling, chemistry and so on) and human, social and economic dimensions of the issues around the causes and solutions to climate change, including individual and collective behaviour, are being brought together and used to inform policy-making in this area. As one of the great philosophers of science, Karl Popper, put it, 'We are not students of some subject matter, but students of problems. And problems may cut right across the borders of any subject matter or discipline' (1963: 88). It is clear that the greening of social theory requires an explicit interdisciplinary perspective that is still in its infancy relatively speaking, but will be one of the hallmarks of a 'greened' social theory in the coming century.

Greening social theory 1: time and future generations

One of the central claims of green social theory is its concern to extend the temporal dimension of social theory to include a sense of futurity. This reorientation of social theory towards the future is at the heart of the idea of sustainable development, with its explicit recognition that current decisions about how we treat and use the environment cannot be taken without considering the likely impact of these decisions on the type of environment that we leave to future generations, and thus the effect the latter may have on the welfare of the future or on the ability of the future to meet its needs. Thus the greening of social theory requires lengthening its temporal frame, as Adam suggests in her work on the the relationship between environment and social theory (Adam, 1990, 1995, 1998).

A sense of the future dimension involved is the Native American Indian saying that 'We don't inherit the Earth from our parents, but borrow it from our children'. How far into the future is of course an open and important question, the answer to which will have important implications for society. However, the key issue is that greening social theory implies extending its temporal range into the future, in particular making the likely ecological impacts of present courses of action an explicit object of analysis. As Adam puts it:

> We learn about and relate knowledgeably to a multidimensional space, but our understanding of the temporal dimension of socio-environmental life is pretty much exhausted with knowledge about the time of calenders and clocks. Nature, the environment and sustainability, however, are not merely matters of space but fundamentally temporal realms, processes and concepts . . . without a knowledge of this temporal complextity . . . environmental action and policy is bound to run aground, unable to lift itself from the spatial dead-end of its own making.
>
> (Adam, 1998: 9)

She goes on to develop the centrality of adopting new modes of conceptualising time and their associated modes and patterns of living and interacting with the environment. She suggests that 'industrial time' based on a simplistic linear perspective and Newtonian physics is at the heart of environmental problems and our lack of concern for future generations. For her:

> industrial time is centrally implicated in the construction of environmental degradation and hazards; second, as a panacea it worsens the damage.

Industrial time, in other words, is both part of the problem and applied as the solution. As long as time is taken for granted as the mere framework within which action takes place and is used in a pre-conscious, pre-theoretical way . . . it will continue to form a central part of the deep structure of environmental damage wrought by the industrial way of life.

(Adam, 1998: 9)

This suggests not only the need for social theoretical research into the industrial conception of time and how it is connected to environmental destruction, but also non-industrial and non-Western conceptions of time and futurity.

Greening social theory 2: beyond the nation-state and globalisation

The greening of social theory requires the adoption of an international and global perspective, since social–environmental relations are not always contained within a specific geographical area and the effects of social intervention on the environment do not always respect national boundaries. In this way the greening of social theory requires that social theory be self-consciously international and global in its outlook and analysis. As Yearley points out, 'environmental dangers pose supranational problems; these need solutions to which national governments are not well suited' (1991: 45).

Another important dimension of the greening of social theory is the process of globalisation. While social theory has always had an international dimension (though often the various links between societies were assumed rather than made an explicit object of study), the main focus of social theory was on the internal dynamics of society. With the advent of global and transnational environmental problems, highlighting the ecological interconnections between geographically separated parts of the Earth, the study of the origins and effects of and alternatives to these various transnational environmental problems necessarily requires that one must look beyond (and sometimes below) the nation-state. The interconnectedness of the world's economic and ecological systems means that what happens within a society increasingly has its origins in processes outside that society. This means that one must look not just beyond the boundaries of the nation-state but also at other international actors other than the nation-state. These include: powerful global non-government actors such as transnational corporations (TNCs); weaker, but still significant environmental non-government organisations (NGOs); and transnational political and economic institutions such as the United Nations and the World Bank.

The work of Giddens and Beck discussed in Chapters 4 and 9 is concerned with the way in which theorising about the environment necessitates a global or international perspective. Put another way, social theorising about the environment is a particular aspect of social theorising about globalisation. This link between globalisation and the environment may be seen in Paterson's view that, 'if globalisation is environmentally problematic . . . the politics of those concerned with environmental problems lies in resisting globalising forces, such as multinationals, banks and governments, in their attempts to negotiate new international regimes on the environment' (1996: 403–4).

The greening of social theory here requires incorporating the ideas of green thinkers and activists who argue that the global political and economic system maintains an exploitative relationship between the affluent Northern countries (the so-called 'developed' world) and the poor Southern countries (the so-called 'developing' world). Here the greening of social theory suggests analysing the cultural, political, economic and ecological effects of globalisation within the context of North–South relations. For example, while global environmental threats such as global warming and climate change affect everyone on the planet, this does not mean that humanity as a whole is in the same boat. First, it is not 'humanity' as a whole that is to blame for global environmental problems; those with economic and political power, mainly 'developed' industrialised nations, are mostly to blame for causing environmental problems, by, for example, being responsible for the vast bulk of global carbon pollution from the use of fossil fuels. Second, environmental problems do not have the same effects on everyone. Thus, for example, affluent countries and groups are better able to protect themselves from environmental problems than are poor countries and groups. As Seabrook notes:

> *Globalisation is not an organic growth*, but a carefully wrought ideological project. . . . If the perpetuation of privilege is to be the guiding force of the world, let it be identified as such. . . . If the abuse of the resource-base of the earth and the intensifying exploitation of its people is the supreme civilisational pursuit of the culture of globalisation, no matter what the consequences, why shrink from saying so?
>
> (1998: 27; emphasis added)

Rejecting the claim that globalisation is not something natural or organic, that is, something which is both 'given', 'inevitable' and beyond human control, and also somehow 'good', has recently played a major part in critiques of parties and governments such as Tony Blair's New Labour administration. According to a recent edition of *Marxism Today* (November 1998), the New Labour government views globalisation as 'a force of nature' (Jacques, 1998), an inevitable and irresistible economic process that the power of the nation-state cannot affect. Against such a 'natural' and inevitable phenomenon all we can do is adapt ourselves, our economies and societies to this current phase of the 'evolution' of the world economy. The result of seeing globalisation as a force of nature can explain why New Labour has much in common with the previous Conservative administration with regard to the economy. As Stuart Hall points out, '*Since globalisation is a fact of life* to which There Is No Alternative, and national governments cannot hope to regulate or impose any order on its processes or effects, New Labour has accordingly largely withdrawn from the active management of the economy' (1998: 11; emphasis added). In this way viewing or describing something as a force of nature, a 'fact of life' or 'natural' has tremendous ideological power in prescribing particular forms of action, or in the case of New Labour, inaction.

Since environmental degradation is caused by poverty and socio-economic inequality, stopping environmental destruction requires alleviating poverty, which in turn requires the creation of a fairer global economic system. The latter, according to Gray (1998), in turn requires regulating the global economy. While many greens would go beyond Gray, and argue for the restructuring or dismantling of the global economy, his argument for the political regulation of the global economy, for ecological and social reasons, is something that fits with the main thrust of green social theory in respect to globalisation. According to Gray:

> What is beyond doubt is that organising the world economy as a single global free market promotes instability. It forces workers to bear the costs of new technologies and unrestricted free trade. It contains no means whereby activities which endanger the global ecological balance can be curbed. Organising the world economy as a universal free market is, in effect, staking the planet's future on the supposition that these vast dangers will be resolved as an unintended consequence of the unfettered pursuit of profits. It is hard to think of a more reckless danger.
>
> (1998: 32)

Greening social theory 3: beyond the species barrier

According to Brian Baxter, 'it is now intellectually unacceptable to develop political theories in which the sole focus of concern is human well-being and values, ignoring the issues which greens have pushed to the fore concerning the well-being of other species, and the biosphere in general' (1996: 68). Baxter makes this claim on the basis that there are compelling moral arguments (from within green theory) which mean that it is illegitimate not to extend moral considerability and concern beyond humans to include (at least parts of) the nonhuman environment (animals, plants, ecosystems). This normative approach to including the natural environment and its interests in human social theory has much to commend it, even though it goes against many settled assumptions about our attitudes towards and treatment of the nonhuman world.

However, one can also advance reasons why the nonhuman world should be part of the agenda of social theory on the more factual grounds that social–environmental relations are constitutive of human societies. That is, we cannot fully grasp or understand any human society without also understanding the types and meanings attached to the ways in which society views, values, treats and uses its natural environment. Of particular importance here is the relationship between domesticated animals and humans, where it is true to say that human societies and cultures in which relations with domesticated animals were absent would be unintelligible to us in ways that societies which have these relations would not present the same problem.

At the same time, there are those like Milton (1996) whose ideas suggest that social theory, at least that part of social theory which is based on reflecting on culture, cannot remain concerned with human culture alone. As she puts it, 'As we learn more about both human and nonhuman animals, *it becomes increasingly difficult to sustain the view that culture is uniquely human*' (1996: 64; emphasis added). In this effort she is joined by some recent moral philosophers who hold that morality, in the sense of acting or behaving in accordance with moral principles, like 'culture', is not something that is unique to humans (see Singer, 1994: part B).

Social theory can be (indeed it must be) *human-based* but does not have to be *human-centred*. That is, while social theory must begin from an analysis of human society, it does not necessarily have to be exclusively concerned with human affairs, interests and events. For example, given both the co-evolutionary history of humans and animals, as well as the various material ways in which human societies use and have relationships with animals, social theory can reorientate its aims and objects of study to include these nonhumans as fellow members of 'society'.

Conclusion

It is rather paradoxical that what started out as an examination of the ways in which social theory has viewed, used and abused the external environment should at various points (in different ways, and for different reasons) oblige us to return inwards towards an examination of 'human nature' and its place in social theory. This was chartered in numerous ways in the last chapter where we saw how the introduction of environment to social theory (naturalising social theory) also leads us in the direction of integrating biological concerns, while in Chapter 3 we saw how arguments based on the 'state of nature' also included assumptions about 'human nature'.

Andrew Brennan puts this nicely when he notes that, 'in order to discover what sort of human life is valuable we must first consider what kind of thing a human being is. Although there is, in my view, no complete answer to this question, we can . . . grasp one important aspect of human nature by reflecting on what are essentially ecological considerations' (1988: xiii). In this way one could say that by its very nature (excuse the pun) when we use the word 'environment' we almost automatically move in the direction of talking about 'external nature' and from there it is difficult to prevent this discussion spilling over into debates about 'internal nature'.

We begin by analysing the natural world around us (that which environs us) and our relations to and with it, and find that this cannot be done without reflecting on 'us' as embodied beings, a particular species of animal evolved from social primates, and our own 'nature'. And the ultimate paradox of humanity is that the 'human condition' (which is also at the same time our 'natural condition') is one in which humanity is both a part of and apart from the environment.

Summary points

- The main principles of green theory are a rejection of the separation of 'humanity' and 'environment'; a stress on the biological embodiedness and ecological embeddedness of humans; viewing social–environmental relations as not only important in human society, but also constitutive of human society; and a claim that social–environment relations are of moral concern.
- The ideas behind 'sustainable development' may be used as a way to explore the main ways in which green social theory can and does contribute to the 'greening' of social theory in general.
- The greening of social theory involves overcoming a strict separation between 'environment' and 'society', and stresses the ecological embeddedness and biologically embodiedness of human beings and human society.
- Ecology and the 'ecological paradigm' has been used to suggest the integration of the natural and social sciences, which for some may lead to a 'unified science of humanity and nature'.
- While not perhaps going so far as a unified science of humanity and nature, the greening of social theory requires the adoption of a more multi- or interdisciplinary approach.

- Three dimensions of the greening of social theory include: integrating a concern for future generations and time; extending social theory beyond the nation-state and focusing on globalisation; and finally, extending social theory beyond the species barrier, such that while social theory must be human-based it does not necessarily have to be human-centred.
- The greening of social theory and the social theorising about the environment may be seen as focusing on the paradox of the 'human condition', in which humans are both a part of yet apart from the natural environment.

Further reading

On green social and political theory: good introductory overviews include Stephen Young's *The Politics of the Environment*, Manchester: Baseline Books, 1992; Robert Garner's *Environmental Politics*, Hemel Hempstead: Harvester Wheatsheaf, 1995; and ch. 2 of of Erika Cudworth's *Environment and Society*, London: Routledge, 2003; and my own 'Green Political Theory', in J. Barry and E.G. Frankland (eds), *International Encyclopedia of Environmental Politics*, London: Routledge, 2001.

More detailed accounts are my own *Rethinking Green Politics: Nature, Virtue and Progress*, London: Sage, 1999; J. Barry, 'Green Political Theory', in A. Lent (ed.), *New Political Thought*, London: Macmillan, 1998; J. Barry, 'Environmentalism', in R. Axtmann (ed.), *Understanding Democratic Politics: Concepts, Institutions, Movements*, London: Sage, 2002; J. Barry and A. Dobson, 'Green Political Theory: A Report', in C. Kukathas and G. Gaus, (eds), *Handbook of Political Theory*, London: Sage, 2003.

More advanced texts include: Andy Dobson's *Green Political Thought: An Introduction* (3rd edn), London: Routledge, 2000; Luke Martell's *Nature and Society: An Introduction*, Cambridge: Polity Press, 1994; Robert Goodin's *Green Political Theory*, Cambridge: Polity Press, 1992; Robyn Eckersley's *Environmentalism and Political Theory: Towards an Ecocentric Approach*, London: UCL Press, 1992; Tim Hayward's *Ecological Thought: An Introduction*, Cambridge: Polity Press, 1995; Brian Baxter's *Ecologism: An Introduction*, Edinburgh: Edinburgh University Press, 1999; Joel Kassiola's *The Death of Industrial Civilization: The Limits of Economic Growth and the Repoliticization of Advanced Industrial Society*, Albany, NY: The State University of New York Press, 1990; James Radcliffe's *Green Politics: Dictatorship or Democracy?*, Basingstoke: Macmillan, 2000; Mike Woodin and Caroline Lucas' *Green Alternatives to Globalisation: A Manifesto*, London: Pluto Press, 2004.

On the greening of social theory see William Catton and Riley Dunlap's 'A New Ecological Paradigm for a Post-exuberant Sociology', *The American Behavioral Scientist*, 24: 1 (1980); and for book-length treatments, see the following for excellent works: Peter Dickens, *Society and Nature: Towards a Green Social Theory*, Hemel Hempstead:

Harvester Wheatsheaf, 1992; Ted Benton, *Natural Relations: Ecology, Animal Rights and Social Justice*, London: Verso, 1993; and Keekok Lee, *Social Philosophy and Ecological Scarcity*, London: Routledge, 1989. On 're-embodying' social theory see Chris Shilling, *The Body and Social Theory*, London: Sage, 1993.

On green social theory and globalisation, see *The New Internationalist*, 'Globalization: Pealing Back the Layers', November 1997, and 'Currencies of Desire', October 1998; Tim Lang and Colin Hines, *The New Protectionism*, London: Earthscan, 1993; ch. 5 of Kay Milton's *Environmentalism and Cultural Theory*, London: Routledge, 1996; ch. 5 of Erika Cudworth's *Environment and Society*, London: Routledge, 2003.

Glossary

acid rain rain, snow or mist in which water droplets have combined with acidic gases such as sulphur dioxide, usually as a result of the burning of fossil fuels, such as coal, gas or petroleum.

anthropocentrism thinking or acting that is predominantly concerned with humans. It can take 'weaker' or 'stronger' forms. The strong version holds that human interests or purposes are the only issue in making moral judgements, while the weaker version holds that while human interests are important, they are not the only ones to be considered.

biodiversity shorthand for biological diversity, referring to the variety and quantity of species of plant and animal life.

classical liberalisman early form of liberal theory the central principles of which were a *laissez-faire* or free market view of the economy, a rejection of state interference in economic and social relations, all with the aim of defending a particularly individualistic view of human liberty. Modern versions of classical liberal themes may be found in contemporary libertarianism.

climate change sometimes termed 'global warming', human-caused changes in the Earth's climate as a result of the emission of greenhouse gases such as carbon dioxide and methane. The main causes of climate change are the burning of fossil fuels such as coal, oil and gas. Globally, the ten hottest years on record have all occurred since the beginning of the 1990s.

ecocentrism thinking or acting that is predominantly concerned with both humans and nonhumans.

ecology has a few different meanings. It can mean the science of ecology, the branch of biology concerned with the relations of living things (mainly animals and plants) and their environments, or it can mean the 'ecological paradigm' that is a summary of green ideas which contain both factual and normative claims.

the Enlightenment sometimes called 'modernity'. Denotes the radical series of changes in European thought and action which occurred towards the end of the eighteenth century.

fact/value distinction refers to an important distinction within social theory between statements about the way things are, or descriptions about the way things are (facts), and statements about how things ought to be or normative judgement about things (values).

fossil fuels carbon-based non-renewable forms of energy such as coal, oil, gas.

genetic engineering and biotechnology human manipulation of genetic information and material to create new forms of organisms.

globalisation denotes the series of cultural, political, economic, environmental and social changes and developments which are turning the world into a single market, creating webs of relations bringing different parts of the world closer together into one unified system.

instrumental value something has instrumental value if its value (or worth) is given by reference to some other purpose or entity.

intrinsic value something which has value in itself as opposed to being valuable only for some other purpose or entity.

libertarianism modern heir of the classical liberal view of society which argues for minimal state interference in the economy, coupled with a belief in the efficiency and liberty-enhancing effects of the unfettered free market, all aimed at realising a particular negative form of individual freedom as 'freedom from' state and social interference.

mode of production in Marxist political economy this refers to the particular mode or ways in which a society produces its means of subsistence (it includes the level and type of technology, property relations, and the division of labour in a society); each society, according to the Marxist theory of historical materialism, will have a different mode of production.

modernisation a highly complex and contentious term, which is sometimes used as another word for 'development'. It describes a particular social development path, largely based on the historical experience of European societies. A modernising society displays a series of changes in its economy, politics and culture, such as a move from an agricultural, rural economy to an industrial, urban one, a shift from non-democratic forms of government to (liberal) democratic forms, and the secularisation of values and culture and the decline of traditional or religious values.

natural order has its roots in historical accounts of the relationship between humans and nature, referring to the 'given' (often God-given) and 'proper' organisation of the world including human society.

NGOs/Non-Governmental Organisations voluntary groups or organisations which operate independently of national and international government agencies and institutions. They are often international in focus and often lobby governments while monitoring government activity in a variety of fields including environmental and economic policy.

political economy a term pre-dating modern economics as a discipline but which deals with much the same subject matter: how to deal with the 'economic problem' of limited resources and infinite human wants, and the efficient allocation of resources with competing uses. Where political economy differs from modern economics is in its explicit recognition of the political context of economic activity, and the links between politics and economics.

social construction refers to a particular approach within social theory which holds that either there is no 'objective' reality for humans which has not been constructed via language, or that there is always a social or discursive dimension to human reality.

steady-state economy a term first used by ecological economist Herman Daly to describe the following: a constant population level, a constant capital stock, and minimising energy and matter 'throughput' in the economy.

sustainability has a variety of meanings, but the main thrust of the concept conveys a sense of something continuing (lasting, enduring, sustaining) into the future. Usually taken to mean ecological sustainability in the sense of 'maximum yield', a sustainable use of resources which does not undermine regeneration rates.

sustainable development in the words of the Brundtland Commission, sustainable development means the ability of the current generation to meet its needs without undermining the ability of the future to meet its own needs.

technocentric used to describe approaches to social and environmental problems as 'technical' matters which can be solved by technological or technical means, usually the application of scientific or expert knowledge and/or some technological innovation.

utilitarianism the moral theory in which the 'greatest happiness' of the 'greatest number', the balance of pleasure over pain, or utility over disutility, is the criterion by which one ought to judge different courses of action.

Internet resources and sites

There are thousands of environmental and green sites. Some are academic, some are government-related and many are maintained by environmental groups, green parties, movements and activists.

Some good places to start your search are:

1 **Fundamentally Green:** http://www.barnsdle.demon.co.uk/pol/fundi.html
2 **EnviroLink:** http://www.envirolink.org
3 **GreenNet Home Page:** http://www.gn.apc.org
4 **Political Science Resources:** http://www.psr.keele.ac.uk/

The latter also has links to a wide variety of sites on political and social theory.

On social theory see also http://www.geocities.com/~sociorealm/ socialth2.htm.

Environmental groups/movements

1 **Green Parties of North America:** www.greenparty.org
2 **Green Party of New Jersey:** http://www.gpnj.org/
3 **Green Party of England and Wales:** www.greenparty.org.uk
4 **The European Federation of Green Parties:** www.europeangreens.org
5 **Green Parties World:** http://www.greens.org/
6 **Worldwide Fund for Nature:** www.panda.org/
7 **Earthaction:** www.oneworld.org/earthaction/
8 **Greenpeace:** www.greenpeace.org/
9 **Eco – The Campaign for Political Ecology:** www.gn.apc.org/eco/
10 **The Land Is Ours:** www.envirolink.org/orgs/tlio/
11 **Centre for Alternative Technology:** www.foe.co.uk/CAT/
12 **Earthfirst!:** www.earthfirstjournal.org
13 **Rainforest Action Network:** www.ran.org
14 **Environmental Protest in Britain:** www.keele.ac.uk/depts/po/pol/courses/m307/britprot.htm

15 **The New Economics Foundation:** www.neweconomics.org
16 **Demos:** www.demos.org
17 **Adbusters:** www.adbusters.org
18 **Sierra Club:** http://www.sierraclub.org/
19 **Corporate Watch:** http://www.corpwatch.org/
20 **Friends of the Earth Publications:**
    http://www.foe.co.uk/pubsinfo/infosyst/other_services.html

Journals, magazines and bibliographical sites

1 **Bibliography on Biodiversity:** http://osu.orst.edu/dept/ag_resrc_econ/
   biodiv/biblio.html
2 **Electronic Green Journal:** http://www.lib.uidaho.edu:70/docs/egj.html
3 **Red Pepper Magazine:** http://www.netlink.co.uk/users/editoria/
4 **Environmental Newsletter E-Zine:** http://www.geocities.com/Eureka/
   Plaza/1697/newsletter.html
5 **EcoSocialist Review:** http://www.dsausa.org/dsa/ESR/index.html
6 **National Library for the Environment from CNIE:** http://www.cnie.
   org/nle/
7 **International Society for Environmental Ethics:** http://www.cep.unt.edu/
   ISEE/html
8 **Green Politics Newsletter:** http://www.keele.ac.uk/depts/po/pol/green/
   march98.htm
9 **New Internationalist:** http://www.oneworld.org/ni/
10 **Grist Magazine**: www.grist.org
11 **Green TV**: www.green.tv/
12 **Green Web Publishing:** http://home.ca.inter.net/~greenweb/gw-other.html
13 **Electronic Green Journal:** http://egj.lib.uidaho.edu/egj07/schrode.htm

Government and academic sites on the environment and environmental issues

1 **European Environment Agency**: http://www.eea.dk
2 **Centre for Social and Economic Research on the Global Environment**:
   http://www.uea.ac.uk/env/cserge/noframe.htm
3 **Department of Environment, Transport and the Regions (UK)**:
   http://www.detr.gov.uk/itwp/index.htm
4 **Southampton Library Environmental Resources:** http://www.southampton.
   liunet.edu/library/environ.htm

5 **Global Environmental Change Site:** http://www.sussex.sc.uk/Units/gec/

6 **National Centre for Sustainability (US):** http://www.islandnet.com/~ncfs/ncsf/homemenu.htm

7 **Sustainable Development:** http://www.ulb.ac.be/ceese/sustul.htm

8 **Institute for Bioregional Studies:** http://www.cycor.ca/IBS/

9 **Systematic Work on Environmental Ethics:** http://www.cep.unt.edu/theo.html

10 **Centre for Study of Social Movements (Cantebury):** http://snipe.ukc.ac.uk/sociology/polsoc.html

11 **US Global Change Research Program:** http://www.usgcrp.gov

12 **Biodiversity and Ecosystem Network (BENE):** http://straylight.tamu.edu/bene/bene.html

13 **Centre for World Indigenous Studies:** http://www.halcyon.com/FWDP/cwisinfo.html

14 **Environmental Ethics:** http://ethics.acusd.edu/Applied/Environment/

15 **Envirolink:** http://www.envirolink.org/

16 **Clearinghouse for Subject-oriented Internet Resource Guides at the University of Michigan:** http://www.clearinghouse.net/cgi-bin/chadmin/viewcat/Environment? kywd

17 **School of Social Ecology:** http://www.seweb.uci.edu/index.uci

18 **Internet Resources for Environment and Economics:** http://sorrel.humboldt.edu/~envecon/internet.html

19 **International Society for Ecological Economics:** http://www.ecoeco.org/index.html

20 **Ecological Economics Research Institutes:** http://www.ecoeco.org/links/resear_insts/linx_inst.htm

21 **Environmental History Timeline:** http://www.radford.edu/~wkovarik/envhist/

22 **Environmental History Web Resources:** http://www.cnr.berkeley.edu/departments/espm/env-hist/

23 **Berkeley Workshop on Environmental Politics:** http://globetrotter.berkeley.edu/EnvirPol/

24 **Science, Technology and Environmental Politics section of American Political Science Association:** http://www.apsanet.org/section_320.cfm

25 **MIT Project on Environmental Politics and Policy:** http://web.mit.edu/polisci/mpepp/

Environmental justice

1 **Working Group on Environmental Justice:** http://ecojustice.net/
2 **Environmental Justice resources on the web:** http://ecojustice.net/document/ejlinks.htm
3 **Environmental Justice Resource Centre:** http://www.ejrc.cau.edu/
4 **Environmental Justice Program:**http://www-personal.umich.edu/~b bryant/envjustice.html
5 **Environmental Justice Database:** http://web1.msue.msu.edu/msue/imp/modej/masterej.html

Gender and environment, eco-feminism

1 **Gender and Urban Planning:** http://www-rcf.usc.edu/~harwood/fem&plan.htm
2 **Women in Natural Resources:** http://www.ets.uidaho.edu/winr/index
3 **Women and Environments:** http://www.web.net/~weed
4 **Eco-feminism:** http://www2.infoseek.com/Titles?qt=eco-feminism

Sustainability, sustainable development

Sustainability Web ring: http://n.webring.com/hub?ring=sustainability

Government and academic sites on the environment and environmental issues

1 **European Environment Agency:** http://www.eea.dk
2 **Centre for Social and Economic Research on the Global Environment:** http://www.uea.ac.uk/env/cserge/noframe.htm
3 **Department of Environment, Transport and the Regions (UK):** http://www.detr.gov.uk/itwp/index.htm
4 **Southampton Library Environmental Resources:** http://www. southampton.liunet.edu/library/environ.htm
5 **Global Environmental Change Site:** http://www.sussex.ac.uk/ Units/gec/
6 **National Centre for Sustainability (US):** http://www.islandnet. com/~ncsf/ncsf/homemenu.htm
7 **Sustainable Development:** http://www.ulb.ac.be/ceese/sustul.htm
8 **Institue for Bioregional Studies:** http://www.cycor.ca/IBS/
9 **Systematic Work on Environmental Ethics:** http://www.cep. unt.edu/theo.html

10 **Centre for Study of Social Movements (Cantebury):** http://snipe. ukc.ac. uk/sociology/polsoc.html

11 **US Global Change Research Program:** http://www.usgcrp.gov

12 **Biodiversity and Ecosystem Network (BENE):** http://straylight. tamu.edu/ bene/bene.html

13 **Centre for World Indigenous Studies:** http://www.halcyon.com/ FWDP/ cwisinfo.html

Bibliography

Abbey, E. (1988) *One Life at a Time Please*, New York: Henry Holt.

Adam, B. (1990) *Time and Social Theory*, Cambridge: Polity Press.

Adam, B. (1995) *Timewatch: The Social Analysis of Time*, Cambridge: Polity Press.

Adam, B.(1998) *Timescapes of Modernity. The Environment and Invisible Hazards,* London: Routledge.

Alper, J. (1978) 'Ethical and Social Implications', in M. Gregory *et al.* (eds), *Sociobiology and Human Nature*, San Francisco, CA: Jossey-Bass.

Andermatt Conley, V. (1997) *Ecopolitics: The Environment in Poststructuralist Thought*, London: Routledge.

Anderson, J. (2006) 'Only Sustain. . . . The Environment, "Anti-globalization" and the Runaway Bicycle', in J. Johnston, M. Gismondi and J. Goodman (eds), *Nature's Revenge: Reclaiming Sustainability in an Age of Corporate Globalization*, Ontario: Broadview Press.

Anderson, T. and Leal, D. (1991) *Free Market Environmentalism*, Boulder, CO: Westview Publications.

Aristotle (1948) *Politics*, translated by E. Barker, Oxford: Oxford University Press.

Arnold, D. (1996) *The Problem of Nature: Environment, Culture and European Expansion*, Oxford: Blackwell.

Arnold, R. (1997) *Ecoterror: The Violent Agenda to Save Nature, the World of the Unabomber*, Washington, DC: Free Enterprise Press.

Bahro, R. (1994) *Avoiding Social and Ecological Disaster: The Politics of World Transformation*, Bath: Gateway Books.

Bandarage, A. (1997) *Women, Population and Global Crisis: A Political-economic Analysis*, London: Zed Books.

Barnes, M. (ed.) (1994) *An Ecology of the Spirit*, Lanham, Maryland: University Press of America.

Barnett, J. (2001) *The Meaning of Environmental Security: Ecological Politics and Policy in the New Security Era*, London: Zed Books.

Barnett, J., Ellemore, H. and Dovers, S. (2003) 'Sustainability and Interdisciplinarity', in S. Dovers, D. Stern and M. Young (eds), *New Dimensions in Ecological Economics: Integrated Appoaches to People and Nature*, Chelthenham: Edward Elgar.

Barry, J. (1990) 'Limits to Growth', unpublished MA dissertation.

Barry, J. (1993) 'Deep Ecology and the Undermining of Green Politics', in J. Holder *et al.* (eds), *Perspectives on the Environment*, Aldershot: Avebury Press.

Barry, J. (1994) 'Beyond the Shallow and the Deep: Green Politics, Philosophy and Praxis', *Environmental Politics*, 3:3.

Barry, J. (1995a) 'Towards a Theory of the Green State', in S. Elworthy *et al.* (eds), *Perspectives on the Environment Two*, Aldershot: Avebury Press.

Barry, J. (1995b) 'Nature in Question: What is the Question?', *Environmental Politics*, 4:1.

Barry, J. (1995c) 'Deep Ecology, Socialism and Human "Being" in the World: A Part of Yet Apart from Nature', *Capitalism, Nature, Socialism*, 6:3.

Barry, J. (1996a) 'Democracy, Judgement and Sustainability', in B. Doherty and M. Geus (eds), *Democracy and Green Political Theory: Sustainability, Justice and Citizenship*, London: Routledge.

Barry, J. (1998a) 'Green Political Theory', in A. Lent (ed.), *New Political Thought: An Introduction*, London: Macmillan.

Barry, J. (1998b) 'Social Policy and Social Movements: Ecology and Social Policy' in N. Ellison and C. Pierson (eds), *Developments in British Social Policy*, London: Macmillan.

Barry, J. (1998c) 'Marxism and Ecology', in A. Gamble *et al.* (eds), *Marxism and Social Science*, Basingstoke: Macmillan.

Barry, J. (1999a) *Rethinking Green Politics: Nature, Virtue and Progress*, London: Sage.

Barry, J. (1999b) 'Marxism and Ecology', in A. Gamble et al (eds), *Marxism and Social Science*. Basingstoke: Macmillan.

Barry, J (2001), 'Greening Liberal Democracy: Theory, Practice and Political Economy', in J. Barry and M. Wissenburg (eds), *Sustaining Liberal Democracy: Ecological Challenges and Opportunities*, Basingstoke: Palgrave.

Barry, J. (2003) 'Holding Tender Views in Tough Ways: Political Economy and Strategies of Resistance in Green Politics', *British Journal of Politics and International Relations*, 5:4.

Barry, J. (2005) 'Resistance is Fertile: From Environmental to Sustainability Citizenship', in D. Bell and A. Dobson (eds), *Environmental Citizenship*, Cambridge, MA: MIT Press.

Barry, J. (2006a) 'Towards a Concrete Utopian Model of Green Political Economy: From Economic Growth and Ecological Modernisation to Economic Security', *Post-Autistic Economics Review*, 38.

Barry, J. (2006b) 'Towards a Model of Green Political Economy: From Ecological Modernisation to Economic Security', *International Journal of Green Economics*, 1:3.

Barry, J. and Doherty, B. (2001) 'The Greens and Social Policy: Movements, Politics and Practice?', *Social Policy and Administration*, 35:5.

Barry, J. and Frankland, E.G. (eds) (2001) *International Encyclopedia of Environmental Politics*, London: Routledge.

Bate, R. (2003) 'The European Union's Confused Position on GM Food', *Economic Affairs*, 23:1.

Bauman, Z. (1993) *Postmodern Ethics*, Oxford: Blackwell.

Baxter, B. (1996) 'Must Political Theory Now Be Green?', in I. Hampsher-Monk and

J. Stanyer (eds), *Contemporary Political Studies 1996*, vols 1–3, Belfast: Political Studies Association.

Baxter, B. (1999) *Ecologism: A Defence*, Edinburgh: Edinburgh University Press.

Beder, S. (1997) *Global Spin: The Corporate Assault on Environmentalism*, Devon: Green Books.

Beder, S. (2001a) 'Neoliberal Think-tanks and Free Market Environmentalism', *Environmental Politics*, 10:2.

Beder, S. (2001b) 'Anti-environmentalism' and 'Green backlash', in J. Barry and E.G. Frankland (eds), *International Encyclopedia of Environmental Politics*, London: Routledge.

Beck, U. (1992a) *Risk Society: Towards a New Modernity*, London: Sage.

Beck, U. (1992b) 'From Industrial Society to Risk Society: Questions of Survival, Social Structure and Ecological Enlightenment', *Theory, Culture and Society*, 9:1.

Beck, U. (1995a) *Ecological Enlightenment: Essays on the Politics of the Risk Society*, Princeton, NJ: Humanities Press International.

Beck, U. (1995b) *Ecological Politics in an Age of Risk*, Cambridge: Polity Press.

Beck, U. (2000) *What is Globalization?*, Oxford: Blackwell.

Beckerman, W. (1974) *In Defence of Economic Growth*, London: Cape.

Beckerman, W. (1995) *Small is Stupid: Blowing the Whistle on the Greens*, London: Duckworth.

Beckerman, W. (2002) *A Poverty of Reason: Sustainable Development and Economic Growth,* Oakland, CA: The Independent Institute.

Bellemy-Foster, J. (2000) *Marx's Ecology: Materialism and Nature*, New York: Monthly Review Press.

Bellemy-Foster, J. (2002) *Ecology against Capitalism*, New York: Monthly Review Press.

Bennholdt-Thompsen, V. and Mies, M. (1999) *The Subsistence Perspective: Beyond the Globalised Economy*, London: Zed Books.

Bentham, J. (1823/1970) *The Principles of Morals and Legislation*, Darien, CT: Hafner Publishing.

Benton, T. (1989) 'Marxism and Natural Limits: An Ecological Critique and Reconstruction', *New Left Review*, 178.

Benton, T. (1991) 'Biology and Social Science: Why the Return of the Repressed Should be Given a (Cautious) Welcome', *Sociology*, 25:1.

Benton, T. (1993) *Natural Relations: Ecology, Animals and Social Justice*, London: Verso.

Benton, T. (1994) 'Biology and Social Theory in the Environmental Debate', in M. Redclift and T. Benton (eds).

Benton, T. (ed.) (1996) *The Greening of Marxism*, New York: The Guilford Press.

Benton, T. (1999) 'Radical Politics – Neither Left nor Right?', in M. O'Brien, S. Penna and C. Hay (eds), *Theorising Modernity: Reflexivity, Environment and Identity in Giddens' Social Theory*, Harlow, Essex: Addison Wesley Longman.

Benton, T. (2000) 'An Ecological Historical Materialism', in F. Gale and M. M'Gonigle (eds), *Nature, Production, Power: Towards an Ecological Political Economy*, Cheltenham: Edward Elgar.

Benton, T. and Redclift, M. (1994) 'Introduction', in M. Redclift and T. Benton (eds), *Social Theory and the Global Environment*, London: Routledge.

Berry, C. (1986) *Human Nature*, London: Macmillan.

Biro, A. (2005) *Denaturalizing Ecological Politics: Alienation from Nature from Rousseau to the Frankfurt School and Beyond*, Toronto: University of Toronto Press.

Blaug, M. (1997) 'Ugly Currents in Modern Economics', *Policy Options*, spring, http://www.irpp.org/po/archive/sep97/blaug.pdf (accessed 20 May 2006).

Blühdorn, I. (2000) *Post-ecologist Politics: Social Theory and the Abdication of the Ecologist Paradigm*, London: Routledge.

Blühdorn, I. (2002) 'Unsustainability as a Frame of Mind – and How We Disguise It: The Silent Counter-revolution and the Politics of Simulation', *The Trumpeter*, 18:1.

Boardman, B. with Bullock, S. and McLaren, D. (1999), *Equity and the Environment: Guidelines for Green and Socially Just Government*, London: Catalyst and Friends of the Earth.

Bocking, S. (2004) *Nature's Experts: Science, Politics and the Environment*, New Brunswick, NJ: Rutgers University Press.

Bookchin, M. (1980) *Towards an Ecological Society*, Montreal and Buffalo: Black Rose Books.

Bookchin, M. (1986) *The Modern Crisis*, Philadelphia, PA: New Society Publishers.

Bookchin, M. (1990) *Remaking Society*, Montreal and New York: Black Rose Books.

Bookchin, M. (1992) 'Libertarian Municipalism', *Society and Nature*, 1:1.

Bookchin, M. (1993) 'What is Social Ecology?', in M. Zimmerman (ed.), *Environmental Philosophy: From Animal Rights to Radical Ecology*, Englewood Cliffs, NJ: Prentice Hall.

Boulding, K. (1966) 'The Economics of the Coming Spaceship Earth', in H. Jarrett (ed.), *Environmental Quality in a Growing Economy*, Baltimore, MD: Johns Hopkins University Press.

Boyle, D. (ed.) (2002) *The Money Changers: Currency Reform from Aristotle to E-cash*, London: Earthscan.

Bradley, R. (2004) *Climate Alarmism Reconsidered*, London: Institute of Economic Affairs.

Bramwell, A. (1989) *Ecology in the 20th Century: An Introduction*, London, and New Haven, CT: Yale University Press.

Brennan, A. (1988) *Thinking about Nature: An Investigation of Nature, Value and Ecology*, London: Routledge.

Brockman, J. (1995) *The Third Culture: Beyond the Scientific Revolution*, New York: Simon & Schuster.

Bromley, D. (1991) *Environment and Economy: Property Rights and Public Policy*, Oxford: Blackwell.

Brown, L. (1989) *Envisioning a Sustainable Society: Learning Our Way Out*, New York: State University of New York Press.

Bruelle, R. (2000) *Agency, Democracy and Nature: The U.S. Environmental Movement from a Critical Theory Perspective*, Cambridge, MA: MIT Press.

Bruelle, R. (2002) 'Habermas and Green Political Thought: Two Roads Converging', *Environmental Politics*, 11:4.

Buckingham, S. (2004) 'Ecofeminism in the Twenty-first Century', *The Geographical Journal*, 170:2.

Buckingham-Hatfield, S. (2000) *Gender and Environment*, London: Routledge.

Buell, F. (2003) *From Apocalypse to Way of Life: Environmental Crisis in the American Century*, New York: Routledge.

Buell, L. (1996), *The Environmental Imagination: Thoreau, Nature Writing, and the Formation of American Culture*, Cambridge, MA: Harvard University Press.

Buell, L. (2004) *Emerson*, Cambridge MA: Harvard University Press.

Bullard, R. (2005) 'Environmental Justice in the 21st Century', in J. Dryzek and D. Schlosberg (eds), *Debating the Earth: The Environmental Politics Reader* (2nd edn), Oxford: Oxford University Press.

Burke, E. (1790/1969) *Reflections on the Revolution in France*, ed. Conor Cruise O'Brien, Harmondsworth: Penguin Books.

Burke, W. (1993) 'The Wise Use Movement: Right-wing Anti-environmentalism', *The Public Eye* (June).

Cahalan, J.A. (2001) *Edward Abbey: A Life*, Tucson: University of Arizona Press.

Cahn, M. and O'Brien, R. (1996) *Thinking About the Environment: Readings on Politics, Property and the Physical World*, New York and London: M.E. Sharpe.

Cakebread, A. (1996) 'Guess Who's Coming to Dinner', *New Internationalist*, September.

Callicott, J.B. (1994) *Earth's Insights: A Multicultural Survey of Ecological Ethics from the Mediterranean Basin to the Australian Outback*, Berkeley: University of California Press.

Carson, R. (1962) *Silent Spring*, Greenwich, CT: Fawcett Premier.

Carter, A. (1993) 'Towards a Green Political Theory', in A. Dobson and P. Lucardie (eds), *The Politics of Nature: Explorations in Green Political Theory*, London: Routledge.

Carter, A. (1999), *A Radical Green Political Theory*, London: Routledge.

Cassell, P. (ed.) (1993) *The Giddens Reader*, London: Macmillan.

Castree, N. (forthcoming) 'Nature', in K.J. Johnston *et al.* (eds), *The Dictionary of Human Geography* (4th edn), Oxford: Blackwell.

Catton, W. and Dunlap, R. (1980) 'A New Ecological Paradigm for a Post-exhuberant Sociology', *American Behavioral Scientist*, 24:1.

Chesterton, G.K. (1959) Orthodoxy, Garden City, NY: Doubleday Image Books.

The Children's Dictionary (1969) London: The Grolier Society.

Christie, I. and Nash, L. (eds) (1998) *The Good Life*, London: Demos.

Clark, J. (2004) 'Municipal Dreams : A Social Ecological Critique of Bookchin's Politics', *Research on Anarchism Forum*, http://raforum.apinc.org/article.php3?id_article=1038 (accessed 11 May 2006).

Clayre, A. (ed.) (1977) *Nature and Industrialization*, Oxford: Oxford University Press.

Coates, I. (1993) 'A Cuckoo in the Nest: The National Front and Green Ideology', in J. Holder *et al.* (eds), *Perspectives on the Environment: Interdisciplinary Reseach in Action,* Aldershot: Avebury Press.

Collard, A. (1988) *Rape of the Wild: Man's Violence against Animals and the Earth*, Indianapolis: Indiana University Press.

Common, M. (1996) 'What is Ecological Economics?', in R. Gill (ed.), *R&D Priorities for Ecological Economics*, Canberra: Land and Waters Research and Development Corporation.

Commoner, B. (1971) *The Closing Circle: Nature, Man and Technology*, New York: Bantam Books.

Condorcet, M. (1995) 'The Future Progress of the Human Mind', in I. Kramnick (ed.), *The Portable Enlightenment Reader*, London: Penguin.

Cooper, D. (1992) 'The Idea of Environment', in D. Cooper and J. Palmer (eds), *The Environment in Question: Ethics and Global Issues*, London: Routledge.

Cooper, D. and Palmer, J. (eds) (1992) *The Environment in Question: Ethics and Global Issues*, London: Routledge.

Copenhagen Consensus (2004) 'The Basic Idea', http://www.copenhagenconsensus.com/Default.aspx?ID=161 (accessed 10 May 2006).

Costanza, R. (1989) 'What is Ecological Economics?', *Ecological Economics*, 1:1.

Costello, A. (1991) 'The Ecology of Failure', *Analysis,* winter.

Croall, S. and Rankin, W. (1981) *Ecology for Beginners*, New York: Pantheon Books.

Croll, E. and Parkin, D. (1992) *Bush Base, Forest Farm: Culture, Environment and Development*, London: Routledge.

Cronon, W. (1995) 'The Trouble with Wilderness; or, Getting Back to the Wrong Nature', in Cronon, W. (ed.), *Uncommon Ground: Rethinking the Human Place in Nature*, New York: W. W. Norton.

Cudworth, E. (2003) *Environment and Society*, London: Routledge.

Daly, H. (ed.) (1973) *Toward a Steady-state Economy*, San Francisco, CA: W.H. Freeman.

Daly, H. (1987) 'The Economic Growth Debate: What some Economists have Learned but Many have not', *Journal of Environmental Economics and Management*, 14:4.

Daly, H. and Cobb, J.B. (1989) *For the Common Good*, Boston, MA: Beacon Press.

Danish Committees on Scientific Dishonesty (2003) http://www.forsk.dk/uvvu/nyt/udtaldebat/bl_decision.htm.

Dasmann, R. (1993) 'Biosphere versus Ecosphere People', in G. Sessions (ed.), *Deep Ecology for the Twenty-first Century*, Convelo, CA: Island Press.

Davies, J. (1976) 'Ions of Emotion and Political Behavior: A Prototheory', in A. Somit (ed.), *Biology and Politics: Recent Explorations*, The Hague and Paris: Mouton.

de Geus, M. (1999) *Ecological Utopias: Envisioning the Sustainable Society*, Utrecht: International Books.

de-Shalit, A. (1997) 'Is Liberalism Environment-Friendly?', in R. Gottlieb (ed.), *The Ecological Community: Environmental Challenges for Philosophy, Politics and Morality*, London: Routledge.

Diamond, C. (1978) 'Eating Meat and Eating People', *Philosophy*, 53.

Diamond, J. (1991) *The Rise and Fall of the Third Chimpanzee*, London: Vintage.

Diamond, J. (1997) *Guns, Germs and Steel: The Fates of Human Societies*, London: Jonathan Cape.

Dickens, P. (1992) *Society and Nature: Towards a Green Social Theory*, London: Harvester Wheatsheaf.

Dickens, P. (1996) *Reconstructing Nature: Alienation, Emancipation and the Division of Labour*, London: Routledge.

Dickens, P. (1997) 'Beyond Sociology: Marxism and the Environment', in M. Redclift and G. Woodgate (eds), *The International Handbook of Environmental Sociology*, Cheltenham: Edward Elgar.

Dickens, P. (2002) 'A Green Marxism? Labor Processes, Alienation, and the Division of Labor', in R. Dunlap, F. Buttel, P. Dickens and A. Gijswijt (eds), *Sociological Theory and the Environment. Classical Foundations, Contemporary Insights*, New York: Rowman & Littlefield.

Dixon-Mueller, H. (1993) *Population Policy and Women's Rights*, Westport, CT: Praeger.

Dobson, A. (1993) 'Critical Theory and Green Politics', in A. Dobson and P. Lucardie (eds), *The Politics of Nature: Explorations in Green Political Theory*, London: Routledge.

Dobson, A. (2000) *Green Political Thought* (3rd edn), London: Routledge.

Dobson, A. (2003) *Citizenship and the Environment*, Oxford: Oxford University Press.

Dobson, A. and Lucardie, P. (eds) (1993) *The Politics of Nature: Explorations in Green Political Theory*, London: Routledge.

Docherty, T. (1993) 'Postmodernism: An Introduction', in T. Docherty (ed.), *Postmodernism: A Reader*, Hemel Hempstead: Harvester Wheatsheaf.

Docherty, T. (ed.) (1993) *Postmodernism: A Reader*, Hemel Hempstead: Harvester Wheatsheaf.

Doherty, B. and de Geus, M. (eds) (1996) *Democracy and Green Political Theory: Sustainability, Justice and Citizenship*, London: Routledge.

Doran, P. (2006) 'Street Wise Provocations: The Global Justice Movement's Take on Sustainable Development', *International Journal of Green Economics,* 1:1/2.

Dovers, S., Stern, D. and Young, M. (eds) (2003) *New Dimensions in Ecological Economics: Integrated Appoaches to People and Nature*, Chelthenham: Edward Elgar.

Doyle, T. and McEachern, D. (1998) *Environment and Politics*, London: Routledge.

Dryzek, J. (1987) *Rational Ecology: Environment and Political Economy*, Oxford: Blackwell.

Dryzek, J. (1990) 'Green Reason: Communicative Ethics for the Biosphere', *Environmental Ethics*, 12:3.

Dryzek, J. (1996) 'Foundations for Environmental Political Economy: The Search for *Homo Ecologicus*?, *New Political Economy*, 1:1.

Dryzek, J. (2005) *The Politics of the Earth* (2nd edn), Oxford: Oxford University Press.

Dryzek, J. and Schlosberg, D. (eds) (2004) *Debating the Earth* (2nd edn), Oxford: Oxford University Press.

Easterbrook, G. (1995) *A Moment on the Earth: The Coming Age of Environmental Optimism*, New York: Penguin.

Eccleshall, R. (2003), 'Conservatism', in R. Eccleshall *et al.* (eds), *Political Ideologies: An Introduction*, London: Routledge.

Eckersley, R. (1989) 'Green Politics: Selfishness or Virtue?', *Political Studies*, 37:2.

Eckersley, R. (1990) 'Habermas and Green Political Theory: Two Roads Diverging', *Theory and Society*, 19:6.

Eckersley, R. (1992) *Environmentalism and Political Theory: Toward an Ecocentric Approach*, London: UCL Press.

Eckersley, R. (1993) 'Free Market Environmentalism: Friend or Foe?, *Environmental Politics*, 2:1.

Eckersley, R. (2005) *The Green State*, Boston, MA: MIT Press.

Eder, K. (1996) *The Social Construction of Nature: A Sociology of Ecological Enlightenment*, London: Sage.

Ehrlich, P. and Ehrlich, A. (1996) *Betrayal of Science and Reason: How Anti-environmental Rhetoric Threatens our Future*, Washington, DC: Island Press.

Ekins, P. (ed.) (1986) *The Living Economy: A New Economics in the Making*, London: Greenprint.

Ekins, P. (ed.) (1992) *A New World Order: Grassroots Movements for Global Change*, London: Routledge.

Eklund, R. and Herbert, R. (1975) *A History of Economic Theory and Method*, New York: McGraw-Hill.

Elliot, D. (1997) *Energy, Society and Environment*, London: Routledge.

Elliott, R. (1997) *Faking Nature*, London: Routledge.

Ellison, N. and Pierson, C. (eds) (1998) *Developments in British Social Policy*, London: Macmillan.

Elvin, J. (1998) 'The Environmental Legacy of Imperial China', *The China Quarterly*, 156.

Engel, J. and Engel, J. (eds) (1990) *The Ethics of Environmental Development*, London: Belhaven Press.

Enhrenfeld, D. (1980) *The Arrogance of Humanism*, Oxford: Oxford University Press.

Environmental Assessment Institute (2003) 'Lomborg Decision Overturned by Danish Ministry of Science', http://www.imv.dk/Default.aspx?ID=233 (accessed 10 May 2006).

Environmental Justice – Principles of Environmental Justice (1991) *Environmental Justice/Environmental Racism*, http://www.ejnet.org/ej/principles.html (accessed 11 May 2006).

Enzensberger, H.M. (1974) 'A Critique of Political Ecology', *New Left Review*, 84.

Evans, J. (1993) 'Ecofeminism and the Politics of the Gendered Self', in A. Dobson and P. Lucardie (eds), *The Politics of Nature: Explorations in Green Political Theory*, London: Routledge.

Fardell, J. (1992) 'Caricature of Greens', *Viz Magazine*.

Feenberg, A. (1992) 'Marcuse: Obstinacy as a Theoretical Virtue', *Capitalism, Nature, Socialism*, September, pp. 38–40.

Feenberg, A. (1996) 'Marcuse or Habermas: Two Critiques of Technology', *Inquiry*, 39.

Feenberg, A. (1999) 'A Fresh Look at Lukács: On Steven Vogel's Against Nature', *Rethinking Marxism,* winter, pp. 84–92, also at http://www-rohan.sdsu.edu/faculty/feenberg/vogel3.htm (accessed 28 April 2005).

Fischer, F. (1990) *Technology and the Politics of Expertise*, London: Sage.

Fischer, F. (2000) *Citizens, Experts and the Environment: The Politics of Local Knowledge*, Durham, NC: Duke University Press.

Foster, J. (1997) 'Introduction', in J. Foster (ed.), *Valuing Nature? Economics, Ethics and Environment*, London: Routledge.

Foster, J. (ed.) (1997) *Valuing Nature? Economics, Ethics and Environment*, London: Routledge.

Foucault, M. (1978) *The History of Sexuality: An Introduction*, translated by R. Hurley, Harmondsworth: Penguin.

Foucault, M. (1980) *Michel Foucault, Power/Knowledge: Selected Interviews and Other Writings 1972–1977*, ed. Colin Gordon, New York: Pantheon.

Freeman, C. (1973) 'Malthus with a Computer', *World Issues*, 25:4.

Freud, S. (1950) '"Civilized" Sexual Morality and Modern Nervousness', *Collected Papers*, Vol. II, London: Hogarth Press.

Fromm, E. (1976) *To Have or to Be?*, New York: Harper & Row.

Fullbrook, E. (ed.) (2003) *The Crisis in Economics*, London: Routledge.

Fullbrook, E. (ed.) (2004) *A Guide To What's Wrong With Economics*, London: Anthem Press.

Fullbrook, E. (2005) 'Post-autistic Economics', *Soundings: A Journal of Politics and Culture*, spring.

Gandy, M. (1996) 'Crumbling Land: The Postmodernity Debate and the Analysis of Environmental Problems', *Progress in Human Geography*, 20:1.

Gare, A. (1995) *Postmodernism and the Environmental Crisis*, London: Routledge.

Gasman, D. (1971) The Scientific Origins of National Socialism: Social Darwinism in Ernst Haeckel and the German Monist League, New York: Transaction Publishers.

Ghate, O. (2002) 'Defend Industry Against Terrorism – Before It's Too Late', http://www.enterstageright.com/archive/articles/0202/0202hometerror.htm (accessed 10 May 2006).

Giarini, O. (1980) *Dialogue on Wealth and Welfare*, London: Pergamon Press.

Giddens, A. (1985) *The Nation-state and Violence*, Cambridge: Polity Press.

Giddens, A. (1990) *The Consequences of Modernity*, Cambridge: Polity Press.

Giddens, A. (1991) *Modernity and Self-identity: Self and Society in the Late Modern Age*, Cambridge: Polity Press.

Giddens, A. (1994) *Beyond Left and Right: The Future of Radical Politics*, Cambridge: Polity Press.

Giddens, A. (1996) 'Affluence, Poverty and the Idea of a Post-scarcity Society', *Development and Change*, 27:2.

Giddens, A. (2000a) 'Interview', in L.B. Kaspersen, *Anthony Giddens: An Introduction to a Social Theorist,* Oxford: Blackwell.

Giddens, A. (2000b) *The Third Way and its Critics,* Cambridge: Polity Press.

Giddens, A. *et al.* (1994) 'Introduction', in Giddens, A. *et al.* (eds), *The Polity Reader in Social Theory*, Cambridge: Polity Press.

Glacken, C. (1967) *Traces on the Rhodian Shore*, Berkeley: University of California Press.

Glotfelty, C. (2001) 'Susan Griffin', in J. Palmer (ed.), *Fifty Key Thinkers on the Environment*, London: Routledge.

Goldblatt, D. (1996) *Social Theory and the Environment*, Cambridge: Polity Press.

Goldsmith, E. *et al.* (1992) 'Whose Common Future?', *The Ecologist*, 22:4.

Goodin, R. (1992) *Green Political Theory*, Cambridge: Polity Press.

Gorz, A. (1989) *Critique of Economic Reason*, London: Verso.

Gorz, A. (1993) 'Political Ecology: Expertocracy Versus Self-limitation', *New Left Review*, 202.

Gorz, A. (1994) *Capitalism, Socialism, Ecology*, London: Verso.

Gottlieb, R. (ed.) (1996) *This Sacred Earth: Religion, Nature, Environment*, London: Routledge.

Gottlieb, R. (ed.) (1997) *The Ecological Community: Environmental Challenges for Philosophy, Politics and Morality*, London: Routledge.

Gould, P. (1988) *Early Green Politics*, Brighton: Harvester Wheatsheaf.

Gray, J. (1993) 'An Agenda for Green Conservatism', in his *Beyond the New Right: Markets, Government and the Common Environment*, London: Routledge.

Gray, J. (1998a) 'Globalisation: The Dark Side', *New Statesman*, 13 March.

Gray, J. (1998b) *False Dawn: The Delusion of Global Capitalism*, London: Granta.

Gray, J. (2002) *Straw Dogs: Thoughts on Humans and Other Animals*, London: Granta.

Green, T.H. (1974) 'Lecture on Liberal Legislation and Freedom of Contract', reprinted in J.B. Diggs (ed.), *The State, Justice and the Common Good*, Glenview, IL: Scott, Foresman.

Gregory, M. (1978) 'Epilogue', in M. Gregory *et al.* (eds), *Sociobiology and Human Nature*, San Francisco: Jossey-Bass.

Griffin, S. (2000) *Women and Nature: The Roaring Inside Her* (revised edn), San Francisco, CA: Sierra Club Books.

Griffiths, I. (2006) *Social Theory and Sustainability: Deep Ecology, Eco-Marxism, Anthony Giddens and a New Progressive Policy Framework for Sustainable Development*, unpublished Ph.D. Queens University of Belfast, Northern Ireland.

Grove, R. (1995) *Green Imperialism: Colonial Expansion, Tropical Island Edens and the Origins of Environmentalism, 1600–1860*, Cambidge: Cambridge University Press.

Guha, R. (1989) 'Radical American Environmentalism and Wilderness Preservation: A Third World Critique', *Environmental Ethics*, 11:1.

Guha, R. (1996) 'Lewis Mumford, The Forgotten American Environmentalist: An Essay in Rehabilitation', in D. Macauley (ed.), *Minding Nature: The Philosophers of Ecology*, New York: Guilford Press.

Guha, R. and Martinez-Alier, J. (1997) *Varieties of Environmentalism: Essays North and South*, London: Zed Books.

Habermas, J. (1970) 'Technology and Science as "Ideology"', in J. Shapiro (ed. and trans.), *Toward a Rational Society*, Boston, MA: Beacon Press.

Habermas, J. (1971) *Knowledge and Human Interests*, Boston, MA: Beacon Press.

Habermas, J. (1975) *Legitimation Crisis*, Boston, MA: Beacon Press.

Habermas, J. (1981) 'New Social Movements', *Telos*, 49.

Habermas, J. (1982) 'A Reply to My Critics', in J. Thompson and D. Held (eds), *Habermas: Critical Debates*, London: Macmillan.

Habermas, J. (1989a) 'The New Obscurity', in *The New Conservatism: Cultural Criticism and the Historians' Debate*, Cambridge, MA: MIT Press.

Habermas, J. (1989b) *The Theory of Communicative Action*, Vol. 2, Boston, MA: Beacon Press.

Habermas, J. (2003*) The Future of Human Nature*, London: Polity Press.

Hall, E. (2004) 'The Dilemmas of German Bioethics', *The New Atlantis: A Journal of Technology and Society*, spring.

Hall, S. (1998) 'Globalisation and New Labour', *New Statesman*, 13 March.

Hampsher-Monk, I. and Stanyer, J. (eds) (1996) *Contemporary Political Studies 1996*, vols 1–3, Belfast: Political Studies Association.

Hannigan, J. (1995) *Environmental Sociology: A Social Constructionist Perspective*, London: Routledge.

Haraway, D. (1995) 'Otherwordly Conversations, Terrain Topics, Local Terms', in V. Shiva and I. Moser (eds), *Biopolitics: A Feminist and Ecological Reader in Biotechnology*, London: Zed Books.

Hardin, G. (1968) 'The Tragedy of the Commons', *Science*, 162:1243–1248.

Harding, S. (1990) 'Feminism and Theories of Scientific Knowledge', in M. Evans (ed.), *The Woman Question*, London: Sage.

Harvey, D. (1993) 'The Nature of Environment: The Dialectics of Social and Environmental Change', *Socialist Register 1993*.

Harvey, D. (1996) *Justice, Nature and the Geography of Difference*, Oxford: Blackwell.

Haught, T. (1994) 'Religion and the Origins of the Environmental Crisis', in M. Barnes (ed.), *An Ecology of the Spirit*, Lanham, Maryland: University Press of America.

Hay, C. (1994) 'Environment: Security and State Legitimacy', in M. O'Connor (ed.), *Is Capitalism Sustainable? Political Economy and the Politics of Ecology*, London: Guilford Press.

Hayward, T. (1992) 'Ecology and Human Emancipation', *Radical Philosophy*, 62.

Hayward, T. (1995) *Ecological Thought: An Introduction*, Cambridge: Polity Press.

Hayward, T. (1996) 'What Is Green Political Theory?', in I. Hampsher-Monk and J. Stanyer (eds), *Contemporary Political Studies 1996*, vols 1–3, Belfast: Political Studies Association.

Heilbroner, R. (1967) *The Worldy Philosophers*, New York: Simon & Schuster.

Henderson, H. *et al.* (1986) 'Indicators of No Real Meaning', in P. Ekins (ed.), *The Living Economy: A New Economics in the Making*, London: Greenprint.

Hodgson, G. (1997) 'Economics, Environmental Policy and the Transcendence of Utilitarianism', in J. Foster (ed.), *Valuing Nature? Economics, Ethics and Environment*, London: Routledge.

Holland, A. (1997) 'Substitutability: Or, Why Strong Sustainability is Weak and Absurdly Strong Sustainability Is not Absurd', in J. Foster (ed.), *Valuing Nature? Economics, Ethics and Environment*, London: Routledge.

Homer-Dixon. T. (2001) *Environment, Scarcity, and Violence*, Princeton, NJ: Princeton University Press.

Horkheimer, M. and Adorno, T. (1973) *Dialectic of Enlightenment*, London: Allan Lane.

Huby, M. (1998) *Social Policy and the Environment*, Buckingham: Open University Press.

Hughes, J. (2000) *Ecology and Historical Materialism*, Cambridge: Cambridge University Press.

Hughes, J.D. (1994) 'Ecology in Ancient Mesopotamia', in D. Wall (ed.), *Green History: A Reader in Environmental Literature, Philosophy and Politics*, London: Routledge.

Hutchinson, F., Mellor, M. and Olsen, W. (2002) *The Politics of Money: Towards Sustainability and Economic Democracy*, London: Pluto.

Hutchinson Dictionary of Ideas (1994) Oxford: Helicon.

Illich, I. (1971) *Deschooling Society*, Harmondsworth: Penguin.

Illich, I. (1973) *Celebration of Awareness: A Call for Institutional Revolution*, Harmondsworth: Penguin.

Illich, I. (1975) *Tools for Conviviality*, London: Fontana.

Illich, I. (1978) *The Right to Useful Unemployment and Its Professional Enemies*, London: Marion Boyers.

Illich, I. (1981) *Shadow Work*, London: Marion Boyers.

Inglehart, R. (1977) *The Silent Revolution: Changing Values and Political Styles among Western Publics*, Princeton, NJ: Princeton University Press.

Ingold, T. (1992) 'Culture and the Perception of the Environment', in E. Croll and D. Parkin (eds), *Bush Base, Forest Farm: Culture, Environment and Development*, London: Routledge.

Irwin, A. (1997) 'Risk, the Environment and Environmental Knowledges', in M. Redclift and G. Woodgate (eds), *The International Handbook of Environmental Sociology*, Cheltenham: Edward Elgar.

Ives, Y. (2004) 'Judaism and the Environment', http://www.aish.com/tubshvat/tubshvatdefault/Judaism_and_the_Environment.asp (accessed 12 April 2006).

Jacobs, M. (1994) 'The Limits of Neoclassicalism: Towards an Institutional Environmental Economics', in M. Redclift and T. Benton (eds), *Social Theory and the Environment*, London: Routledge.

Jacobs, M. (1996) *The Politics of the Real World*, London: Earthscan.

Jacques, M. (1998) 'Editorial', *Marxism Today*, November/December.

Jameson, F. (1991) *Postmodernism; or, the Cultural Logic of Late Capitalism*, London: Verso.

Journo, E. (2002), 'Home Grown Terrorism', http://www.aynrand.org/site/News2?page=NewsArticle&id=5255 (accessed 10 May 2006).

Kassiola, J. (1990) *The Death of Industrial Civilization: The Limits of Economic Growth and the Repoliticization of Advanced Industrial Society*, Albany, NY: The State University of New York Press.

Keat, R. (1994) 'Citizens, Consumers and the Environment: Reflections on *The Economy of the Earth*', *Environmental Values*, 3:4.

Kellert, S. and Wilson, E.O. (eds) (1993) *The Biophilia Hypothesis*, Washington, DC: Island Press/Shearwater.

Kenny, M. (2003) 'Ecologism', in R. Eccleshall *et al.* (eds), *Political Ideologies: An Introduction*, London: Routledge.

Kidner, D.W. (1994) 'Why Psychology Is Mute about the Environmental Crisis', *Environmental Ethics*, 16:4.

Kidner, D.W. (2001) *Nature and Psyche: Radical Environmentalism and the Politics of Subjectivity,* Albany, NY: State University of New York Press.

Kinsley, D. (1996) 'Christianity as Ecologically Harmful', in R. Gottlieb (ed.), *This Sacred Earth: Religion, Nature, Environment,* London: Routledge.

Kohák, E. (1984) *The Embers and the Stars: A Philosophical Inquiry into the Moral Sense of Nature,* Chicago, IL, and London: University of Chicago Press.

Kovel, J. (1988) *The Radical Spirit: Essays on Psychoanalysis and Society,* London: Free Association Books.

Kovel, J. (2002) *The Enemy of Nature: The End of Capitalism or the End of the World?,* London: Zed Books.

Kramnick, I. (ed.) (1995) *The Portable Enlightenment Reader,* London: Penguin.

Krieger, M.H. (2000) *What's Wrong with Plastic Trees?: Artifice and Authenticity in Design,* Westport, CT: Praeger.

Lash, S., Szersynski, B. and Wynne, B. (eds) (1996) *Risk, Environment and Modernity: Towards a New Ecology,* London: Sage.

Latouche, S. (1993) *In the Wake of the Affluent Society,* London: Zed Books.

Layder, D. (1994) *Understanding Social Theory,* London: Sage.

Lee, K. (1989) *Social Philosophy and Ecological Scarcity,* London: Routledge.

Lee, K. (1993a) 'To De-industrialize: Is it so Irrational?', in A. Dobson and P. Lucardie (eds), *The Politics of Nature: Explorations in Green Political Theory,* London: Routledge.

Lee, K.N. (1993b) *Compass and Gyroscope: Integrating Science and Politics for the Environment,* Washington, DC: Island Press.

Leeson, S. (1979) 'Philosophic Implications of the Ecological Crisis: The Authoritarian Challenge to Liberalism', *Polity,* 11: 303–18.

Lent, A. (ed.) (1998) *New Political Thought: An Introduction,* London: Macmillan.

Levett, R. (2001) 'What Quality? Whose Lives?', *Green Futures,* 28 (May/June).

Levidow, L. (1995) 'Whose Ethics for Agricultural Biotechnology?', in V. Shiva and I. Moses (eds), *Biopolitics: A Feminist and Ecological Reader on Biotechnology,* London: Zed Books.

Levidow, L. and Tait, J. (1995) 'The Greening of Biotechnology: GMOs and Environment-Friendly Products', in V. Shiva and I. Moses (eds), *Biopolitics: A Feminist and Ecological Reader on Biotechnology,* London: Zed Books.

Lewontin, R. (1982) 'Organism and Environment', in H. Plotkin (ed.), *Learning, Development and Culture,* Chichester: Wiley.

Light, A. (ed.) (2003) *Social Ecology after Bookchin,* New York: Guilford Press.

Locke, J. (1959) *Two Treatises on Government,* Cambridge: Cambridge University Press.

Lomberg, B. (2001) *The Skeptical Environmentalist: The True State of the World,* Cambridge: Cambridge University Press.

Lomborg, B. (2003) 'Smearing a Skeptic: Something is Rotten in the State of Denmark', *Wall Street Journal,* 13 January.

Lovelock, J. (1988) *The Ages of Gaia: A Biography of our Living Earth,* New York: W.W. Norton.

Luke, T. (1997) *Ecocritique: Contesting the Politics of Nature, Economy, and Culture,* Minneapolis: University of Minnesota Press.

Luke, T. (1999a) 'Training Eco-managerialists: Academic Environmental Studies as a Power/Knowledge Formation', in F. Fischer and M. Hajer (eds), *Living with Nature: Environmental Discourse as Cultural Politics*, Oxford: Oxford University Press.

Luke, T. (1999b) 'Environmentality as Green Governmentality', in E. Darier (ed.), *Discourses of the Environment*, Oxford: Blackwell.

Luke, T. (2002) 'The People, Politics, and the Planet: Who Knows, Protects, and Serves Nature Best?', in B. Minteer and B. Pepperman Taylor (eds), *Democracy and the Claims of Nature: Critical Perspectives for a New Century*, Lanham, MD: Rowman & Littlefield.

Lyotard, J. (1984) *The Postmodern Condition: A Report on Knowledge*, Minneapolis: University of Minnesota Press.

Macgregor, S. (2006) *Beyond Mothering Earth: Ecological Citizenship and the Gendered Politics of Care*, Vancouver: University of British Columbia Press.

McKibben, B. (1989) *The End of Nature*, New York: Random House.

McMurtry, J. (1999) *The Cancer Stage of Capitalism*, London: Pluto Press.

Macpherson, C.B. (1973) *Democratic Theory*, Oxford: Clarendon Press.

Marcuse, H. (1955) *Eros and Civilization: A Philosophical Inquiry into Freud*, New York: Vintage Books.

Marcuse, H. (1964) *One Dimensional Man*, Boston, MA: Beacon Press.

Marcuse, H. (1972) *Counterrevolution and Revolt*, London: Penguin.

Marcuse, H. (1992) 'Ecology and the Critique of Modern Society', *Capitalism, Nature, Socialism*, 3:3.

Markovic, M. (1974) *The Contemporary Marx: Essays on Humanist Communism*, Nottingham: Spokesman Books.

Marshall, P. (1995) *Nature's Web: Rethinking Our Place on Earth*, London: Cassell.

Martin, E. (1991) 'The Egg and the Sperm: How Science has Constructed a Romance based on Stereotypical Male–Female Roles', in E.F. Keller and H.E. Longino (eds) (1996) *Feminism and Science*, Oxford: Oxford University Press.

Martin-Brown, J. (1992) 'Women in the Ecological Mainstream', *International Journal*, 47: 4.

Martinez-Alier, J. (1995) 'Political Ecology, Distributional Conflicts, and Economic Incommensurability', *New Left Review*, 211.

Martinez-Alier, J. (2001) *Environmentalism of the Poor*, Cheltenham: Edward Elgar.

Marx, K. (1971) *Theories of Surplus Value*, Moscow: Progress Publishers.

Marx, K. (1975) 'Economic and Philosophical Manuscripts', in L. Colletti (ed.), *Karl Marx: Early Writings*, Harmondsworth: Penguin.

Marx, K. and Engels, F. (1848/1967) *The Communist Manifesto*, Harmondsworth: Penguin.

Marx, K. and Engels, F. (1978) *The Communist Manifesto*, in R. Tucker (ed), *The Marx–Engels Reader*, New York: W.W. Norton.

Masters, R. (1991) 'Jean-Jacques Rousseau', in D. Miller *et al.* (eds), *The Blackwell Encyclopaedia of Political Thought*, Oxford: Blackwell.

May, T. (1996) *Situating Social Theory*, Buckingham: Open University Press.

Meadowcroft, J. (2005) 'From Welfare State to Ecostate', in J. Barry and R. Eckersley (eds) *The Global Ecological Crisis and the Nation-state*, Cambridge, MA.: MIT Press.

Meadows, D. *et al.* (1973) *The Limits to Growth: A Report for the Club of Rome's Project on the Predicament of Mankind*, New York: Universe.

Mellor, M. (1992a) *Breaking the Boundaries: Towards a Feminist, Green Socialism*, London: Virago.

Mellor, M. (1992b) 'Green Politics: Ecofeminist, Ecofeminine or Ecomasculine?', *Environmental Politics*, 1:2.

Mellor, M. (1995) 'Materialist Communal Politics: Getting from There to Here', in J. Lovenduski and J. Stanyer (eds), *Contemporary Political Studies,1995*, Belfast: Political Studies Association.

Mellor, M. (1997) *Feminism and Ecology*, Cambridge: Polity Press.

Mellor, M. (2001) 'Ecofeminism', in J. Barry and E.G. Frankland (eds), *International Encyclopedia of Environmental Politics*, London: Routledge.

Merchant, C. (1990) *The Death of Nature*, New York: Harper & Row.

Meyer, J. (2001) *Political Nature: Environmentalism and the Interpretation of Western Thought*, Boston, MA: MIT Press.

Midgley, M. (2001) *Gaia: The Next Big Idea*, London: Demos.

Milbrath, L. (1984) *Environmentalists: Vanguard for a New Society*, Albany: State University of New York Press.

Mill, J.S. (1848/1900) *Principles of Political Economy*, London: Longmans, Green & Co.

Mill, J.S. (1977) 'On Nature', in A. Clayre (ed.), *Nature and Industrialization*, Oxford: Oxford University Press.

Miller, C. (1988) *Jefferson and Nature: An Interpretation*, Baltimore, MD: Johns Hopkins University Press.

Miller, D. (1991) 'Peter Kropotkin', in D. Miller *et al.* (eds), *The Blackwell Encyclopaedia of Political Thought*, Oxford: Blackwell.

Miller, D. *et al.* (eds) (1991) *The Blackwell Encyclopaedia of Political Thought*, Oxford: Blackwell.

Milton, K. (1996) *Environmentalism and Cultural Theory: Exploring the Role of Anthropology in Environmental Discourse*, London: Routledge.

Milton, K. (2002) *Loving Nature: Towards an Ecology of Emotion*, London: Routledge.

Mirowski, P. (1994) 'Doing What Comes Naturally: Four Metanarratives on What Metaphors Are For', in P. Mirowski (ed.), *Natural Images in Economic Thought: Markets Read in Tooth and Claw*, Cambridge: Cambridge University Press.

Mirowski, P. (ed.) (1994) *Natural Images in Economic Thought: Markets Read in Tooth and Claw*, Cambridge: Cambridge University Press.

Morgan, P. (2001) 'Religions and the Environment', in J. Barry and E.G. Frankland (eds), *International Encyclopedia of of Environmental Politics*, London: Routledge.

Mulberg, J. (1995) *Social Limits to Economic Theory*, London: Routledge.

Nash, R. (1967) *Wilderness and the Ameican Mind*, New Haven, CT, and London: Yale University Press.

Naess, A. (1989), *Community, Ecology, Lifestyle*, Cambridge: Cambridge University Press.

Nisbet, R. (1982) *The Social Philosophers*, New York: Washington Square Press.

North, R. (1995) *Life on a Modern Planet: A Manifesto for Progress*, Manchester: Manchester University Press.

Northcott, M. (1996) *The Environment and Christian Ethics*, Cambridge: Cambridge University Press.

Norton, P. (2003) 'A Critique of Generative Class Theories of Environmentalism and of the Labour–Environmentalist Relationship', *Environmental Politics*, 12:4.

O'Brien, R. and Cahn, M. (1996) 'Thinking About the Environment: What's Theory Got to Do with It?', in M. Cahn and R. O'Brien, *Thinking About the Environment: Readings on Politics, Property and the Physical World*, New York and London: M.E. Sharpe.

O'Conner, J. (1991) 'The Second Contradiction of Capitalism: Causes and Consequences', *Capitalism, Nature, Socialism*, Pamphlet 1.

O'Conner, J. (1992) 'Socialism and Ecology', *Society and Nature*, 1:1.

O'Connor, M. T. (ed.) (1995) *Is Capitalism Sustainable?: Political Economy and the Politics of Ecology*, New York and London: Guilford Press.

O'Neill, J. (1993) *Ecology, Policy and Politics: Human Well-being and the Natural World*, London: Routledge.

O'Riordan, T. (1981) *Environmentalism* (2nd revised edn), London: Pion.

O'Riordan, T. and Jordan, A. (1994) 'The Precautionary Principle in Contemporary Environmental Politics', *Environmental Values*, 4:3.

Odum, E. (1971) *Fundamentals of Ecology* (3rd edn), Philadelphia, PA: Saunders & Co.

Oelschlaeger, M. (1992) 'Wilderness, Civilization and Language', in Oelschlaeger, M. (ed.), *The Wilderness Condition: Essays on Environment and Civilization*, Washington, DC, and Covelo, CA: Island Press.

Oelschlaeger, M. (ed.) (1992) *The Wilderness Condition: Essays on Environment and Civilization*, Washington, DC, and Covelo, CA: Island Press.

Offe, C. (1984) *Contradictions of the Welfare State*, London: Hutchinson.

Ophuls, W. (1977) *Ecology and the Politics of Scarcity*, San Francisco, CA: W.H. Freeman.

Outhwaite, W. (ed.) (1996) *The Habermas Reader*, Oxford: Polity Press.

Paehlke, R. (1989) *Environmentalism and the Future of Progressive Politics*, New Haven, CT, and London: Yale University Press.

Parsons, H. (1977) *Marx and Engels on Ecology*, Westport, CT: Greenwood Press.

Passmore, J. (1980) *Man's Responsibility for Nature* (2nd edn), London: Duckworth.

Paterson, M. (1996) 'UNCED in the Context of Globalisation', *New Political Economy*, 1: 3.

Paterson, M. (1999) 'Understanding the Green Backlash', *Environmental Politics*, 8:2.

Paterson, M. (2001) *Understanding Global Environmental Politics: Accumulation, Domination, Resistance*, Basingstoke: Palgrave.

Pavitt, K. (1974) 'Malthus and Other Economists: Some Doomsdays Revisited', in H. Cole *et al.*, *Thinking About the Future: A Critique of the Limits to Growth*, London: Chatto & Windus.

Pearce, D. (1992) 'Green Economics', *Environmental Values*, 1:1.

Pearce, D., Markandya, A. and Barbier, E. (1989) *Blueprint for a Green Economy*, London: Earthscan.

Pepper, D. (1984) *The Roots of Modern Environmentalism*, London: Croom Helm.

Pepper, D. (1993) *Eco-Socialism: From Deep Ecology to Social Justice*, London: Routledge.

Pepper, D. (1996) *Modern Environmentalism: An Introduction*, London: Routledge.

Perkins, P. (2000) 'Equity, Economic Scale and the Role of Exchange in a Sustainable Economy', in F. Gale and M. M'Gonigle (eds), *Nature, Production, Power: Towards an Ecological Political Economy*, Cheltenham: Edward Elgar.

Pietilä, H. (1990) 'The Daughters of Earth: Women's Culture as a Basis for Sustainable Development', in J. Engel and J. Engel (eds), *The Ethics of Environmental Development*, London: Belhaven Press.

Pinter, H. (2005) 'Nobel Lecture: Art, Truth and Politics', http://nobelprize.org/literature/laureates/2005/pinter-lecture-e.html (accessed 14 May 2006).

Plant, J. (1989) 'Toward a New World Order: An Introduction', in J. Plant (ed.), *Healing the Wounds: The Promise of Ecofeminism*, Philadelphia, PA: New Society Publishers.

Plant, J. (ed.) (1989) *Healing the Wounds: The Promise of Ecofeminism*, Philadelphia, PA: New Society Publishers.

Plotkin, H. (ed.) (1982) *Learning, Development and Culture*, Chichester: Wiley.

Plumwood, V. (1993) *Feminism and the Mastery of Nature*, London: Routledge.

Podobnik, B. (2002) 'The Globalisation Protest Movement: An Analysis of Broad Trends and the Impact of September 11', Paper presented at the American Sociological Association Annual Meeting, Chicago, IL, 16–19 August.

Polanyi, K. (1947) *The Great Transformation: The Political and Economic Origins of Our Time*, Boston, MA: Beacon Press.

Polyp, P.J. (1996a) 'The Economic Totem Pole', *New Internationalist*, April.

Polyp, P.J. (1996b) 'Economic Growth', *New Internationalist*, June.

Polyp, P.J. (1996c) 'We're All Gonna Die', *New Internationalist*, September.

Ponting, C. (1991) *A Green History of the World*, London: Sinclair-Stevenson.

Popper, K. (1963) *Conjectures and Refutations: The Growth of Scientific Knowledge*, New York: Routledge & Kegan Paul.

Porritt, J. (1984) *Seeing Green: The Politics of Ecology Explained*, Oxford: Blackwell.

Porteous, J.D. (1997) *Environmental Aesthetics: Ideas, Politics and Planning*, London: Routledge.

Porter, R. (1994) 'The Enlightenment', in *The Hutchinson Dictionary of Ideas*, Oxford: Helicon.

Pratt, V. with Howarth, J. and Brady, E. (2000) *Environment and Philosophy*, London: Routledge.

Purdue, D. (1995) 'Hegemonic Trips: World Trade, Intellectual Property Rights and Biodiversity', *Environmental Politics* 4:1.

Quinby, L. (1990) 'Ecofeminism and the Politics of Resistance', in I. Diamond and G. Orenstein (eds), *Reweaving the World*, San Francisco, CA: Sierra Club Books.

Redclift, M. and Benton, T. (eds) (1994) *Social Theory and the Global Environment*, London: Routledge.

Rennie-Short, J. (1991) *Imagined Country: Society, Culture and Environment*, London: Routledge.

Ridley, M. (1995), *Down to Earth: A Contrarian View of Environmental Problems*, London: Institute of Economic Affairs in association with the *Sunday Telegraph*.

Robertson, G., Mash, M., Tickner, L., Bird, J., Curtis, B. and Putnam, T. (eds) (1996) *FutureNatural: Nature, Science, Culture*, London: Routledge.

Robertson, J. (1976) *Power, Money and Sex: Towards a New Social Balance*, London: Marion Boyars.

Robertson, J. (1983) *The Sane Alternative*, published by the author.

Rose, S., Lewontin, R. and Kamin, L. (1984) *Not in Our Genes*, Harmondsworth: Penguin.

Rousseau, J.J. (1995) 'A Critique of Progress', in I. Kramnick (ed.), *The Portable Enlightenment Reader*, London: Penguin.

Rousseau, J.J. (1997) *The Discourses and Other Early Political Writings*, edited by V. Gourevitch, Cambridge: Cambridge University Press.

Rowell, A. (1996) *Green Backlash: Global Subversion of the Environmental Movement*, London: Routledge.

Roy, A. (1999) *The Cost of Living*, London: Flamingo.

Rudy, A. and Light, A. (1996) 'Social Ecology and Social Labor: A Consideration and Critique of Murray Bookchin', in D. Macauley (ed.), *Minding Nature: The Philosophers of Ecology*, New York: Guilford Press.

Sachs, W. (1995) 'Global Ecology and the Shadow of Development', in G. Sessions (ed.), *Deep Ecology for the Twenty-first Century*, Boston, MA, and London: Shambala Press.

Sachs, W., Loske, R. and Linz, M. (1998) *Greening the North: A Post-industrial Blueprint for Ecology and Equity*, London: Zed Books.

Sallah, A. (1992) 'The Ecofeminist/Deep Ecology Debate: A Reply to Patriarchal Reason', *Environmental Ethics*, 14:3.

Sallah, A. (1995) 'Nature, Woman, Labor, Capital: Living the Deepest Contradiction', in M. T. O'Connor (ed.), *Is Capitalism Sustainable?: Political Economy and the Politics of Ecology*, New York and London: Guilford Press.

Salleh, A. (1997) *Ecofeminism as Politics: Nature, Marx and the Postmodern*, London: Zed Books.

Salleh, A. (2005) 'Moving to an Embodied Materialism', *Capitalism, Nature, Socialism*, 16:2.

Samuels, A. (1993) '"I am a Place": Depth Psychology and Environmentalism', *British Journal of Psychotherapy*, 10:2.

Sandilands, C. (1999) *The Good-natured Feminist: Ecofeminism and the Quest for Democracy*, Minneapolis: University of Minnesota Press.

Sarkar, S. (1999) *Eco-socialism or Eco-capitalism?*, London: Zed Books.

Schlosberg, D. (1999) *Environmental Justice and the New Pluralism*, Oxford: Oxford University Press.

Schumacher, E.F. (1973) *Small is Beautiful: Economics as if People Mattered*, London: Blond & Briggs.

Scott, J. (2000) 'Rational Choice Theory', in G. Browning, A. Halcli and F. Webster (eds), *Understanding Contemporary Society: Theories of The Present*, London: Sage.

Seabrook, J. (1998) 'A Global Market for All', *New Statesman*, 26 June.

Seager, J. (1993) *Earth Follies: Feminism, Politics and the Environment*, London: Earthscan.

Selin, H. (ed.) (2003) *Nature Across Cultures: Views of Nature and the Environment in Non-Western Cultures*, Boston, MA: Kluwer Academic.

Sessions, G. (ed.) (1995) *Deep Ecology for the Twenty-first Century*, Boston, MA, and London: Shambala Press.

Shiva, V. (1988) *Staying Alive: Women, Ecology and Development*, London: Zed Books.

Shiva, V. (1991) *Minding our Lives: Women from the South and North Reconnect Ecology and Health*, New Delhi: Kali for Women.

Shiva, V. (1992) 'Overview', *Ms Magazine*, 11:6.

Shiva, V. and Moses, I. (eds) (1995) *Biopolitics: A Feminist and Ecological Reader on Biotechnology*, London: Zed Books.

Simon, J. (1996) *The Ultimate Resource 2*, http://www.juliansimon.org/writings/Ultimate_Resource/TCONCLUS.txt (accessed 15 May 2006).

Simon, J. and Kahn, H. (1984) *The Resourceful Earth: A Response to Global 2000*, Oxford: Blackwell.

Singer, P. (1990) *Animal Liberation* (2nd edn), London: Jonathoan Cape.

Singer, P. (ed.) (1994) *Ethics*, Oxford: Oxford University Press.

Smith, G. (2003) *Deliberative Democracy and the Environment*, London: Routledge.

Smith, N. (1996) 'The Production of Nature', in G. Robertson *et al.* (eds), *FutureNatural: Nature, Science, Culture*, London: Routledge.

Snow, C.P. (1959) *The Two Cultures and the Scientific Revolution*, Cambridge: Cambridge University Press.

Soper, K. (1995) *What Is Nature?*, Oxford: Blackwell.

Soper, K. (1996) 'Greening Prometheus: Marxism and Ecology', in T. Benton (ed), *The Greening of Marxism*, New York: The Guilford Press.

Spencer, H. (1884/1982) *The Man Versus the State*, Indianapolis, MS: Liberty Press.

Spretnak, C. (1996) 'Beyond Humanism, Modernity, and Patriarchy', in R. Gottlieb (ed.), *This Sacred Earth: Religion, Nature, Environment*, London: Routledge.

Stephens, P. (1996) 'Plural Pluralisms: Towards a More Liberal Green Political Theory', in I. Hampsher-Monk and J. Stanyer (eds), *Contemporary Political Studies 1996*, vols 1–3, Belfast: Political Studies Association.

Sumner, W. (1992) *On Liberty, Society and Politics: The Essential Essays of William Graham Sumner*, Indianapolis, MS: Liberty Press.

Swartz, D. (1996) 'Jews, Jewish Texts, and Nature: A Brief History', in R. Gottlieb (ed.), *This Sacred Earth: Religion, Nature, Environment*, London: Routledge.

Synnott, A. (1992) 'Tomb, Temple, Machine and Self: The Social Construction of the Body', *British Journal of Sociology*, 43:1.

Thiele, L.P. (2000) Book Review: M. de Geus (1999) *Ecological Utopias: Envisioning the Sustainable Society, Conservation Ecology*, 4:1, http://www.consecol.org/vol4/iss1/art18/ (accessed 20 April 2006).

Thomas, K. (1983) *Man and the Natural World: Changing Attitudes in England 1500–1800*, Harmondsworth: Penguin.

Thompson, J. and Held, D. (eds) (1982) *Habermas: Critical Debates*, London: Macmillan.

Thompson, P. (1995) *The Spirit of the Soil: Agriculture and Environmental Ethics*, London: Routledge.

Thropy, Miss Ann (1986) 'AIDS and Africa', http://www.off-road.com/green/ef_aids_wish.html, (accesssed 5 November 2004).

Tocqueville, A. (1956) *Democracy in America*, edited by R. Heffner, New York: Mentor Books.

Tokar, B. (1995) 'The "Wise Use" Backlash: Responding to Militant Anti-environmentalism', *The Ecologist*, 25:4.

Torgerson, D. (1999) *The Promise of Green Politics: Environmentalism and the Public Sphere*, Durham, NC: Duke University Press.

van den Bergh, J. (1996) *Ecological Economics and Sustainable Development: Theory, Methods and Application*, Cheltenham: Edward Elgar.

Vincent. A. (1992) 'Ecologism', in his *Modern Political Ideologies*, Oxford: Blackwell.

Vogel, S. (1996) *Against Nature: The Concept of Nature in Critical Theory*, New York: State University of New York Press.

Vogel, S. (1997) 'Habermas and the Ethics of Nature', in R. Gottlieb (ed.), *The Ecological Community: Environmental Challenges for Philosophy, Politics and Morality*, London: Routledge.

von Bulow, G. (1962) 'Die Sudwalder von Reichenhall', *Mitt. aus der Staatsforstverwaltung Bayerns*, vol. 33.

Wall, D. (1994a) *Green History: A Reader in Environmental Literature, Philosophy and Politics*, London: Routledge.

Wall, D. (1994b) 'Towards a Green Political Theory: In Defence of the Commons?', in J. Stanyer and P. Dunleavy (eds), *Contemporary Political Studies, 1994*, Belfast: Political Studies Association.

Wall, D. (2005) *Babylon and Beyond: The Economics of Anti-capitalist, Anti-globalist and Radical Green Movements*, London: Pluto Press.

Wall, D. (2006) 'Green Economics: An Introduction and Research Agenda', *International Journal of Green Economics*, 1:1/2.

Warren, K. (1987) 'Feminism and Ecology: Making Connections', *Environmental Ethics*, 9:1.

Watson, D. (1996) *Beyond Bookchin: Preface for a Future Social Ecology*, New York: Autonomedia.

Wells, D. (1982) 'Resurrecting the Dismal Parson: Malthus, Ecology and Political Thought', *Political Studies*, 30:1.

Wells, D. and Lynch, D. (2000) *The Political Ecologist*, Aldershot: Ashgate.

Weston, I. (ed.) (1986) *Red and Green*, London: Pluto Press.

White, D. (2003) 'Hierarchy, Domination, Nature: Considering Bookchin's Critical Social Theory', *Organization and Environment*, 16:1.

White, L. (1967) 'The Historical Roots of Our Ecologic Crisis', *Science*, 155.

White, S.K. (1988) *The Recent Work of Jürgen Habermas: Reason, Justice and Modernity*, Cambridge: Cambridge University Press.

Whitebrook, J. (1996) 'The Problem of Nature in Habermas', in D. Macauley (ed.), *Minding Nature: The Philosophers of Ecology*, New York and London: Guilford Press.

Wickramasinghe, A. (2003) 'Women and Environmental Justice in South Asia', in J. Agyeman, R. Bullard and B. Evans, *Just Sustainabilities: Development in an Unequal World*, London: Earthscan.

Williams, R. (1988) *Keywords: A Vocabulary of Culture and Society*, London: Fontana.

Wilson, E.O. (1978) 'What Is Sociobiology?', in M. Gregory *et al.* (eds), *Sociobiology and Human Nature*, San Francisco: Jossey-Bass.

Wilson, E.O. (1984) *Biophilia*, Harvard, MA: Harvard University Press.

Wilson, E.O. (1993) 'Biophilia and the Conservation Ethic', in S. Kellert and E.O. Wilson (eds), *The Biophilia Hypothesis*, Washington, DC: Island Press/Shearwater.

Wilson, E.O. (1997) *In Search of Nature*, London: Allen Lane, the Penguin Press.

Wingrove, E.R. (2000) *Rousseau's Republican Romance*, Princeton: Princeton University Press.

Wissenburg, M. (1998) *The Free and the Green Society: Green Liberalism*, London: UCL Press.

Wollheim, R. (1976) *Freud*, London: Fontana.

Woodgate, G. and Redclift, M. (1998) 'From a "Sociology of Nature" to Environmental Sociology: Beyond Social Construction', *Environmental Values*, 7:1.

World Commission on Environment and Development (WCED) (1987) *Our Common Future*, Oxford: Oxford University Press.

World Social Forum (2001) *Charter of Principles*, http://www.wsfindia.org/charter.php (accessed 11 May 2006).

World Wildlife Fund (WWF) (2005) *Living Planet Report 2004*, http://assets.panda.org/downloads/lpr2004.pdf (accessed 5 October 2005).

Worster, D. (1994) *Nature's Economy: A History of Ecological Ideas* (2nd edn), Cambridge: Cambridge University Press.

Wynne, B. (2002) 'Risk and Environment as Legitimatory Discourses of Technology', *Current Sociology*, 50:3.

Yearley, S. (1991) *The Green Case: A Sociology of Environmental Issues, Arguments and Politics*, London: HarperCollins.

Zimmerman, M. (1987) 'Feminism, Deep Ecology and Environmental Ethics', *Environmental Ethics*, 9:1.

Zimmerman, M. (1992) 'The Blessing of Otherness: Wilderness and the Human Condition', in M. Oelschlaeger (ed.), *The Wilderness Condition: Essays on Environment and Civilization*, Washington, DC, and Covelo, CA: Island Press.

Zimmerman, M. (1994) *Contesting Earth's Future: Radical Ecology and Postmodernity*, Berkeley: University of California Press.

Index